Fundamentals of Power System Economics

Fundamentals of Power System Economics

Second Edition

Daniel S. Kirschen
University of Washington, United States

Goran Strbac
Imperial College London, United Kingdom

Edition History
1e: © 2004 John Wiley & Sons Ltd

Registered Office(s)
John Wiley & Sons, Inc., 111 River Street, Hoboken, NJ 07030, USA
John Wiley & Sons Ltd, The Atrium, Southern Gate, Chichester, West Sussex, PO19 8SQ, UK

Editorial Office
The Atrium, Southern Gate, Chichester, West Sussex, PO19 8SQ, UK

For details of our global editorial offices, customer services, and more information about Wiley products visit us at www.wiley.com.

Wiley also publishes its books in a variety of electronic formats and by print-on-demand. Some content that appears in standard print versions of this book may not be available in other formats.

Library of Congress Cataloging-in-Publication Data

Names: Kirschen, Daniel Sadi, author. | Strbac, Goran, author.
Title: Fundamentals of power system economics / by Daniel S. Kirschen, University of Washington, United States, Goran Strbac, Imperial College London, United Kingdom.
Description: Second Edition. | Hoboken : Wiley, [2019] | Revised edition of the authors' Fundamentals of power system economics, c2004. | Includes index. |
Identifiers: LCCN 2018015513 (print) | LCCN 2018017093 (ebook) | ISBN 9781119309888 (pdf) | ISBN 9781119213253 (epub) | ISBN 9781119213246 (cloth)
Subjects: LCSH: Electric utilities. | Electric power distribution. | Interconnected electric utility systems.
Classification: LCC HD9685.A2 (ebook) | LCC HD9685.A2 K57 2019 (print) | DDC 333.793/2–dc23
LC record available at https://lccn.loc.gov/2018015513

Cover Design: Wiley
Cover Image: © onurdongel/iStockphoto

Set in 10/12 pt WarnockPro-Regular by Thomson Digital, Noida, India

Printed in the UK

For Penny and Philippe

For Dragana, Jelena, Anna, and Emily

Contents

Preface to the First Edition

For about a hundred years, the electricity supply industry was in the hands of vertically integrated monopoly utilities. During that time, engineers treated the management of this industry as a set of challenging optimization problems. Over the years, these optimization problems grew in size, complexity, and scope. New algorithms were developed, and ever more powerful computers were deployed to refine the planning and the operation of the power systems. With the introduction of competition in the electricity supply industry, a single organization is no longer in charge. Multiple actors with divergent or competing interests must interact to deliver electrical energy and keep the lights on. Conventional optimization problems are often no longer relevant. Instead, dozens of new questions are being asked about a physical system that has not changed. To deliver the promised benefits of competition, old issues must be addressed in radically new ways. To stay in business, new companies must maximize the value of the service they provide. Understanding the physics of the system is no longer enough. We must understand how the economics affect the physics and how the physics constrain the economics.

An environment with many independent participants evolves very rapidly. Over the last two decades, hundreds of technical papers, thousands of reports, and a few books have been written to discuss these new issues and to propose solutions. The objective of this book is not to summarize or repeat what is in these documents. Instead, we have chosen to concentrate on delivering a clear and in-depth explanation of the fundamental issues. Our aim is to give the readers a solid understanding of the basics and help them develop innovative solutions to problems that vary in subtle ways from country to country, from market to market, and from company to company. Therefore, we do not discuss the organization of specific markets. Neither do we attempt to describe all the solution techniques that have been proposed.

The plan of this book is simple. After introducing the participants to a restructured electricity supply industry, we discuss the concepts from microeconomics that are essential for the understanding of electricity markets. We then move on to the analysis of the operation of power systems in a competitive environment. To keep matters simple, we begin by ignoring the transmission network and we consider the operation of pure energy markets. We then discuss power system security and the effects that networks have on electricity prices. Finally, in the last two chapters, we address the issue of investments in power generation and transmission equipment in a competitive environment.

The typical reader we had in mind while writing this book was a first-year graduate student or a final-year undergraduate student specializing in power engineering. We have assumed that these students know the physical structure of power systems, understand the purpose and principles of a power flow calculation, and are familiar with basic optimization theory. We believe that this book will also be valuable to engineers who are working on deregulation or competition issues and who want to acquire a broader perspective on these questions. Finally, this book might also be useful to economists and other professionals who want to understand the engineering perspective on these multidisciplinary issues.

Except when a specific source is cited, we have made no attempt to use or produce realistic numbers in the problems and examples. We have used $ as a unit for money because it is probably the best-known symbol for a currency. We could have used €, £, or ¥ instead without any change in meaning. Some of our examples refer to the fictitious countries of Syldavia and Borduria, which are the product of the fertile imagination of the Belgian cartoonist Hergé, creator of the character Tintin.

This book stems from our research and teaching activities in power system economics at UMIST. We are grateful to our colleagues Ron Allan and Nick Jenkins for fostering an environment in which this work was able to flourish. We also thank Fiona Woolf for fascinating interdisciplinary discussions on transmission expansion. A few of our students spent considerable time proofreading drafts of this book and checking answers to the problems. In particular, we thank Tan Yun Tiam, Miguel Ortega Vazquez, Su Chua Liang, Mmeli Fipaza, Irene Charalambous, Li Zhang, Jaime Maldonado Moniet, Danny Pudjianto, and Joseph Mutale. Any remaining errors are our sole responsibility.

Manchester, England
February 2004

Daniel Kirschen

Manchester, England
February 2004

Goran Strbac

Preface to the Second Edition

Since the publication of this book's first edition in 2004, competitive electricity markets have become increasingly sophisticated. While their fundamental principles have not changed, we felt that our text needed to be updated to more closely reflect current practice. In particular, the phenomenal increase in the amount of energy produced from intermittent and stochastic renewable energy sources significantly increases the uncertainty that power system operators and market participants have to manage. This second edition, therefore, includes a number of new sections devoted to analyzing the effect of uncertainty and the need for technical flexibility (e.g. from storage and the demand side) as well as for more flexible market rules.

The chapter on system security and ancillary service has been expanded into a chapter on power system operation and placed after the chapter on the effect of the transmission network. We have also carefully revised the text throughout the book to reflect current terminology and our deeper understanding of the workings of electricity markets.

We would like to thank the students and postdocs who helped us by developing some examples and pointing out mistakes in a draft of this second edition: Bolun Xu, Muhammad Danish Farooq, Yujie Zhou, Linyue Qiao, Mareldi Ahumada Paras, Namit Chauhan, Ben Walborn, Dimitrios Papadaskalopoulos, Rodrigo Moreno, Yujian Ye and Yang Yang. Any remaining errors are our sole responsibility.

Seattle, WA
March 2018

Daniel Kirschen

London, England
March 2018

Goran Strbac

1 Introduction

1.1 Why Competition?

For most of the twentieth century, when consumers wanted to buy electrical energy, they had no choice. They had to buy it from the utility that held the monopoly for the supply of electricity in the area where these consumers were located. Some of these utilities were vertically integrated, which means that they generated the electrical energy, transmitted it from the power plants to the load centers, and distributed it to individual consumers. In other cases, the utility from which consumers purchased electricity was responsible only for its sale and distribution in a local area. This distribution utility in turn had to purchase electrical energy from a generation and transmission utility that had a monopoly over a wider geographical area. In some parts of the world, these utilities were regulated private companies, while in others they were public companies or government agencies. Irrespective of ownership and level of vertical integration, geographical monopolies were the norm.

Electric utilities operating under this model made truly remarkable contributions to economic activity and quality of life. Most people living in the industrialized world have access to an electricity distribution network. For several decades, the amount of energy delivered by these networks doubled about every 8 years. At the same time, advances in engineering improved the reliability of the electricity supply to the point that in many parts of the world the average consumer is deprived of electricity for less than 2 min per year. These achievements were made possible by ceaseless technological advances. Among these, let us mention only the development and erection of transmission lines operating at over 1 000 000 V and spanning thousands of kilometers, the construction of power plants capable of generating more than 1000 MW and the on-line control of the networks connecting these plants to the consumers. Some readers will undoubtedly feel that on the basis of this record, it may have been premature to write the first paragraph of this book in the past tense.

In the 1980s, some economists started arguing that this model had run its course. They said that the monopoly status of the electric utilities removed the incentive to operate efficiently and encouraged unnecessary investments. They also argued that the cost of the mistakes that private utilities made should not be passed on to the consumers. Public utilities, on the other hand, were often too closely linked to the government. Politics could then interfere with good economics. For example, some public utilities were treated as cash cows, others were prevented from setting rates at a level that reflected costs or were deprived of the capital that they needed for essential investments.

These economists suggested that prices would be lower and the overall economy more efficient if the supply of electricity was subjected to market discipline rather than monopoly regulation or government policy. This proposal was made in the context of a general deregulation of Western economies that had started in the late seventies. Before attention turned toward electricity, this movement had already affected airlines, transportation and the supply of natural gas. In all these sectors, a regulated market or monopolies had previously been deemed the most efficient way of delivering the "products" to the consumers. It was felt that their special characteristics made them unsuitable for trading on free markets. Advocates of deregulation argued that the special characteristics of these products were not insurmountable obstacles and that they could and should be treated like all other commodities. If companies were allowed to compete freely for the provision of electricity, the efficiency gains arising from competition would ultimately benefit the consumers. In addition, competing companies would probably choose different technologies. It was therefore less likely that the consumers would be saddled with the consequences of unwise investments.

If kilowatt-hours could be stacked on a shelf – like kilograms of flour or television sets – ready to be used as soon as the consumer turns on the light or starts the industrial process, electricity would be a simple commodity, and there would be no need for this book. However, despite recent technological advances in electricity storage and micro-generation, this concept is not yet technically or commercially feasible. The reliable and continuous delivery of significant amounts of electrical energy still requires large generating plants connected to the consumer through transmission and distribution networks and careful attention must be paid to reliability.

In this book, we explore how various aspects of the supply of electricity can be packaged into products that can be bought and sold on open markets. Because these products cannot be fully separated from the supply infrastructure, we also discuss how their trading affects the operation of the power system and, in turn, how operational constraints impinge on the electricity markets.

In the long run, the need always arises to invest in new facilities, either because a new technology holds the promise of greater profits or simply because equipment age and need to be replaced. Here again we will need to examine the interplay between market-driven behavior, physical constraints, and the need for reliability.

1.2 Market Structures and Participants

Before we delve into the analysis of electricity markets, it is useful to consider the various ways in which they can be structured and to introduce the types of companies and organizations that play a role in these markets. In the following chapters, we will discuss in much more detail the function and motivations of each of these participants. Since markets have evolved at different rates and in somewhat different directions in each country or region, not all these entities will be found in each market.

1.2.1 Traditional Model

In the traditional market model (Figure 1.1), trading is limited to consumers purchasing electricity from their local electric utility. This utility has two main characteristics. First, it

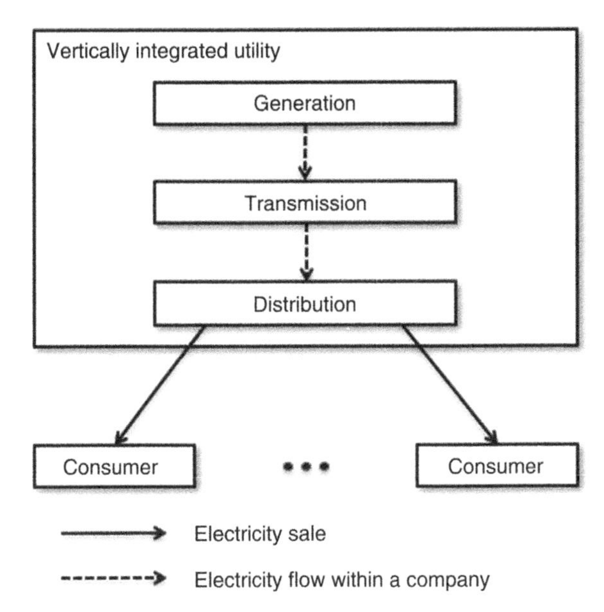

Figure 1.1 Traditional model of electricity supply.

has a monopoly for the supply of electricity over its service territory. If consumers want to purchase electricity, they do not have a choice: they have to buy it from this utility. Second, the utility is vertically integrated. This means that it performs all the functions required to supply electricity: building generating plants, transmission lines and distribution networks, operating these assets in a reliable manner, and billing the consumers for the service provided.

In a fairly common variant of the traditional model (Figure 1.2), the vertically integrated utility is split in two parts. One organization generates and transmits electricity over a fairly wide area and sells it to several distribution companies (*Discos*), each of which has a local monopoly for the sale of electricity to consumers.

Because monopolies could take advantage of the fact that their customers do not have a choice to charge them extortionate prices, they must either be government entities or be subject to oversight by a government department, which we shall call the regulator. In the traditional model, the regulator enforces what is called the regulatory compact. This is an agreement that gives a utility a monopoly for the supply of electricity over a given geographical area. In exchange, the utility agrees that its prices will be set by the regulator, that it will supply all the consumers in that area, and that it will maintain a certain quality of service.

This model does not preclude bilateral energy trades between utilities operating in different geographical areas. Such trades take place at the wholesale level, i.e. through interconnections between transmission networks.

The problem with the traditional model and its variant is that monopolies tend to be inefficient because they do not have to compete with others in order to survive. Furthermore, because their operations are rather opaque, regulators have difficulties assessing where improvements could be made.

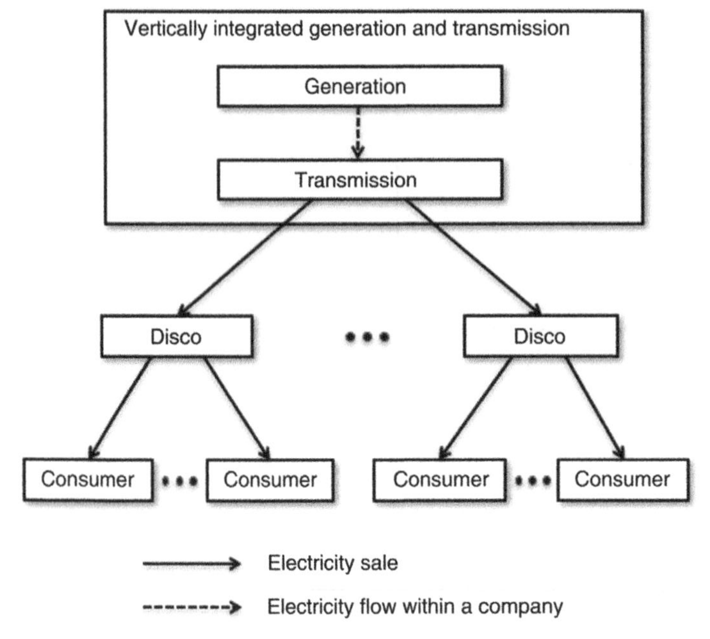

Figure 1.2 Variant on the traditional model of electricity supply.

1.2.2 Introducing Independent Power Producers

A first step toward a more competitive industry structure consists in allowing other companies (called *independent power producers* or *IPPs*) to produce part of the electrical energy that the incumbent vertically integrated utility must supply to its customers. Figure 1.3 illustrates this arrangement. While this model introduces a degree of competition at the generation level, it does not provide a mechanism for discovering cost-reflective prices in the same way that a free market does (see Chapter 2). The incumbent

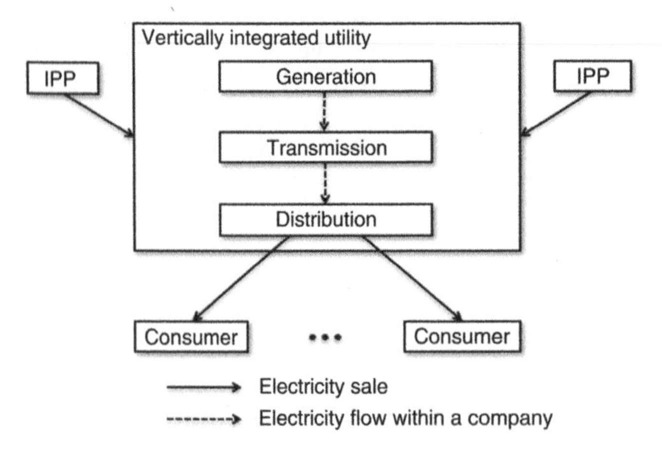

Figure 1.3 Incumbent vertically integrated utility with independent power producers (IPPs).

utility would like to pay as little as possible for the energy produced by the IPPs to discourage them from expanding their generation capacity. It must therefore be forced by law to buy the power produced by the IPPs. Given this guarantee that their production will be purchased, the IPPs will try to get as high a price as they can. This leaves the regulator with the task of deciding what an equitable price would be. In the absence of detailed and reliable information, the result will often be economically inefficient.

1.2.3 Wholesale Competition

A further step toward competitive electricity markets consists in getting rid of the incumbent utility. As illustrated in Figure 1.4, all the companies that own large generating plants (*Gencos*) then compete on an equal basis to sell electrical energy.

Distribution companies purchase the electrical energy consumed by their customers on this wholesale electricity market. The largest consumers are often allowed to participate directly in this market. As we will discuss in Chapter 3, this wholesale market can operate in a centralized manner or can be based on bilateral transactions. In this model, the wholesale price of electricity is determined by the interplay of supply and demand. On the other hand, the retail price of electrical energy must remain regulated because each distribution company retains a local monopoly over the sale of electrical energy flowing through its network.

When the wholesale market is operated in a centralized manner, an organization called *independent system operator* (*ISO*) must be created. This ISO has two main functions. First, it must manage the market in an impartial and efficient manner. Second, it is responsible for the reliable operation of the transmission system. As its name implies, to ensure the fairness of the market, the ISO has to be institutionally independent from all market participants.

In a bilateral wholesale market, these functions are often split between one or more *market operators* (*MOs*), whose role is to facilitate commercial transactions between buyers and sellers of electrical energy, and a *transmission system operator* (*TSO*), who keeps the system in balance and operationally reliable. While TSOs often own the transmission assets (lines, transformers, substations, etc.), ISOs usually do not.

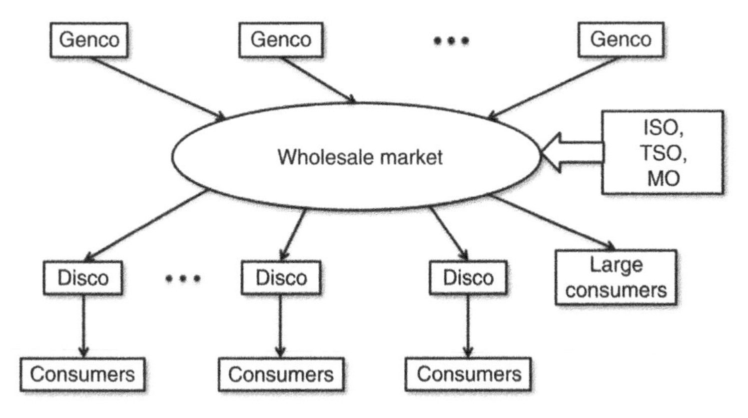

Figure 1.4 Wholesale electricity market structure.

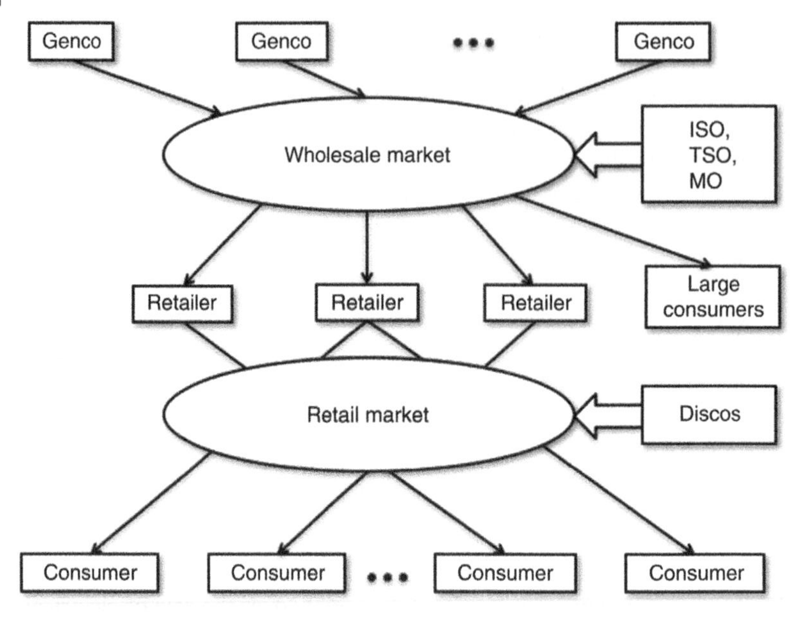

Figure 1.5 Market with retail competition.

1.2.4 Retail Competition

Competition can also be introduced in the retail market. This leads to the structure illustrated in Figure 1.5, where *retailers* purchase electrical energy in bulk on the wholesale market and resell it to small- and medium-size consumers. In this model, the "wires" activities of the distribution companies are normally separated from their retail activities because they no longer have a local monopoly for the supply of electrical energy in the area covered by their network. One can view the wholesale market as operating over the transmission network while the retail market takes place over the distribution network. Building and operating the transmission and distribution networks remain monopoly activities because it is generally agreed that building competing sets of wires would be wasteful. The regulator thus has to decide what investments in network assets are justified and how the cost of these investments should be allocated to the users of the networks.

Once sufficiently competitive markets have been established, the retail price no longer has to be regulated because small consumers have the option to change retailer if they are offered a better price or better service. As we will see in Chapter 2, from an economics perspective, this is desirable because market interactions lead to the discovery of economically efficient prices.

1.2.5 Renewable and Distributed Energy Resources

Over the last two decades, public policies aimed at reducing carbon emissions to mitigate climate change have significantly altered the mix of generation technologies in many parts of the world. Because wind and solar generation now contribute a substantial fraction of the overall production of electrical energy, electricity markets have had to adapt to their intermittent and stochastic nature. To deal more efficiently with the larger imbalances between generation and load that renewable generation causes, markets operate on a much shorter time frame than before. Another adaptation is the increasing reliance on flexibility from the demand side to help maintain this balance. Marshaling demand-side resources is challenging because they tend to be small and distributed throughout the system. Direct participation in the wholesale electricity markets by distributed energy resources (such as demand response, small-scale energy storage, and photovoltaic generation) is not possible because it would vastly increase the number of market participants and render these markets unmanageable. In addition, the rules of the wholesale markets are complex and the requirements for participation are strict, making the transaction costs prohibitively expensive for small participants. To overcome this problem, new entities called *aggregators* are emerging. Their role is to serve as a commercial and technical intermediary between the wholesale markets and the owners of distributed energy resources who could contribute to the economic efficiency of the overall system.

1.3 Dramatis Personae

This section summarizes the roles of the different types of organizations that operate in the various market structures. Some of these organizations were already introduced in the previous section. In some markets, a single commercial entity or subsidiaries of this entity may be allowed to perform the functions of two or more of the organizations listed below. The names given to different types of organizations may also differ from country to country.

Vertically integrated utilities own and operate all the assets needed to supply electrical energy: generating plants, transmission networks, and distribution networks. In a traditional regulated environment, such a company has a monopoly for the supply of electricity over a given geographical area. Once a wholesale electricity market has been established, the functions and assets of the vertically integrated utilities are divided between other types of organizations.

Generating companies (*gencos*) own generating plants and sell electrical energy. They may also sell services such as regulation, voltage control and reserve that the system operator needs to maintain the quality and operational reliability of the electricity supply. A generating company can own a single plant or a portfolio of plants of different technologies. Generating companies that coexist with vertically integrated utilities are called IPPs.

Distribution companies (*discos*) own and operate distribution networks. Unless a retail market has been organized, discos have a monopoly for the sale of electrical energy to all consumers connected to their network. When a retail market is in operation, discos are no longer responsible for the sale of energy to consumers and their role is limited to the operation and development of the distribution network.

Retailers buy electrical energy on the wholesale market and resell it to consumers who do not wish or are not allowed to participate in this wholesale market. Retailers do not have to own any power generation, transmission or distribution assets. Some retailers are subsidiaries of generation or distribution companies. All the customers of a retailer do not have to be connected to the network of the same distribution company.

Market Operators (*MOs*) run computer systems to match the bids and offers that buyers and sellers of electrical energy submit. They also take care of the settlement of the accepted bids and offers, i.e. they forward payments from buyers to sellers following delivery of the energy. Independent for-profit MOs often manage electricity markets that close some time ahead of real time. On the other hand, the ISO runs the market of last resort, i.e. the market where load and generation are balanced in real time.

The primary responsibility of an **Independent System Operator** (*ISO*) is to maintain the stability and operational reliability of the power system. It is called independent because in a competitive environment, the system must be operated in a manner that does not favor or penalize one market participant over another. ISOs normally own only the computing and communications assets required to monitor and control the power system. An ISO usually combines its system operation responsibility with the role of operator of the market of last resort. ISOs are also called **regional transmission organizations** (*RTOs*).

Transmission companies (*transcos*) own transmission assets such as lines, cables, transformers, and reactive compensation devices. They operate this equipment according to the instructions of the ISO.

Transmission System Operators (*TSOs*) combine the function of ISOs with the ownership of transmission assets.

Small consumers buy electrical energy from a retailer and lease a connection to the power system from their local distribution company. Their participation in the electricity market usually amounts to no more than choosing one retailer among others when they have this option. **Aggregators** contract with a number of small consumers to reduce or shift their demand in time on request. The combined effect is then sufficiently large to be sold on the wholesale market.

On the other hand, **large consumers** often have the skills and technical resources needed to trade directly on the wholesale electricity markets.

A **regulator** is a governmental body responsible for ensuring the fair and efficient operation of the electricity sector. It determines or approves the rules of the electricity market and investigates suspected cases of abuse of market power. The regulator also sets the prices for the products and services that are provided by monopolies. Regulatory functions are sometimes divided between two levels of government. For example, in the United States, the Federal Energy Regulatory Commission regulates interstate transmission and wholesale electricity markets, while the public utilities commission of each state regulates the retail markets and the distribution networks. In addition to their purely economic function, regulators also set the rules required to ensure the reliability and quality of the electricity supply. In some cases, the technical details of these rules are administered by a specialist organization, such as NERC in North America or ENTSO-E in Europe.

1.4 Competition and Privatization

In many countries, the introduction of competition in the supply of electricity has been accompanied by the privatization of some or all components of the industry. Privatization is the process by which publicly owned utilities are sold by the government to private investors. These utilities then become private, for-profit companies. Privatization is not, however, a prerequisite for the introduction of competition. None of the models of competition described above implies a certain form of ownership. Public utilities can coexist with private companies in a competitive environment.

1.5 Experience and Open Questions

In the monopoly utility model, all technical decisions regarding the operation and the development of the power system are taken within a single organization. In the short term, this means that, at least in theory, the operation of all the components of the system can be coordinated to achieve least cost operation. For example, the maintenance of the transmission system can be scheduled jointly with the maintenance of the generation units to co-optimize reliability and economy. Similarly, the long-term development of the system can be planned to ensure that the transmission capacity and topology match the generation capacity and location.

Introducing competition implies renouncing centralized control and coordinated planning. A single integrated utility is replaced by a constellation of independent companies, none of which has the responsibility to supply electrical energy to all the consumers. Each of these companies decides independently what it will do to maximize its private objectives. When the idea of competitive electricity markets was first mooted, it was rejected by many on the grounds that such a disaggregated system could not keep the lights on. There is now ample evidence to demonstrate that separating the operation of generation from that of the transmission system does not necessarily reduce the operational reliability of the overall system.

Having a separate, independent organization in charge of operating the power system has the significant advantage that it makes processes more open and transparent. Buyers and sellers have an interest in exploring how market rules and operational procedures can be improved to reduce costs and improve the profitability of their assets. This attitude has led to markets operating much closer to real time and to the development of "products" to accommodate the increasing amount of renewable energy sources as well as new technologies such as demand-side participation and energy storage.

Electricity markets have also grown geographically because bigger markets provide more trading opportunities and are thus more liquid and efficient. This growth happened either through additional participants joining an existing market or through the establishment of market coupling mechanisms. Increased trading opportunities result in more frequent and larger transactions between distant generators and loads. Such power flows increase the physical interdependence between parts of the grid that used to be loosely connected. Maintaining the stability and operational reliability of large interconnections under these conditions has forced system operators to enhance the scope and functionality of their data acquisition and analysis capabilities.

As we will discuss in Chapter 4, electricity markets have some unique characteristics that facilitate the abuse of market power. Many, if not most, electricity markets therefore have had to deal with the fact that they were often less than perfectly competitive. This has led to a number of inquiries by regulators, the creation of market monitoring bodies, the implementation of price caps, and other less controversial market power mitigation measures.

In terms of long-term development, the argument in favor of competition is that central planners always get their forecasts wrong. In particular, monopoly utilities have a tendency to overestimate the amount of generation capacity that will be needed. Their captive consumers are then obliged to pay for unnecessary investments. With the introduction of competition, it is hoped that the sum of the independent investment decisions of several profit-seeking companies will match the actual evolution of the demand more closely than the recommendations of a single planning department. In addition, underutilized investments by a company operating in a free market represent a loss for its owners and not a liability for its customers. Some markets rely entirely on the profits that power plants can obtain from the sale of energy and services to motivate investments in generation capacity. In other jurisdictions, market designers have introduced additional revenue streams to ensure that enough generation capacity is available to supply the load in a reliable manner. We will discuss this issue in more detail in Chapter 7.

Vertically integrated utilities can plan the development of their transmission network to suit the construction of new generating plants. In a competitive environment, the transmission company does not know years in advance where and when generating companies will build new plants. This uncertainty makes the transmission planning process much more difficult. Conversely, generating companies are not guaranteed that enough transmission capacity will become or will remain available for the output of their plants. Other companies may indeed build new plants in the vicinity and compete for the available transmission capacity.

The transmission and distribution networks have so far been treated as natural monopolies. Having two separate and competing sets of transmission lines or distribution feeders clearly does not make sense. From the economic and the reliability points of view, all lines, feeders, and other components should be connected to the same system. On the other hand, some economists and some entrepreneurs argue that not all these components must be owned by the same company. They believe that independent investors should have the opportunity to build new transmission facilities to satisfy specific needs that they have identified. Taken individually, such opportunities could be lucrative for the investors. However, the prevalent view is that such investments must take place within a framework that maximizes the overall benefits derived by all users of the network while minimizing their environmental impact.

Electricity is not a simple commodity whose trading is governed by the principles of classical economics. In addition to the need to maintain reliability, electricity markets are also affected by policy decisions driven by a desire to promote renewable energy sources and protect the environment, concerns about energy security and independence, as well as subsidies aimed at spurring the development of new technologies or helping a national industry.

1.6 Problems

1.1 Determine the electricity market structure that exists in your region or country or in another area where you have access to sufficient information. Discuss any difference that you observe between the basic model and the electricity market implementation in this area.

1.2 Identify the companies that participate in the electricity market in the area that you chose for Problem 1.1. Map the basic functions defined in this chapter with these companies and discuss any difference that you observe. Identify the companies that enjoy a monopoly status in some or all their activities.

1.3 Identify the regulatory agencies that oversee the electricity supply industry in the area that you chose for Problem 1.1.

1.4 Identify the organizations that fulfill the functions of market operator and system operator in the area that you chose for Problem 1.1.

1.5 Identify policies that have been implemented to promote the development of renewable energy sources in the area that you chose for Problem 1.1.

Further Reading

European Commission (1999). Opening up to choice – the single electricity market. https://bookshop.europa.eu/en/opening-up-to-choice-pbCS1798782 (accessed 28 February 2018).

Federal Energy Commission (2015). Energy primer: a handbook of energy market basics. www.ferc.gov/market-oversight/guide/energy-primer.pdf (accessed 28 February 2018).

Hunt, S. and Shuttleworth, G. (1996). *Competition and Choice in Electricity*. Chichester: Wiley.

2

Basic Concepts from Economics

2.1 Introduction

In this chapter, we introduce the concepts from the theory of microeconomics that are needed to understand electricity markets. We also take this opportunity to explain some of the economics terminology that has become common in the power system engineering literature. This chapter has a limited and utilitarian scope and does not pretend to provide a complete or rigorous course in microeconomics. The reader who feels the need or the inclination to study this subject in more depth is encouraged to consult a microeconomics textbook.

As we will see in the following chapters, electricity is not a simple commodity and electricity markets are more complex than markets for other products. To avoid unnecessary complications, we therefore introduce the basic concepts of microeconomics using examples that have nothing to do with electricity.

2.2 Fundamentals of Markets

Markets are a very old invention that can be found in most civilizations. Over the years, they have evolved from being simply a location where a few people would occasionally gather to barter goods to virtual environments where information circulates electronically and millions of dollars change hands at the click of a mouse. Despite these technological changes, the fundamental principle has not changed: a market is a place where buyers and sellers meet to see if deals can be made.

To explain how markets function, we will first develop a model that describes the behavior of the consumers. Then, we will develop a model explaining the activities of the producers. By combining these two models, we will be able to show under what conditions deals can be struck.

2.2.1 Modeling the Consumers

2.2.1.1 Individual Demand

Let us begin with a simple example: suppose that you work close enough to a farmers' market to be able to walk there during your mid-morning break. While the farmers sell different types of fruit and vegetables on this market, today you are looking at the apples.

Fundamentals of Power System Economics, Second Edition. Daniel S. Kirschen and Goran Strbac.

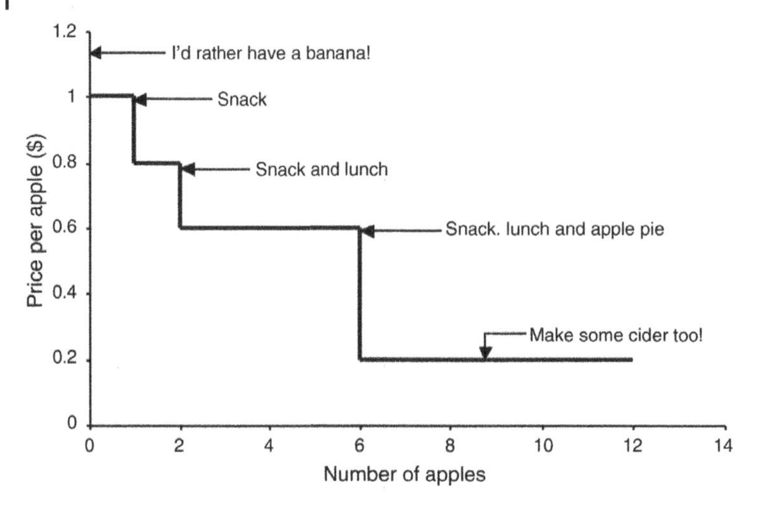

Figure 2.1 Typical relation between the price of apples and the demand of a particular customer.

The number of apples you purchase depends on their current price. There is certainly a price above which you will decide to forego your daily snack or buy another type of fruit instead. If the price is below that threshold but still quite high, you will probably buy only one apple to eat on your way back to work. If the price is lower still, you may buy one for now and another for lunch. At even lower prices, you may decide to purchase apples to make a pie for dinner. Finally, if the price is lower than you have ever seen it before, this may be the opportunity to experiment with the cider-making kit that your brother-in-law gave you for your last birthday. Figure 2.1 summarizes how your demand for apples varies with price. In other words, this curve represents the value that you place on each apple. You might argue that your decision to buy apples would also be influenced by the quality of those that are for sale. This is an important point and we will take care of it by assuming that all the non-price characteristics of the commodity considered (type, size, and quality) are precisely defined.

More generally, such curves show what the price should be for a consumer to purchase a certain amount of a particular *good* or *commodity*. Traditionally (and, at first, counter-intuitively), they are plotted with the price on the vertical axis and drawn assuming that the consumer's income and the price of other commodities remain constant.

2.2.1.2 Surplus

Let us suppose that when you get to the market, the price is $0.40 per apple. At that price, as Figure 2.2 shows, you decide to buy six apples. We can calculate the *gross consumer's surplus* that you, as a consumer, achieve by buying these apples. This represents the total value that you attach to the apples that you decide to purchase. The calculation goes as follows:

Value of the first apple:	1 × $1.00	=	$1.00
Value of the second apple:	1 × $0.80	=	$0.80
Value of the next four apples:	4 × $0.60	=	$2.40
Gross surplus:			**$4.20**

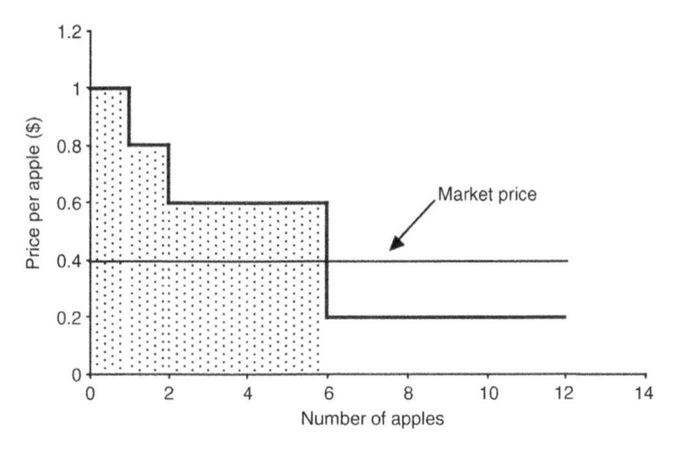

Figure 2.2 Gross surplus of purchasing apples.

As Figure 2.2 shows, your gross consumer's surplus is equal to the area under the curve. However, you have had to pay $6 \times \$0.40 = \$2.40$ to purchase these apples, and this represents money that you no longer have to purchase other goods. We define the *net consumer's surplus* as the difference between the gross consumer's surplus and the expense of purchasing the goods. Graphically, as Figure 2.3 illustrates, the net consumer's surplus is equal to the area between the inverse demand curve and the horizontal line at the market price. The net consumer's surplus represents the "extra value" that you get from being able to buy all the apples at the same market price, even though the value you attach to some of them is higher than the market price.

2.2.1.3 Demand and Inverse Demand Functions

It is very unlikely that all the consumers going to the market have exactly the same appetite for apples as you do. Some consumers would pay much more for the same number of apples while others buy apples only when they are cheap. If we aggregate the demand characteristics of a sufficiently large number of consumers, the discontinuities introduced by the individual decisions are smoothed away, leading to a curve like the one shown in

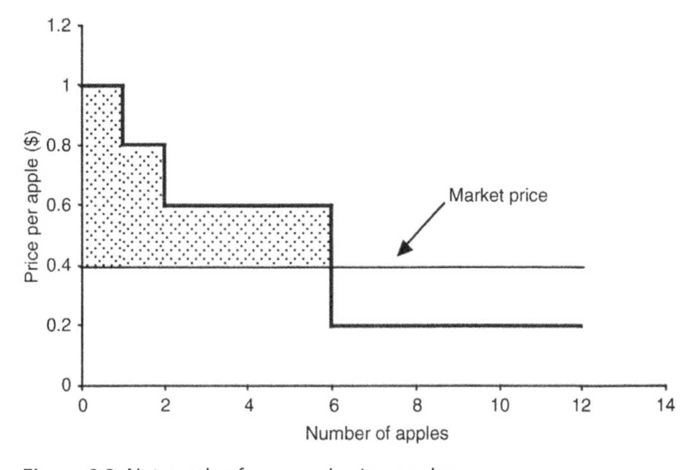

Figure 2.3 Net surplus from purchasing apples.

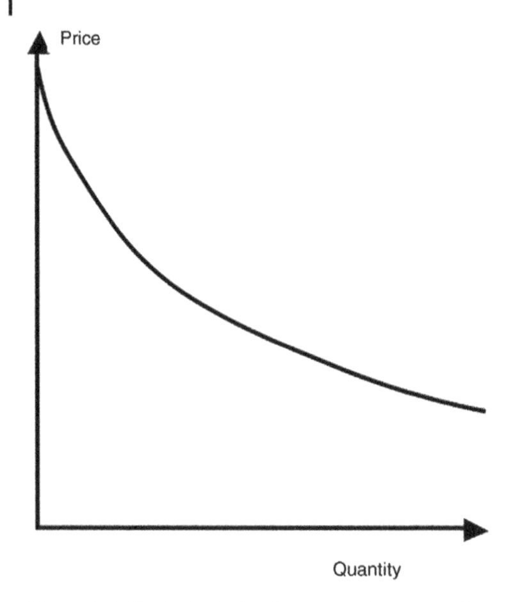

Figure 2.4 Typical relation between the price of a commodity and the demand for this commodity by a group of consumers. This curve is called the inverse demand function or the demand function depending on the perspective adopted.

Figure 2.4. This curve represents the *inverse demand function* for this good by this group of consumers. If q denotes the quantity purchased and π the price of the commodity, we can write:

$$\pi = D^{-1}(q) \tag{2.1}$$

If we look at the same curve from the other direction, we have the *demand function* for this commodity:

$$q = D(\pi) \tag{2.2}$$

For most, if not all, practical commodities, the demand function is downward sloping, i.e. the amount consumed decreases as the price increases. The inverse demand function has an important economic interpretation. For a given consumption level, it measures how much money the consumers are willing to pay to purchase a small additional amount of the good considered. Not spending this amount of money on this commodity would allow them to purchase more of another commodity or save it for purchasing something else at a later date. In other words, the demand curve gives the *marginal value* that consumers attach to the commodity. The typical downward-sloping shape of the curve indicates that consumers are usually willing to pay more for additional quantities when they have only a small amount of a commodity, i.e. their marginal willingness to pay decreases as their consumption increases.

The concepts of gross and net consumer's surplus that we defined for a single consumer can be extended to the gross and net surpluses of a group of consumers. As Figure 2.5 illustrates, the gross surplus is represented graphically by the area below the inverse demand function up to the quantity that the consumers purchase at the current market price. The net surplus corresponds to the area between the inverse demand function and the horizontal line at the market price.

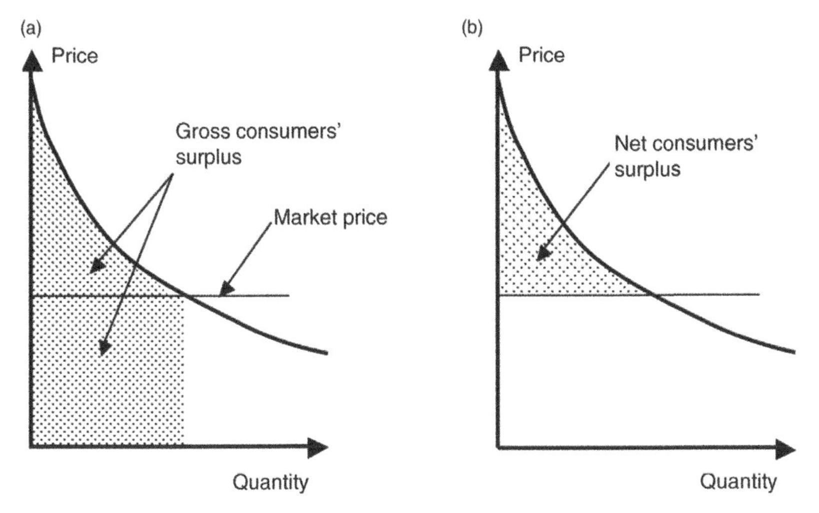

Figure 2.5 Gross consumers' surplus (a) and net consumers' surplus (b).

The concept of net surplus is much more important than the calculation of an absolute value for this quantity. Calculating the absolute value of the net surplus is quite difficult because the inverse demand function is not known accurately. Examining how this net surplus varies with the market price is much more interesting. Figure 2.6 illustrates the change in net surplus when the market price increases. If the market price is π_1, the consumers purchase a quantity q_1 and the net surplus is equal to the shaded area. If the price increases to π_2, the consumption level decreases to q_2 and the consumers' net surplus is reduced to the roughly triangular area labeled A. Two effects contribute to this reduction in net surplus. First, because the price is higher, consumption decreases from q_1 to q_2. This loss of net surplus or welfare is equal to the area labeled C. Second, because consumers have to pay a higher price for the quantity q_2 that they still purchase, they lose an additional amount of welfare represented by the area labeled B.

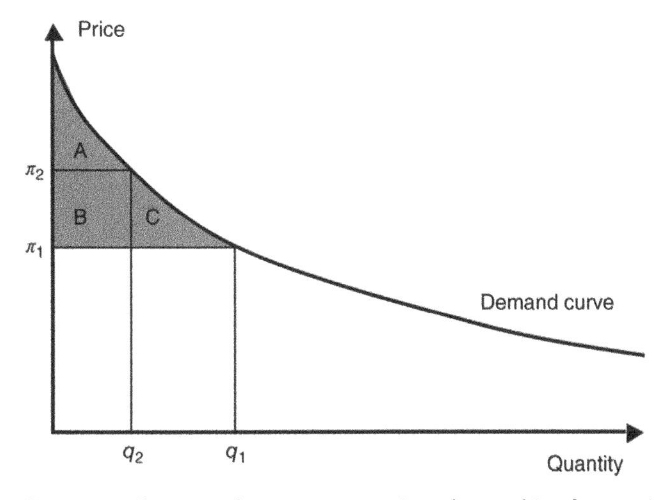

Figure 2.6 Change in the net consumers' surplus resulting from an increase in the market price.

2.2.1.4 Elasticity of Demand

Increasing the price of a commodity even by a small amount will clearly decrease the demand. But by how much? To answer this question, we could use the derivative ${}^{dq}/_{d\pi}$ of the demand curve. However, if we use this slope directly, the numerical value depends on the units that we use to measure the quantity and the price. Comparing the demand's response to price changes for various commodities would then be impossible. Similarly, comparing this response to changes in prices expressed in different currencies would also be impossible. To get around these difficulties, we define the *price elasticity of the demand* as the ratio of the relative change in demand to the relative change in price:

$$\varepsilon = \frac{\frac{dq}{q}}{\frac{d\pi}{\pi}} = \frac{\pi}{q}\frac{dq}{d\pi} \tag{2.3}$$

The demand for a commodity is said to be *elastic* if a given percentage change in price produces a larger percentage change in demand. On the other hand, if the relative change in demand is smaller than the relative change in price, the demand is said to be *inelastic*. Finally, if the elasticity is equal to −1, the demand is *unit elastic*.

The elasticity of the demand for a commodity depends, in large part, on the availability of substitutes. For example, the elasticity of the demand for coffee would be much smaller if consumers did not have the option to drink tea. When discussing elasticities and substitutes, one has to be clear about the time scale for substitutions. Suppose that electric heating is widespread in a region. In the short run, the price elasticity of the demand for electricity will be very low because consumers do not have a choice if they want to stay warm. In the long run, however, they can install gas-fired heating and the price elasticity of the demand for electricity will be much higher.

The concept of substitute products can be quantified by defining the *cross-elasticity* between the demand for commodity i and the price of commodity j:

$$\varepsilon_{ij} = \frac{\frac{dq_i}{q_i}}{\frac{d\pi_j}{\pi_j}} = \frac{\pi_j}{q_i}\frac{dq_i}{d\pi_j} \tag{2.4}$$

While the elasticity of a commodity to its own price (its *self-elasticity*) is always negative, cross-elasticities between substitute products are positive because an increase in the price of one will spur the demand for the other. If two commodities are *complements*, a change in the demand for one will be accompanied by a similar change in the demand for the other. Electricity and electric heaters are clearly complements. The cross-elasticities of complementary commodities are negative.

2.2.2 Modeling the Producers

2.2.2.1 Opportunity Cost

Our model of the consumers' behavior is based on the assumption that consumers can choose how much of a commodity they purchase. We also argued that the consumption level is such that the marginal benefit that consumers get from this commodity is equal to the price that they have to pay to obtain it. A similar argument can be used to develop our model of the producers.

Let us consider one of the apple growers who bring their products to the market that we visited earlier. There is a price below which she will decide that selling apples is not worthwhile. There are several reasons why she could conclude that this revenue is insufficient. First, it might be less than the cost of producing the apples. Second, it might be less than the revenue she could get by using these apples for some other purposes, such as selling them to a cider-making factory. Finally, she could decide that she would rather devote the resources needed to produce apples (money, land, machinery, and her own time) into some other activity, such as growing pears or opening a bed-and-breakfast. One can summarize these possibilities by saying she will not produce apples if the revenue from their sale is less than the *opportunity cost* associated with their production.

2.2.2.2 Supply and Inverse Supply Functions

On the other hand, if the market price for apples is higher, our producer may decide that it is worthwhile to increase the quantity of apples that she brings to the market. Other producers have different opportunity costs and will therefore decide to adjust the amount they supply at different price thresholds. If we aggregate the amounts supplied by a sufficiently large number of producers, we get a smooth, upward-sloping curve such as the one shown in Figure 2.7. This curve represents the *inverse supply function* for this commodity:

$$\pi = S^{-1}(q) \tag{2.5}$$

This function indicates the value that the market price should take to make it worthwhile for the aggregated producers to supply a certain quantity of the commodity to the market. We can, of course, look at the same curve from the other direction and define the *supply function*, which gives us the quantity supplied as a function of the market price:

$$q = S(\pi) \tag{2.6}$$

Goods produced by different producers (or by the same producer but using different means of production) are located on different parts of the supply curve. The marginal

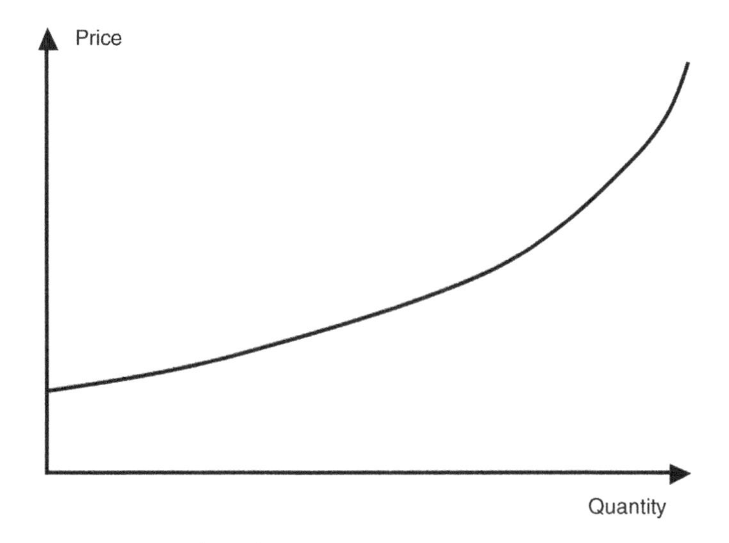

Figure 2.7 Typical supply curve.

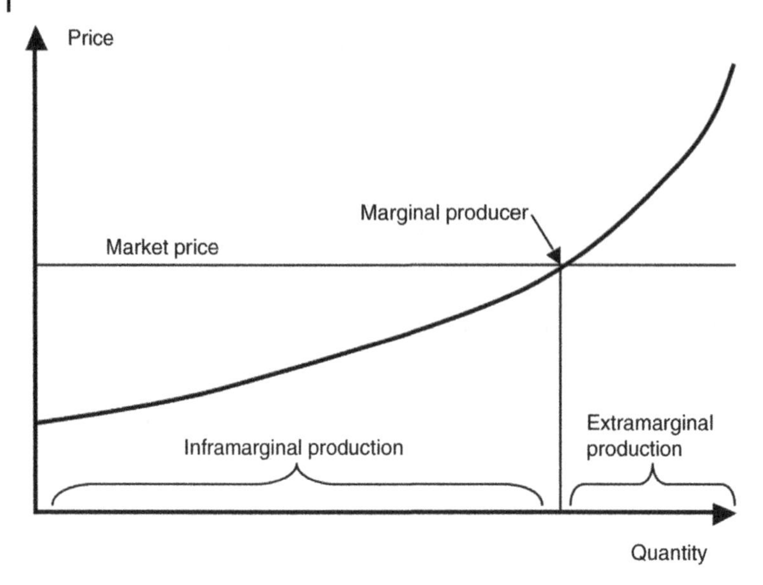

Figure 2.8 Marginal production is such that its opportunity cost is equal to the market price.

producer is the producer whose opportunity cost is equal to the market price. As Figure 2.8 illustrates, if this market price decreases even by a small amount, this producer would decide that it is not worthwhile to continue producing this good. Extramarginal production refers to production that could become worthwhile if the market price were to increase. On the other hand, the opportunity cost of the inframarginal producers is below the market price. These producers are thus able to sell at a price that is higher than the lowest price at which they would find it worthwhile to produce.

2.2.2.3 Producers' Revenue

Since the entire supply of the commodity is traded at the same price, the *producers' revenue* is equal to the product of the traded quantity q_1 and the market price π_1. This quantity is thus equal to the shaded area in Figure 2.9. The *producers' net surplus* or *producers' profit* arises from the fact that all the goods (except for the marginal production) are traded at a price that is higher than their opportunity cost. As Figure 2.10 shows, this net surplus or profit is equal to the area between the supply curve and the horizontal line at the market price. Producers with a low opportunity cost capture a proportionately larger share of this profit than those who have a higher opportunity cost. The marginal producer does not reap any profit.

Figure 2.11 shows that an increase in the market price from π_1 to π_2 affects the net producers' surplus in two ways. It increases the quantity that they supply to the market from q_1 to q_2 (area labeled C) and increases the price for the quantity supplied to the market at the original price (area labeled B).

2.2.2.4 Elasticity of Supply

An increase in the price of a commodity encourages suppliers to make available larger quantities of this commodity. The *price elasticity of supply* quantifies this relation. Its

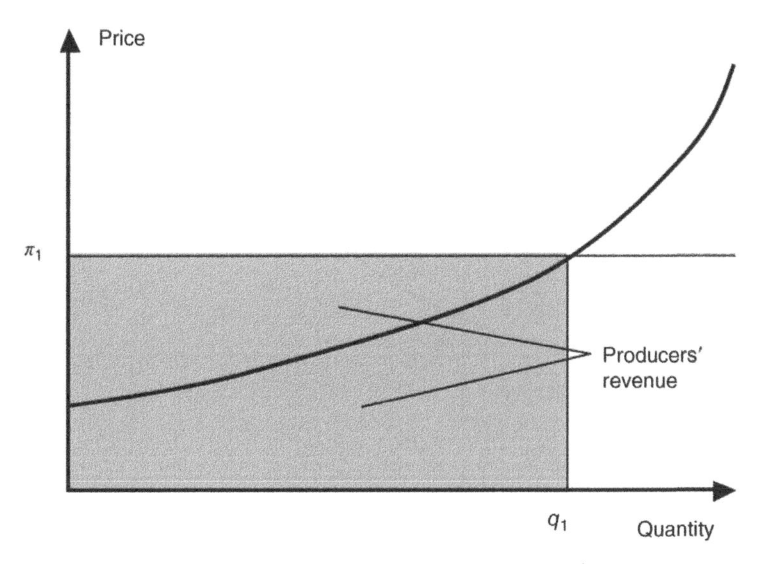

Figure 2.9 The producers' revenue is equal to the product of the market price π_1 and the traded quantity q_1.

definition is similar to that of the price elasticity of the demand, but it involves the derivative of the supply curve rather than that of the demand curve:

$$\varepsilon = \frac{\frac{dq}{q}}{\frac{d\pi}{\pi}} = \frac{\pi}{q}\frac{dq}{d\pi} \tag{2.7}$$

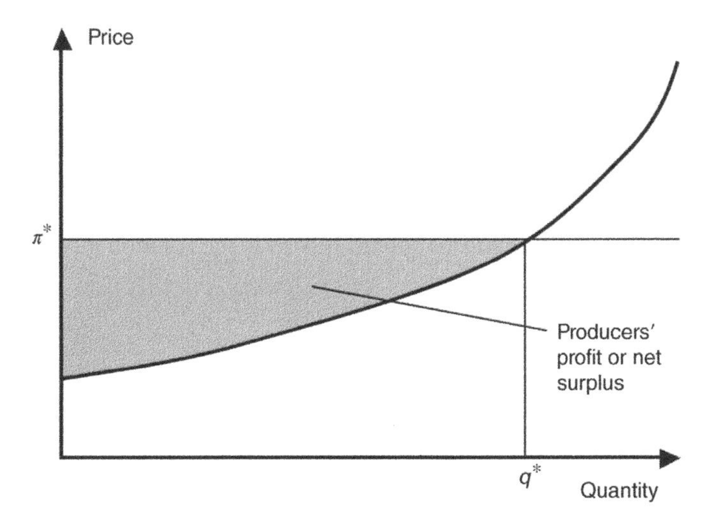

Figure 2.10 The producers' profit or net surplus arises because inframarginal producers are able to sell at a price higher than their opportunity cost.

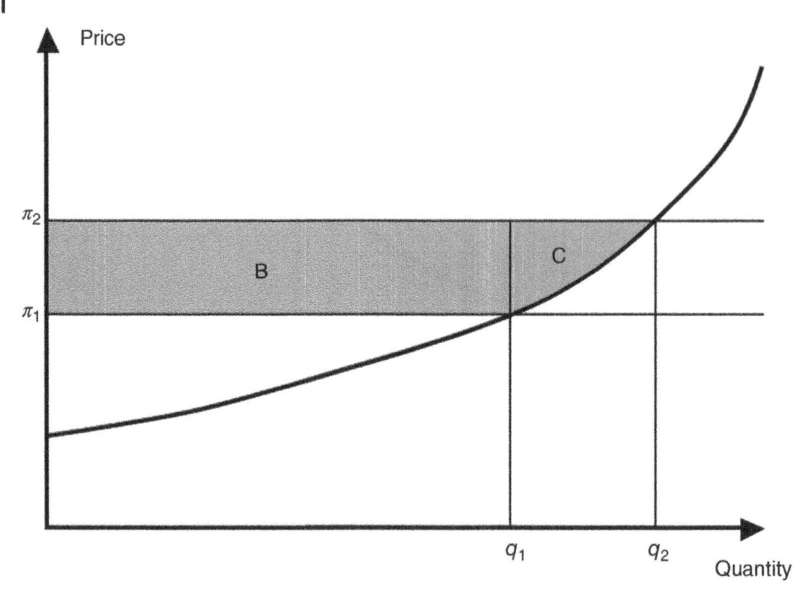

Figure 2.11 Change in the producers' profit or net surplus when the market price changes.

The elasticity of supply is always positive. It will usually be higher in the long run than in the short run because suppliers have more opportunities to increase the means of production.

2.2.3 Market Equilibrium

So far, we have considered producers and consumers separately. It is time to see how they interact in a market. In this section, we will make the assumption that each supplier or consumer cannot affect the price by its individual actions. In other words, all market participants take the price as given. When this assumption is true, the market is said to be a *perfectly competitive market*. However, we should already note that this assumption is usually not true for electricity markets. We will discuss in a later section how markets operate when some participants can influence the price through their actions.

In a competitive market, it is the combined action of all the consumers on one side and of all the suppliers on the other that determines the price. The *equilibrium price* or *market clearing price* π^* is such that the quantity that the suppliers are willing to provide is equal to the quantity that the consumers wish to obtain. It is thus the solution of the following equation:

$$D(\pi^*) = S(\pi^*) \tag{2.8}$$

This equilibrium can also be defined in terms of the inverse demand function and the inverse supply function. The equilibrium quantity q^* is such that the price that the consumers are willing to pay for that quantity is equal to the price that producers must receive to supply that quantity:

$$D^{-1}(q^*) = S^{-1}(q^*) \tag{2.9}$$

Figure 2.12 illustrates these concepts.

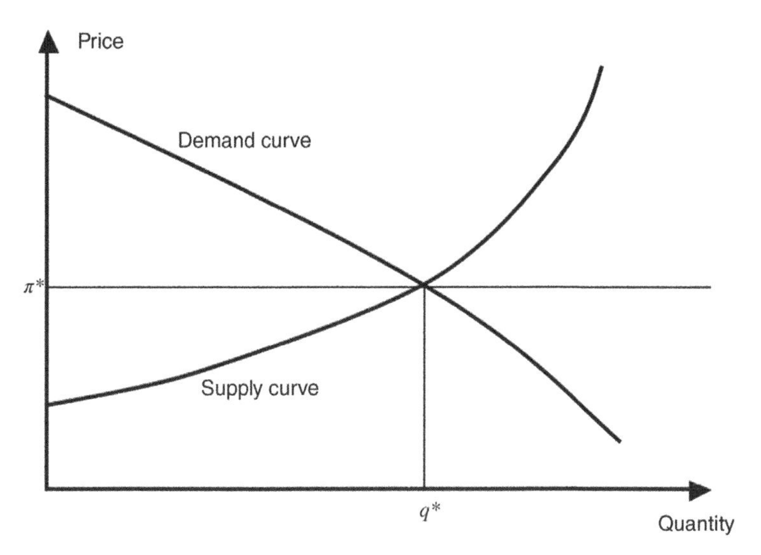

Figure 2.12 Market equilibrium.

So far, we have shown that at the market equilibrium, the behaviors of the consumers and the suppliers are consistent. We have not yet shown, however, that this point represents a stable equilibrium. Let us demonstrate that the market will inevitably settle at that point. Suppose, as shown in Figure 2.13, that the market price is $\pi_1 < \pi^*$ where the demand is greater than the supply. Some suppliers will inevitably realize that there are some unsatisfied customers to whom they could sell their goods at more than the going price. The traded quantity will increase and so will the price until the equilibrium conditions are reached. Similarly, if the market price is $\pi_2 > \pi^*$, the supply exceeds the demand and some suppliers are left with goods for which they cannot find

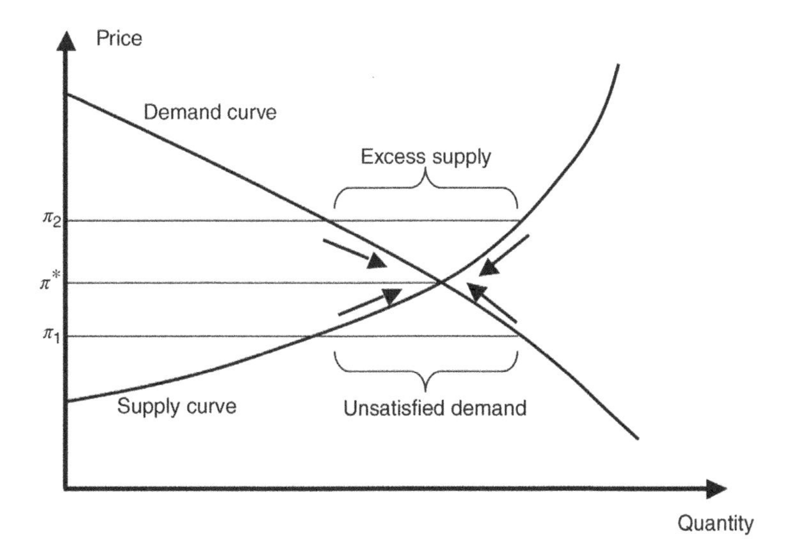

Figure 2.13 Stability of the market equilibrium.

buyers. Since this is unsustainable, they will reduce their production until the amount that producers are willing to sell is equal to the amount that consumers are willing to buy.

2.2.4 Pareto Efficiency

When a system is completely under the control of a single organization, this organization normally attempts to optimize some measure of the benefit it derives from the system. On the other hand, when a system depends on the interactions of various parties with diverging interests, conventional optimization is not applicable and must be replaced by the concept of *Pareto efficiency*. An economic situation is said to be Pareto efficient if the benefit derived by any of the parties can be increased only by reducing the benefit enjoyed by one of the other parties.

The equilibrium situation in a competitive market is Pareto efficient in terms of both the quantity of goods exchanged and the allocation of these goods. Let us first consider the quantity exchanged with the help of Figure 2.14. Suppose that the quantity exchanged is q, which is less than the equilibrium quantity q^*. At that quantity, there is someone willing to sell extra units of the good considered at a price π_1 that is less than the price π_2 that someone else is willing to pay for that extra unit. If a trade can be arranged between these two parties at any price between π_1 and π_2, both parties will be better off. Thus, if the total amount traded is less than the equilibrium q^*, the situation is not Pareto efficient. Similarly, any amount in excess of the equilibrium value is not Pareto efficient because the price that someone would be willing to pay for an extra unit is lower than the price that it would take to get it supplied.

Let us now consider the efficiency of the allocation of goods. In a competitive market, all units of a given commodity are traded at the same price and this price represents the marginal rate of substitution between this good and all other goods. Consumer A's willingness to pay this price means that he values the last unit he purchased of this good more than other goods that he could purchase. On the other hand, consumer B may

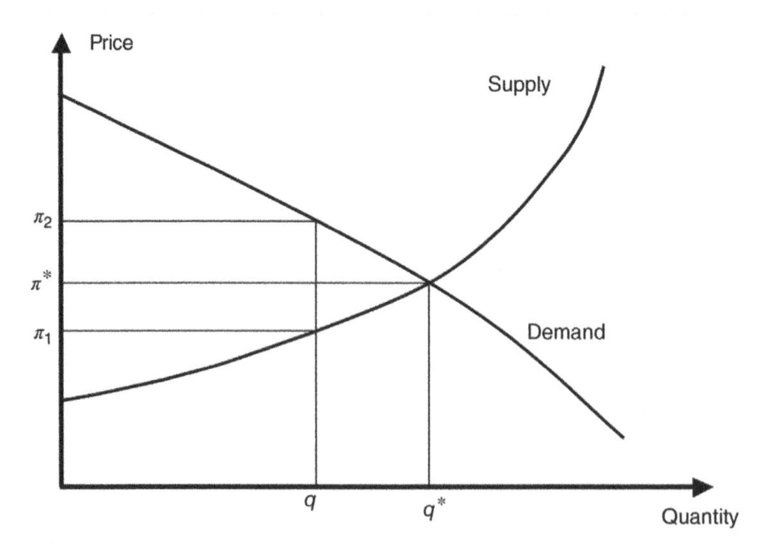

Figure 2.14 Pareto efficiency of the market equilibrium.

decide that at that price, she would rather buy other goods. Suppose now that the goods are not allocated on the basis of the willingness to pay the market price but on some other basis. Consumer A may find himself in a situation where he may be willing to pay $10 to buy an extra unit of the good in addition to those that he has been allocated. On the other hand, consumer B may have received an allocation such that she only values the last unit that she has received at $8. Since these two consumers place different values on one unit of the same good, they would therefore both be better off if they could trade this unit at any price between $8 and $10. It is thus only when goods are allocated on the basis of a single marginal rate of substitution, as happens in a competitive market, that Pareto efficiency is achieved.

2.2.5 Global Welfare and Deadweight Loss

The sum of the net consumers' surplus and of the producers' profit is called the *global welfare*. It quantifies the overall benefit that arises from trading. We will now show that the global welfare is maximum when a competitive market is allowed to operate freely and the price settles at the intersection of the supply and demand curves. Under these conditions, Figure 2.15 shows that the consumers' surplus is equal to the sum of the areas labeled A, B, and E and the producers' profit to the sum of the areas labeled C, D, and F.

External intervention sometimes prevents the price of a good from settling at the equilibrium value that would result from a free and competitive market. For example, in an effort to help producers, the government could set a minimum price for a commodity. If this price is set at a value π_2 that is higher than the competitive market clearing price π^*, this minimum price becomes the market price and consumers reduce their consumption from q^* to q. Under these conditions, the net consumers' surplus shrinks to area A while the net producers' surplus is represented by the sum of the areas B, C, and D.

Similarly, the government could enforce a maximum price for a good. If this price is set at a value π_1 that is lower than the competitive market clearing price π^*, producers will cut

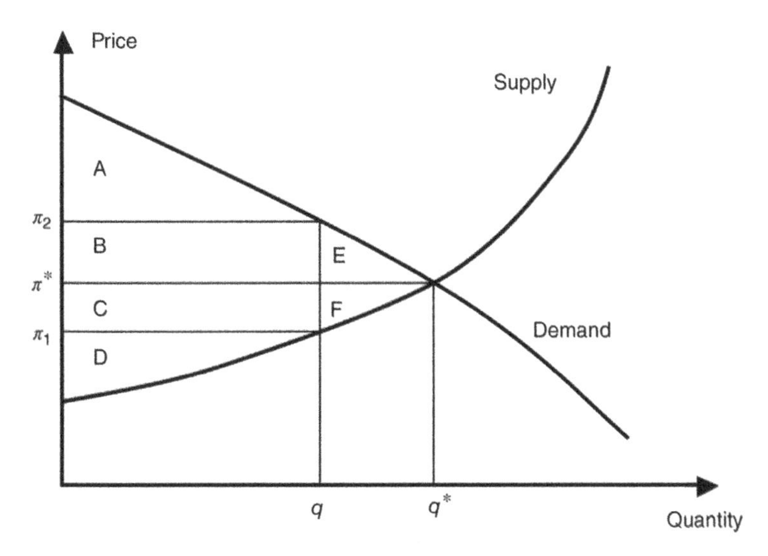

Figure 2.15 Deadweight loss.

their output to q. In this case, the consumers enjoy a net surplus equal to the sum of areas A, B, and C, while the net producers' surplus is only area D.

Finally, the government could decide to tax this commodity. If we assume that the tax is passed on to the consumers in its entirety, it creates a difference between the price paid by the consumers (say π_2) and the price received by the producers (say π_1). The government collects the difference $\pi_2 - \pi_1$ for each unit traded. Under these conditions, the demand again drops from q^* to q, the net consumers' surplus contracts to area A and the net producers' surplus to area D. The total amount collected by the government in taxes is equal to the sum of areas B and C. In each case, the external intervention redistributes the global welfare in favor of the producers, the consumers, or the government, respectively. Unfortunately, all these interventions have the undesirable side effect of reducing the global welfare by an amount equal to the sum of the areas labeled E and F. This drop in global welfare is called the *deadweight loss* and is the result of the reduction in the amount traded caused by the price distortion. Note that for simplicity we have assumed the same drop in demand for all three forms of external intervention. Obviously, this does not have to be the case.

We will see in later chapters that, in some markets, the price of electrical energy is set through a centralized calculation and not through the direct interaction of producers and consumers. To maximize the benefits of trading, this centralized calculation should simulate the operation of a free market and maximize the global welfare.

2.2.6 Time-varying Prices

Demand functions for particular goods go up or down over time because consumers' preferences are affected by fashion or changes in needs. Similarly, supply functions drop if the technology used to produce the goods improves. On the other hand, they rise if the cost of labor or of raw materials increases. These changes affect the market price and the quantity transacted. For example, the marginal cost of supplying strawberries is low in early summer because they ripen naturally at that time of the year. The price of a punnet of strawberries is therefore much lower in summer than when they have to be grown in greenhouses or imported from warmer regions. Letting the market price vary as external conditions change is economically efficient because it encourages the most efficient use of resources, which is another way of saying that we should eat strawberries in summer and apples in winter and not the opposite.

On the other hand, regulatory agencies sometimes determine that the price for a particular good or service should remain constant, irrespective of how the demand or supply for this good or service might vary over time. For example, in most large cities, conventional taxis operate on this basis. The argument in favor of this approach is that it provides certainty for the consumers, i.e. a given trip costs the same today as it did yesterday. The price per kilometer can also be set in such a way that a fixed number of taxi operators can make a living from providing the service. The downside of this approach is that it does not encourage economically efficient behavior by the producers or the consumers. For example, allowing the price to rise when demand is high would encourage more taxis to become available and consumers to use alternative modes of transportation.

The price of electrical energy at the retail level does not typically vary with the supply or the demand. We will discuss in later chapters the benefits of letting this price fluctuate.

2.3 Concepts from the Theory of the Firm

Let us now take a more detailed look at the behavior of firms that produce the goods that are traded on the market.

2.3.1 Inputs and Outputs

For the sake of simplicity, we consider a firm that produces a quantity y of a single good. In order to produce this output, our firm needs some inputs, which are called *factors of production*. Factors of productions vary widely depending on the output produced by the firm. They can be classified into broad categories such as raw materials, labor, land, buildings, or machines. Let us assume that our firm needs only two factors of production. The output is related to the input by a *production function* that reflects the technology used by the firm in the production of the good:

$$y = f(x_1, x_2) \tag{2.10}$$

For example, y might represent the amount of wheat produced by a farmer, with x_1 being the amount of fertilizer and x_2 the surface of land that this farmer uses to raise wheat.

To get some insight into the shape of the production function, let us keep the second factor of production constant and progressively increase the first. At the beginning, the output y increases with x_1. However, for almost all goods and technologies, the rate of increase of y decreases as x_1 gets larger. This phenomenon is called the *law of diminishing marginal product.*

In our example, the yield of a fixed amount of land cultivated by our farmer will go up as he increases the amount of fertilizer. However, above a certain density, the effectiveness of fertilizer declines. Similarly, cultivating more land increases the total amount of wheat produced. However, as the amount of land goes up, the rate of increase in output will inevitably decrease, as the fixed amount of fertilizer must be spread over a larger area.

2.3.2 Long Run and Short Run

Some factors of production can be adjusted faster than others. For example, a horticulturist can increase her production of apples by increasing the amount of organic fertilizer she uses or by hiring more labor to harvest the fruit. The effect of these adjustments will be felt at the next harvest. She could also increase production by planting more trees. In this case, the results will not materialize until these new trees have had time to mature, a process that obviously takes a few years.

There is, however, no specific deadline separating the short and long runs. Economists define the long run as being a period of time sufficiently long as to allow all factors of productions to be adjusted. On the other hand, in the short run, some of the factors of production are fixed. For example, if we assume that the second factor of production has a fixed value $\overline{x_2}$, the production function becomes a function of a single variable:

$$y = f(x_1, \overline{x_2}) \tag{2.11}$$

Figure 2.16 shows the shape of a typical short-run production function.

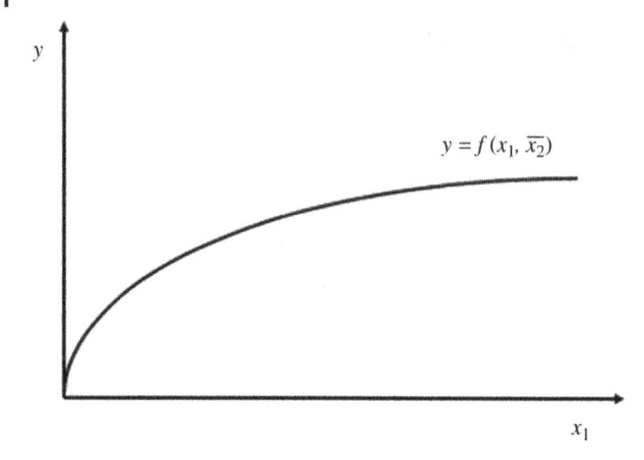

Figure 2.16 Typical short-run production function.

Since in the short run the output often depends on a single production factor, it is convenient to define the input–output function, which is the inverse of the production function:

$$x_1 = g(y) \text{ for } x_2 = \overline{x_2} \tag{2.12}$$

The input–output function indicates how much of the variable production factor is required to produce a specified amount of goods. For example, the input–output curve of a thermal power plant shows how much fuel is required every hour to produce a given amount of power using this plant. We can then define the short-run cost function:

$$c_{SR}(y) = w_1 \cdot x_1 + w_2 \cdot \overline{x_2} = w_1 \cdot g(y) + w_2 \cdot \overline{x_2} \tag{2.13}$$

where w_1 and w_2 are the unit costs of the factors of production x_1 and x_2. Figure 2.17a illustrates a typical short-run cost function. The convexity of this function is due to the law of diminishing marginal products. Because of this convexity, the derivative of the cost function, which is called the *marginal cost function,* is a monotonically increasing function of the quantity produced. Figure 2.17b shows the marginal cost function corresponding to the cost function of Figure 2.17a. Note that if the cost of production is expressed in \$, the marginal cost is expressed in \$ per unit produced. At a given level of production, the numerical value of the marginal cost function is equal to the cost of producing one more unit of the good.

Using these functions, we can determine the short-run behavior of a firm in a perfectly competitive market. In such a market, no firm can influence the market price. Therefore, the only action that a firm can take to maximize its profits is to adjust its output. Since profit is defined as the difference between the firm's revenues and costs, the optimal level of production is given by:

$$\max_{y} \{\pi \cdot y - c_{SR}(y)\} \tag{2.14}$$

At the optimum, we must have:

$$\frac{d\{\pi \cdot y - c_{SR}(y)\}}{dy} = 0$$

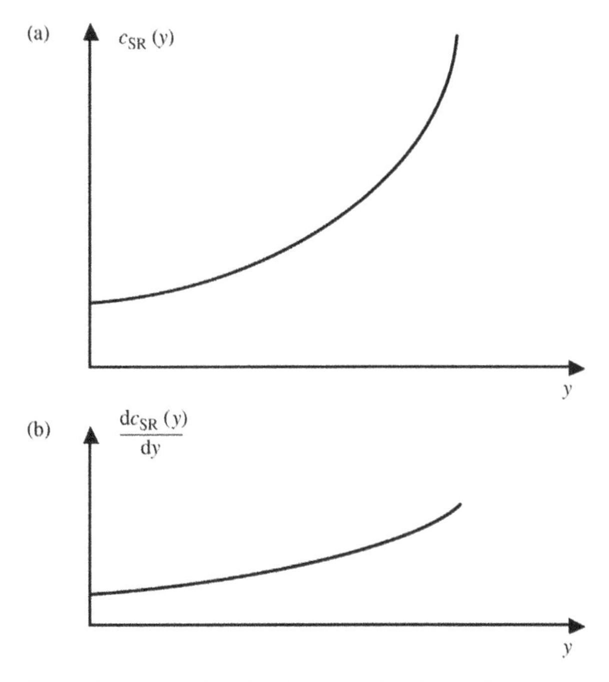

Figure 2.17 (a) Typical short-run cost function and (b) corresponding short-run marginal cost function.

or

$$\pi = \frac{dc_{SR}(y)}{dy} \tag{2.15}$$

The firm will thus increase its production up to the point where its short-term marginal cost of production is equal to the market price. Gaining an intuitive understanding of this relation is useful. If the firm were operating at a point where its marginal cost of production is less than the current market price, it could increase its profits by producing another unit and selling it on the market. Similarly, if the firm's marginal cost of production were higher than the market price, it would save money by not producing the last unit it sold.

Defining a long-run cost function is more complicated because in the long run the firm has more flexibility in deciding how it will produce. For example, a firm could decide to buy more expensive machines and reduce its labor costs or vice versa. The production function therefore cannot be treated as a function of a single variable. However, we can assume that the firm behaves in an optimal manner. By this we mean that in the long run it chooses the combination of factors that produces any quantity of goods at minimum cost. The long-run cost function is thus the solution of an optimization problem and can be expressed as follows:

$$c_{LR}(y) = \min_{x_1, x_2} (w_1 \cdot x_1 + w_2 \cdot x_2) \quad \text{such that} \quad f(x_1, x_2) = y \tag{2.16}$$

where, for the sake of simplicity, we have considered only two factors of production.

In the first part of this book, we will use short-run cost functions because we will be concerned with the operation of an existing power system. Long-run cost functions will be used in later chapters of this book when we consider the expansion of the power system.

2.3.3 Costs

In this section, we define the various components of the production cost and introduce various curves that are used to characterize these costs.

In the short run, some factors of production are fixed. The cost associated with these factors does not depend on the amount produced and is thus a *fixed cost*. For example, if a generating company has bought land and built a power plant on this land, the cost of the land and the plant do not depend on the amount of energy that this plant produces. On the other hand, the quantity of fuel consumed by this plant and, to a certain extent, the manpower required to operate it depend on the amount of energy it produces. Fuel and manpower costs are thus examples of *variable costs*. There is also a third class of costs called *quasi-fixed costs*. These are costs that the firm incurs if the plant produces any amount of output but does not incur if the plant produces nothing. For example, in the case of a generating plant, the cost of the fuel required to startup the plant is fixed in the sense that it does not depend on the amount of energy that the plant produces once it has been synchronized to the grid. However, this startup cost does not need to be paid if the plant stays idle.

In the long run, there are no fixed costs because the firm can decide the amount of money it spends on all production factors. At the limit, the firm's long-run costs can be zero if it decides to produce nothing and go out of business. A *sunk cost* is the difference between the amount of money a firm pays for a production factor and the amount of money it would get back if it sold this asset. For example, in the case of a power plant, the cost of the land on which the plant is built is not a sunk cost because land can be resold. It is thus a *recoverable cost*. On the other hand, if production with this plant is no longer profitable, the difference between the cost of building the plant and its scrap metal value is a sunk cost.

2.3.3.1 Short-run Costs

If we assume that the costs of the production factors are constant, the cost functions defined in the previous section can be expressed as a function of the level of output y:

$$c(y) = c_v(y) + c_f \tag{2.17}$$

where $c_v(y)$ represents the variable costs and c_f represents the fixed costs. The *average cost function* measures the cost per unit of output. It is equal to the sum of the *average variable cost* and the *average fixed cost:*

$$\mathrm{AC}(y) = \frac{c(y)}{y} = \frac{c_v(y)}{y} + \frac{c_f}{y} = \mathrm{AVC}(y) + \mathrm{AFC}(y) \tag{2.18}$$

Let us sketch what these average cost curves might look like. Since the fixed costs do not depend on the production, the average fixed cost will be infinite for a zero output. As production increases, these fixed costs are spread over a larger output. The average fixed cost curve is thus a monotonically decreasing function, as shown in Figure 2.18a. For

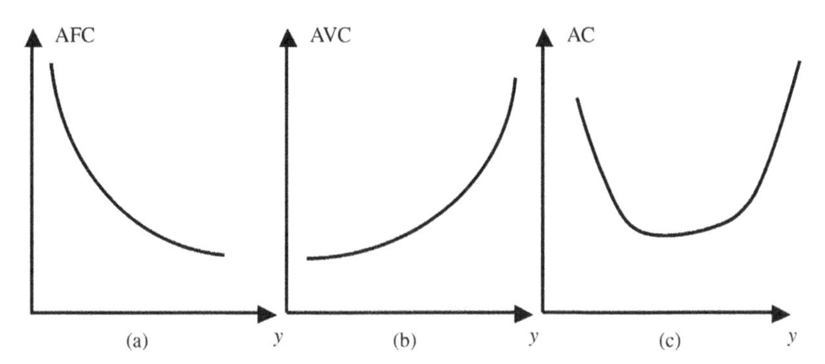

Figure 2.18 Typical shapes of the average fixed cost function (AFC) (a), the average variable cost function (AVC) (b), and the average cost function (AC) (c).

modest production levels, variable costs typically increase linearly with the output. The average variable cost is then constant. However, as the production increases, the fixed factors of production start affecting the efficiency of the production of goods. Figure 2.18b shows that the average variable cost eventually and inevitably rises. For example, the output of a manufacturing plant can often be increased beyond the capacity for which the plant was designed. However, this might require paying the workers higher overtime rates, maintaining the machines more frequently and generally adopting less efficient procedures. Similarly, in the case of generating plants, the maximum efficiency is usually achieved for an output that is somewhat below the maximum capacity of the plant. The average cost curve combines these two effects and has the typical U-shape shown in Figure 2.18c.

It is essential to understand the difference between the average and the marginal costs. Both quantities are expressed in $ per unit produced, but the marginal cost reflects only the cost of the last unit produced. On the other hand, the average cost factors in the cost of all the units already produced. Since the fixed costs are constant, they do not contribute to the marginal cost. Figure 2.19 illustrates the relation between the marginal cost curve and

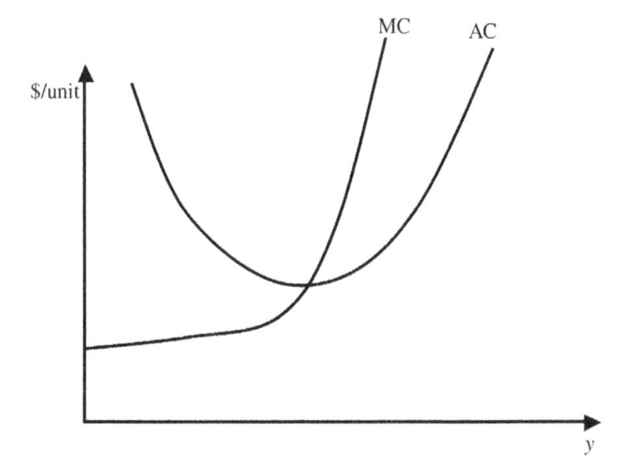

Figure 2.19 Typical relation between the average cost curve (AC) and the marginal cost curve (MC).

the average cost. For low production levels, the marginal cost is smaller than the average cost because of the influence of fixed costs. On the other hand, for high production levels, the marginal cost is higher than the average cost. The marginal cost curve intersects the average cost curve at its minimum.

2.3.3.2 Long-run Costs

We have argued above that, in the long run, there are no fixed costs because all the factors of production can be changed and the firm has the option to produce nothing and get out of business. However, the technology may be such that some costs are incurred independently of the level of production. There may therefore be some quasi-fixed costs in the long run. The long-run average cost curve therefore tends to have a U-shape, as shown in Figure 2.20.

What can we say about the relation between the short-run cost and the long-run cost? In the long run, we can minimize the production cost for any level of output because we can adjust all the factors of production. On the other hand, in the short run, some of the production factors are fixed. The short-run production cost is therefore equal to the long-run production cost only for the value of output y^* for which the fixed production factors were optimized. For other levels of output, the short-run cost is higher than the long-run cost. The short-run average cost curve is therefore above the long-run average cost curve, except for the output for which the fixed production factors have been optimized. At that point, the two curves are tangent, as shown in Figure 2.20. We could, of course, select other sets of fixed production factors that would minimize the production cost for other values of the output $y_1, y_2, \ldots, y_n$. In other words, we could build plants with other capacities. For each plant size, the short-run average cost would be equal to the long-run average cost only for the designed plant capacity. As Figure 2.21 shows, the long-run average cost curve is therefore the lower envelope of the short-run average cost curves.

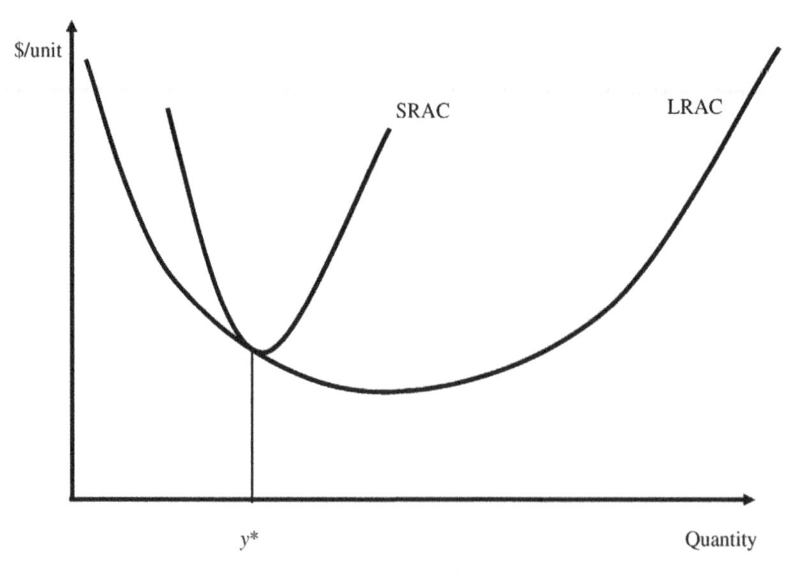

Figure 2.20 Relation between the short-run average cost curve (SRAC) and the long-run average cost curve (LRAC) if the fixed production factors are chosen to minimize the production cost for a value y^* of the output.

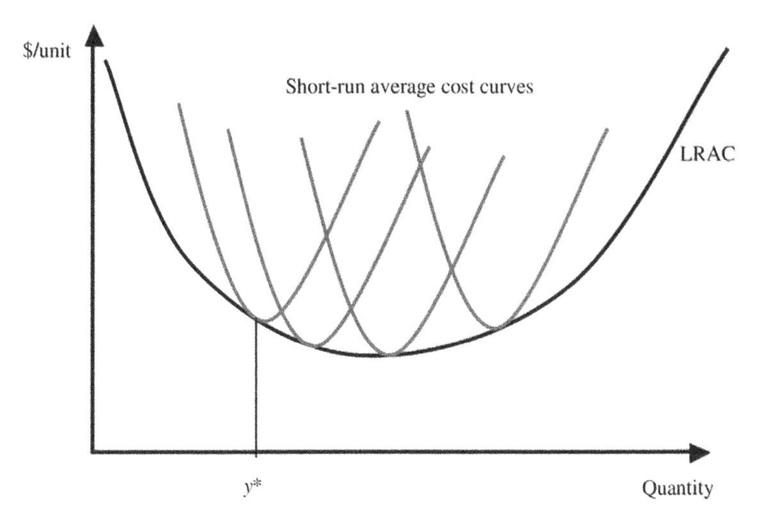

Figure 2.21 The long-run average cost curve (LRAC) is the lower envelope of the short-run average cost curves.

When all factors of production can be adjusted, the cost of a unit increase in production is given by the *long-run marginal cost curve.* Figure 2.22 illustrates two observations about this long-run marginal cost curve. First, the long-run and short-run marginal costs are equal only for the production level y^* for which the fixed production factors have been optimized. Second, the long-run marginal cost is equal to the long-run average cost for the production level that results in the minimum long-run average cost. As long as the long-run marginal cost is smaller than the long-run average cost, this long-run average cost decreases. As long as the average cost decreases, the production is said to exhibit *economies of scale.*

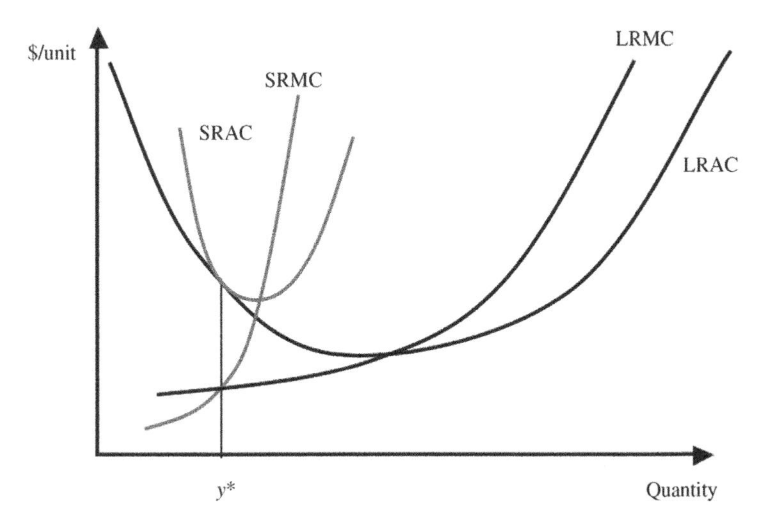

Figure 2.22 Relation between the short-run average cost (SRAC), the short-run marginal cost (SRMC), the long-run average cost curve (LRAC), and the long-run marginal cost (LRMC).

2.4 Risk

An obvious characteristic of the future is that it is uncertain. There is thus always a possibility that things will not turn out the way we expect them to. Businesses are affected by different types of risks, which in economics translates into a loss of revenue. *Technical risks* are associated with failures in the equipment used to produce a commodity. For example, failure of a power plant might prevent a generating company from delivering energy to the market. *External risks* describe the consequences of natural disasters or other catastrophic events. In this chapter, we are mostly concerned about *price risks*, i.e. the risk of having to buy a commodity at a much higher price than anticipated or having to sell a commodity at a much lower price than expected.

While doing business always implies accepting some risks, excessive risk hampers economic activity because the possibility of a large loss often discourages companies or individuals from potentially profitable undertakings. Reducing risks is the obvious way of managing it. For example, technical risks can be reduced by using better quality machinery or performing regular preventive maintenance. Locating production facilities in areas less affected by natural disasters reduces the external risks. Price risks can be reduced by limiting the quantities traded. However, over the centuries, more sophisticated approaches to managing risks have been developed. Individuals or companies can reduce the risks that they face by sharing it with others. For example, most homeowners buy insurance to protect themselves from the consequence of a fire destroying their house. Insurance works because few houses burn each year and most people prefer paying periodically a modest amount of money rather than face catastrophic risks.

Markets provide yet another way of managing risks: relocating it to other parties that are better able to handle it. This is possible because not everybody has the same willingness to accept risks. Some market participants may be willing to accept a somewhat lower profit in exchange for a reduced level of risk, while others seek higher returns because they can tolerate large risks. Furthermore, not all parties have the same ability to control risks. For example, large financial companies are better placed to assess and mitigate the risks associated with a complex investment than a small, novice investor. In the following sections, we introduce various types of markets and discuss how they support the relocation of price risks.

2.5 Types of Markets

So far, we have treated markets only as an abstract mechanism for matching the supply and the demand for a commodity through the discovery of an equilibrium price. We are now going to discuss how markets operate and how different types of markets serve different purposes.

Besides the obvious need to agree on quality, quantity, and price of the goods, three other important matters must be decided when a buyer and a seller arrange a trade:

- The time of delivery of the goods
- The mode of settlement
- Any conditions that might be attached to this transaction.

How buyers and sellers settle these matters defines the type of contracts that they conclude and hence the type of market in which they participate.

2.5.1 Spot Market

In a spot market, the seller delivers the goods immediately and the buyer pays for them "on the spot." No conditions are attached to the delivery. This means that neither party can back out of the deal. A fruit and vegetable market is a good example of a spot market: you inspect the quality of the produce and tell the vendor how many cucumbers you want, she hands them to you, you pay the price indicated, and the transaction is complete. If later on you decide that you would rather eat lettuce, you probably would not even think of trying to return the cucumbers and getting your money back. On the surface, the rules of such markets may appear very informal. In fact, they have behind them the weight of centuries of tradition. Modern spot markets for commodities such as oil, coffee, or barley are superficially more sophisticated because the quantities traded are larger and because traders communicate electronically. However, the principles are exactly the same.

A spot market has the advantage of immediacy. As a producer, I can sell exactly the amount that I have available. As a consumer, I can purchase exactly the amount I need. Unfortunately, prices in a spot market tend to change quickly. A sudden increase in demand (or a drop in production) sends the price soaring because the stock of goods available for immediate delivery is limited. Similarly, a glut in production or a dip in demand depresses the price. Spot markets also react to news about the future availability of a commodity. For example, a forecast about a bumper harvest of an agricultural commodity sends its spot price plunging if enough consumers have the ability to wait until this harvest comes to market. Changes in the price on the spot market (the spot price) are essentially unpredictable because if they were predictable, the market participants would anticipate them.

Large and unpredictable variations in the price of a commodity make life harder for both suppliers and consumers of this commodity. Both are running businesses and are thus facing a variety of risks. Bad weather or a pest can ruin a harvest. The breakdown of a machine can stop production. A strike can stop the shipment of finished goods. While being in business means taking some risks, an excessive amount of risk endangers the survival of a business. Most businesses therefore strive to reduce their exposure to price risks. For example, the producer of a commodity tries to avoid being forced to sell its output at a very low price. Similarly, a consumer does not want to be obliged to buy an essential commodity at a very high price. This desire to avoid being exposed to the wild price fluctuations that are common in spot markets has led to the introduction of the other types of transactions and markets that are described in the following sections.

2.5.2 Forward Contracts and Forward Markets

Imagine that J. McDonald is a farmer who raises wheat. Even though it is early summer, he is very confident that he will be able to deliver 100 tons at harvest time. On the other hand, he is very concerned about price fluctuations. He would very much like to "lock in" an acceptable price now and stop worrying about having to sell at a low price when the wheat is ripe. Will he be able to find someone ready to agree to such a deal? Just like farmers are concerned about having to sell at a low price, the Pretty Good Breakfast food-processing company does not want to have to pay a high price for the wheat it uses to make its well-known pancake mix. If an acceptable price can be agreed, it is ready to sign a contract with farmer McDonald now for the delivery of his wheat harvest in a few months' time. This *forward contract* specifies the following:

- The quantity and quality of the wheat to be delivered
- The date of delivery
- The date of payment following delivery
- The penalties if either party fails to honor its commitment
- The price to be paid.

On what basis can the farmer and the food-processing company agree on a price for the delivery of a commodity in a few months' time when even the spot price is volatile? Both parties start by calculating their best estimate of what the spot price might be at the time of delivery. This estimate takes into account historical data about the spot price and any other information that the farmer and the food-processing company might have about harvest yields, long-term weather forecasts, and demand forecasts. Since a lot of that information is publicly available, the estimates of both parties at any given time are unlikely to be very different. However, the price agreed for the contract may differ from the best estimates because of differences in bargaining positions. If farmer McDonald is concerned about the possibility of a very low price on the spot market, he may agree to a price below his expected value of this spot market price. The difference between his expectation of the spot market price and the price agreed in the forward contract represents a *premium* that he is willing to pay to reduce his exposure to a downward price risk. On the other hand, if the food-processing company is vulnerable to an upward price risk, farmer McDonald might be able to extract a price that reflects a premium above his expectations of the spot market price.

If the spot price at the time of delivery is higher than the agreed price, the forward contract represents a loss for the seller and a profit for the buyer. On the other hand, if the spot price is lower than the agreed price, the forward contract represents a loss for the buyer and a profit for the seller. This profit or loss only reflects the fact that a party could have done better and the other worse by trading on the spot market. They are therefore often called "paper profit" or "paper loss." Nevertheless, a paper loss makes a company less competitive because it means that it has bought or sold a commodity at a worse price than some of its competitors did.

Forward contracts make it possible for parties to trade at a price acceptable to both sides and hence provides a way to share the price risk. Over the years, these two parties could enter into similar forward contracts with a premium over or below the expected spot price. If their estimates of future spot prices are unbiased, in the long run the difference between the average spot price and the average forward price should be equal to the average premium. The party that gets the premium is therefore being remunerated for accepting the price risk.

Going back to our agricultural example, suppose that the Pretty Good Breakfast Company signs every year a forward contract with farmer McDonald at a price that is below the expected spot price for wheat at the time of delivery. In the long run, the company should profit from accepting to shoulder this risk. In the short run, however, it may have to endure a string of large losses if the spot price moves in the wrong direction. To ride through such losses, it must have large financial reserves or demand a substantial premium. If the premium it demands is too large, farmer McDonald may decide that signing a forward contract with the Pretty Good Breakfast Company is not worthwhile. Could other food-processing companies offer him a better deal? Similarly, the Pretty Good Breakfast Company will look for other farmers that might agree to sign forward contracts. If enough farmers and food-processing companies are interested in trading wheat in advance of delivery, a *forward market* for wheat will develop. The establishment of such a market gives all parties access to a larger number of possible trading partners and helps them determine whether the price they are being offered is reasonable.

In some cases, two parties may want to negotiate all the details of a forward contract. This approach is justified if the contract is designed to cover the delivery of a large quantity of a commodity over a long period of time or if special terms need to be discussed. Since such negotiations are expensive, many forward contracts use standardized terms and conditions. This standardization makes possible the resale of forward contracts. For example, let us suppose that the sales of a new Belgian waffle mix manufactured by the Pretty Good Breakfast Company do not meet expectations. Over the summer, the company realizes that it will not need all the wheat for which it has signed forward contracts. Rather than wait until the contracted date of delivery to sell the excess wheat on the spot market, it can resell the forward contracts it holds to other food-processing companies. Other producers will have signed contracts during the spring. As the summer goes by, some of them may realize that they have overestimated the quantity that they will be able to produce. If they cannot deliver the quantities specified in the contracts, they will have to cover the deficit by buying wheat on the spot market. Rather than hope that the spot price will be favorable on the date of delivery, these producers could buy the forward contract from the Pretty Good Breakfast Company to offset their anticipated deficit. The price at which forward contracts are traded will be the current market price for forward contracts with the same delivery date. Depending on the market's view of the evolution of the spot price, this resale price may be higher or lower than the price agreed by the originators of the contract.

2.5.3 Futures Contracts and Futures Markets

The existence of a *secondary market* where producers and consumers of the commodity can buy and sell standardized forward contracts helps these parties manage their exposure to fluctuations in the spot price. Participation in this market does not have to be limited to firms that produce or consume the commodity. Parties that cannot produce or take physical delivery of the commodity may also want to take part in such a market. These parties are speculators who buy a contract for delivery at a future date, in the hope of being able to sell it later at a higher price. Similarly, a speculator can sell a contract first, hoping to buy another one later at a lower price. Since these contracts are not backed by physical delivery, they are called *futures contracts* rather than forwards. As the date of delivery approaches, the speculators must balance their position because they cannot produce, consume, or store the commodity.

At this point we may wonder why any rational person might want to engage in this type of speculation. If the markets are sufficiently competitive and all participants have access to enough information, the forward price should reflect the consensus expectation of the spot price. Hence, buying low in the hope of selling high would seem more like gambling than a sound business strategy. To be successful as a speculator, therefore, one needs an advantage over other parties. This advantage is usually being less risk averse than other market participants. Shareholders in some companies expect stable but not extraordinary returns. The managers of these risk-averse companies therefore try to limit their exposure to risks that might reduce profits significantly below expectations. On the other hand, shareholders in companies that engage in commodity speculation hope for very high returns but should not be surprised by occasional large losses. The management of these risk-loving companies is therefore free to take significant risks in order to secure larger profits. A risk-averse company usually accepts a price somewhat worse than it might be able to get later in exchange for the security of getting a fixed price now. A speculator, on the other hand, demands a better price in exchange for accepting to shoulder the risk of future fluctuations. In essence, risk-averse companies remunerate speculators for their willingness to buy the risk.

As we discussed in Section 2.5.1 on spot markets, producers and consumers of a commodity face other risks besides the price risk. They are therefore usually quite eager to pay another party to reduce their exposure to the price risk. A speculator does not face other risks and has large financial resources that put it in a better position to offset losses against profits over a sufficiently long period of time. In addition, most speculators do not limit themselves to one commodity. By diversifying into markets for different commodities, they further reduce their exposure to risk. Even though speculators make a profit from their trades, as a whole the market benefits from their activities because their presence increases the number and diversity of market participants. *Physical participants* (i.e. those who produce or consume the commodity) thus find counterparties for their trades more easily. This increased *liquidity* helps the market discover the price of a commodity.

2.5.4 Options

Futures and forward contracts are *firm contracts* in the sense that delivery is unconditional. Any seller who is unable to deliver the quantity agreed must buy the missing amount on the spot market. Similarly, any buyer who cannot take full delivery must sell the excess on the spot market. In other words, imbalances are liquidated at the spot price on the date of delivery.

In some cases, participants may prefer contracts with a conditional delivery, which means contracts that are exercised only if the holder of the contract decides that it is in its interest to do so. Such contracts are called *options* and come in two varieties: *calls* and *puts*. A call option gives its holder the right to buy a given amount of a commodity at a price called the *exercise price*. A put option gives its holder the right to sell a given amount of a commodity at the exercise price. Whether the holder of an option decides to exercise its rights under the contract depends on the spot price for the commodity. A European option can be exercised only on its expiry date, while an American option can be exercised at any time before the expiry date. When an option contract is agreed, the seller of the option receives a nonrefundable option fee from the holder of the option.

Example 2.1

On June 1, the Pretty Good Breakfast Company purchased from farmer McDonald a European call option for 100 tons of wheat with an expiry date of September 1 and an exercise price of $50 per ton. On September 1, the spot price for wheat stands at $60 per ton. Buying wheat on the spot market would cost the company $10 per ton more than exercising the option. This call option therefore has a value of $100 \times 10 = \$1000$. The option thus gets exercised: farmer McDonald delivers 100 tons of wheat and the company pays $100 \times 50 = \$5000$.

On the other hand, if the spot price on September 1 is lower than the exercise price of the call, the option is worthless and lapses because it is cheaper for the company to buy wheat on the spot market.

Example 2.2

On July 1, farmer McDonald bought a European put option for 100 tons of wheat from the Great Northern Wheat Trading Company. The exercise price of this contract is $55 per ton and the expiry date is September 1. If on September 1 the spot price for wheat is $60, farmer McDonald does not exercise the option and sells his wheat on the spot market instead. On the other hand, if the spot price is $50 per ton, the option has a value of $100 \times (55 - 50) = \$500$ and should be exercised.

Buying an option contract can therefore be viewed as a way for the holder of the contract to protect itself against the risk of having to trade the commodity at a price less favorable than the spot price. At the same time, it leaves the holder free to trade at a price that is better than the exercise price of the option. The seller of the option assumes the price risk in the place of the holder. In exchange for taking this risk, the seller receives the option fee when the contract is sold. This option fee represents a sunk cost for the buyer and does not affect whether the option is exercised or not.

It is worth noting at this point that option contracts for the delivery of electrical energy are not commonly traded. On the other hand, long-term contracts for the provision of reserve often include both an option fee and an exercise price and thus operate like option contracts.

2.5.5 Contracts for Difference

Producers and consumers of some commodities are sometimes obliged to trade solely through a centralized market. Since they are not allowed to enter into bilateral agreements, they do not have the option to use forward, futures, or option contracts to reduce their exposure to price risks. In such situations, parties often resort to *contracts for difference* that operate in parallel with the centralized market. In a contract for difference, the parties agree on a *strike price* and an amount of the commodity. They then take part in the centralized market like all other participants. Once trading on the centralized market is complete, the contract for difference is settled as follows:

- If the strike price agreed in the contract is higher than the centralized market price, the buyer pays the seller the difference between these two prices times the amount agreed in the contract.
- If the strike price is lower than the market price, the seller pays the buyer the difference between these two prices times the agreed amount.

A contract for difference thus insulates the parties from the price on the centralized market while allowing them to take part in this market. A contract for difference can be described as a combination of a call option and a put option with the same exercise price. Unless the market price is exactly equal to the strike price, one of these options will be exercised.

Example 2.3

The Syldavia Steel Company is required to purchase its electrical energy from the Central Electricity Market of Syldavia. Because the price of energy in that market is highly volatile and it wants to limit its exposure to price risks, Syldavia Steel has signed a contract for difference with the Quality Electrons Generating Company. This contract specifies a uniform quantity of 500 MW and a uniform price of 20 \$/MWh at all hours of the day for a one-year period. Let us suppose that the market price for a specific 1-h trading period is 22 \$/MWh. Syldavia Steel pays to the Central Electricity Market $22 \times 500 = \$11\,000$ for the purchase of 500 MW during that hour. Quality Electrons gets paid $22 \times 500 = \$11\,000$ by the Central Electricity Market for the supply of 500 MW during the same hour. To settle their contract for difference, Quality Electrons pays Syldavia Steel $(22 - 20) \times 500 = \$1000$. Both firms have thus effectively traded 500 MW at 20 \$/MWh. If the market price had been less than 20 \$/MWh, Syldavia Steel would have had to pay Quality Electrons to settle the contract.

2.5.6 Managing the Price Risks

Firms that produce or consume large amounts of a commodity are exposed to other types of risk and will generally try to reduce their exposure to price risks by hedging their positions using a combination of forwards, futures, options, and contracts for difference. Markets for these different types of contracts develop for all major commodities. Firms tend to use the spot market only for the residual volumes that result from unpredictable fluctuations in demand or production. The volume of trades in the spot market therefore typically represents only a small fraction of the volume traded on the other markets.

While the spot market volume may be relatively small, the spot price is the signal that drives all the other markets. Since the spot market is the market of last resort, the spot price represents the alternative against which other opportunities must be measured. A sustained increase in the spot price therefore drives up the prices on the other markets while a continuing reduction forces them lower.

2.5.7 Market Efficiency

The theory that we developed at the beginning of this chapter suggests that if two parties put different values on the same good, a trade should take place. If such transactions are to happen quickly and easily, the market must be *liquid*. This means that there should always be enough participants willing to buy or sell goods. The mechanism through which the market price is discovered should also be reliable. Good mechanisms for disseminating widely comprehensive and unbiased information about the market conditions are indispensable to this price discovery process. Participants also have more confidence in the fairness of the market if its operation is as transparent as possible. Finally, the costs associated with trading (fees, administrative expenses, and the cost of gathering market information) should represent a small fraction of the value of each transaction. These *transaction costs* are considerably smaller if the commodity traded is standardized in terms of quantity and quality. A market that satisfies these criteria is said to be *efficient*.

2.6 Markets with Imperfect Competition

2.6.1 Market Power

So far, we have assumed that no market participant has the ability to influence the market price through its individual actions. This assumption is valid if the number of market participants is large and if none of them controls a large proportion of the production or consumption. Under these circumstances, any supplier who asks more than the market price and any consumer who offers less than the market price will simply be ignored because others can replace their contribution to the market. The price is thus set by the interactions of the buyers and the sellers, taken as groups. A market where all participants act as price takers is said to have *perfect competition*. Achieving or approximating perfect competition is a very desirable goal from a global perspective because it ensures that the marginal cost of production is equal to the marginal value of the goods to the consumers. Such a situation encourages efficient behavior on both sides.

Markets for agricultural commodities are one of the best examples of perfect competition because the number of small producers and consumers of an undifferentiated commodity is very large. For many other goods, some producers and consumers control a large enough share of the market that they are able to exert *market power*. These market participants are called strategic players. As the following example shows, prices can be manipulated either by withholding quantity (physical withholding) or by raising the asking price (economic withholding).

In a perfectly competitive market, the market price is a parameter over which firms have no control. Equation (2.15) led us to the conclusion that under perfect competition, each firm should increase its production up to the point where its marginal cost is equal to the market price. When competition is not perfect, each firm must consider how the quantity it produces might affect the market price. Conversely, it should consider how the price it chooses might affect the quantity it sells. We discuss in more detail how imperfect competition can be modeled in Chapter 4.

Example 2.4

Suppose that a firm sells 10 widgets at a market price of $1800 per widget. Its revenue from the sale of widgets is thus $18 000. If this firm decides to offer only nine widgets for sale and as a consequence the market price for widgets rises, this firm has market power. If the price rises to $2000, the firm achieves the same revenue even though it sells fewer widgets. Furthermore, its profit increases because it incurs the cost of producing only nine widgets instead of 10.

Instead of withholding production, this firm could offer to sell nine widgets at $1800 and one widget at a higher price in the hope that this last widget will sell and boost its profits.

2.6.2 Monopoly

The *minimum efficient size* (MES) of a firm in a particular industry provides a rough indication of the number of competitors that one is likely to find in the market for the product of this industry. This MES is equal to the level of output that minimizes the average cost for a typical firm in that industry. The shape of this curve is determined by the technology used to produce the goods. If, as illustrated in Figure 2.23a, the MES is much smaller than the demand for the goods at this minimum average cost, the market should be able to support a large number of competitors. On the other hand, if, as shown in Figure 2.23b, the MES is comparable to the demand, the market cannot support two profitable firms and a monopoly situation is likely to develop.

Given the opportunity, a monopolist will reduce its output and raise its price above its marginal cost of production to maximize its profit. From a global perspective, this is not satisfactory because consumers purchase less of the goods than if they were sold on a competitive basis. One possible remedy to this problem is to establish a regulatory body whose function is to monitor the activities of the producer and set the price at an acceptable level. Ideally, the regulator should set the price at the marginal cost of production of the monopoly firm. Determining this marginal cost is not an easy task because the regulator does not have access to the same amount of information as the monopoly firm. Even when the regulator succeeds in determining accurately the marginal cost of production, setting the price at that level may not be acceptable because it could bankrupt the monopolist.

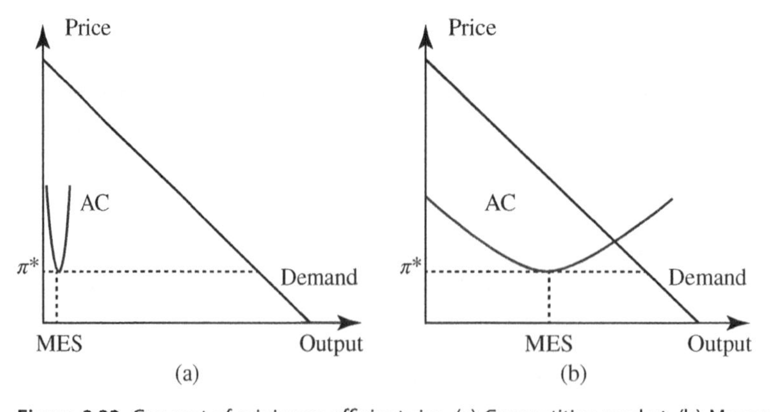

Figure 2.23 Concept of minimum efficient size. (a) Competitive market. (b) Monopoly situation.

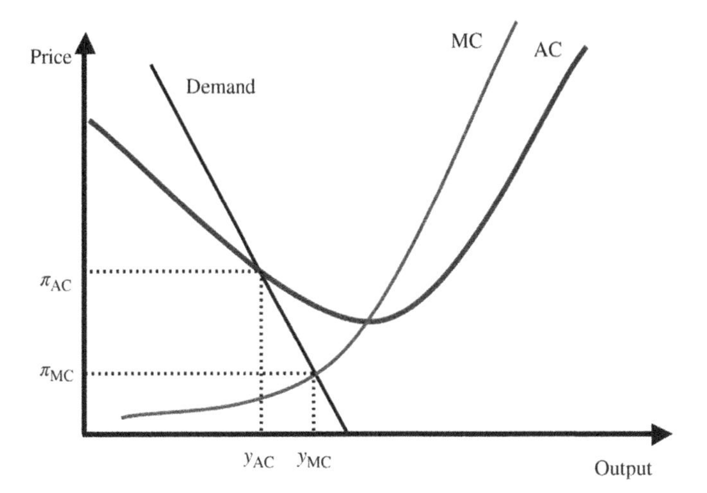

Figure 2.24 Setting a fair price for a natural monopoly.

For example, in the case shown in Figure 2.24, the intersection of the demand and marginal cost curves gives a price π_{MC} that is lower than the average cost of production. To avoid driving the monopolist out of business, the regulator must set the price at least at the value π_{AC} given by the intersection of the demand curve and the average cost curve AC. Such a situation is called a *natural monopoly*. It arises when producing the goods involves large fixed costs and relatively small variable costs. The business of transmitting and distributing electrical energy is a very good example of natural monopoly.

2.7 Problems

2.1 A manufacturer estimates that its variable cost for manufacturing a given product is given by the following expression:

$$C(q) = 25\ q^2 + 2000\ q\ (\$)$$

where C is the total cost and q is the quantity produced.

a Derive an expression for the marginal cost of production.
b Derive expressions for the revenue and the profit when the widgets are sold at marginal cost.

2.2 The inverse demand function of a group of consumers for a given type of widgets is given by the following expression:

$$\pi = -10q + 2000\ (\$)$$

where q is the demand and π is the unit price for this product.

a Determine the maximum consumption of these consumers.
b Determine the price that no consumer is prepared to pay for this product.
c Determine the maximum net consumers' surplus. Explain why the consumers will not be able to realize this surplus.

d For a price π of 1000 \$/unit, calculate the consumption, the consumers' gross surplus, the revenue collected by the producers, and the consumers' net surplus.
e If the price π increases by 20%, calculate the change in consumption and the change in the revenue collected by the producers.
f What is the price elasticity of demand for this product and this group of consumers when the price π is 1000 \$/unit.
g Derive an expression for the gross consumers' surplus and the net consumers' surplus as a function of the demand. Check these expressions using the results of part d.
h Derive an expression for the net consumers' surplus and the gross consumers' surplus as a function of the price. Check these expressions using the results of part d.

2.3 Economists estimate that the supply function for the widget market is given by the following expression:

$$q = 0.2 \cdot \pi - 40$$

a Calculate the demand and price at the market equilibrium if the demand is as defined in Problem 2.2.
b For this equilibrium, calculate the consumers' gross surplus, the consumers' net surplus, the producers' revenue, the producers' profit, and the global welfare.

2.4 Calculate the effect on the market equilibrium of Problem 2.3 of the following interventions:

a A minimum price of \$900 per widget
b A maximum price of \$600 per widget
c A sales tax of \$450 per widget.

In each case, calculate the market price, the quantity transacted, the consumers' net surplus, the producers' profit, and the global welfare. Illustrate your calculations using diagrams. Calculate the deadweight loss compared to the results of Problem 2.3. Summarize your results in a table and discuss briefly.

2.5 The demand curve for a product is estimated to be given by the expression:

$$q = 200 - \pi$$

Calculate the price and the price elasticity of the demand for the following values of the demand: 0, 50, 100, 150, and 200.

Repeat these calculations for the case where the demand curve is given by the expression:

$$q = \frac{10\,000}{\pi}$$

2.6 Vertically integrated utilities often offer two-part tariffs to encourage their consumers to shift demand from on-peak load periods to off-peak periods. Consumption of electrical energy during on-peak and off-peak periods can be viewed as substitute products. The table below summarizes the results of experiments that the

Southern Antarctica Power and Light Company has conducted with its two-part tariff. Use these results to estimate the elasticities and cross-elasticities of the demand for electrical energy during peak and off-peak periods.

	On-peak price π_1 (\$/MWh)	Off-peak price π_2 (\$/MWh)	Average on-peak demand D_1 (MWh)	Average off-peak demand D_2 (MWh)
Base case	0.08	0.06	1000	500
Experiment 1	0.08	0.05	992	509
Experiment 2	0.09	0.06	985	510

2.7 Demonstrate that the marginal production cost is equal to the average production cost for the value of the output that minimizes the average production cost.

2.8 A firm's short-run cost function for the production of gizmos is given by the following expression:

$$C(y) = 10y^2 + 200y + 100\,000$$

a Calculate the range of output over which it would be profitable for this firm to produce gizmos if it can sell each gizmo for \$2400. Calculate the value of the output that maximizes this profit.
b Repeat these calculations and explain your results for the case where the short-run cost function is given by:

$$C(y) = 10y^2 + 200y + 200\,000$$

2.9 The Pretty Good Breakfast Company is going to launch a new line of breakfast drinks. To reduce its exposure to the spot market price, it has bought a call option for 150 000 lb of frozen concentrated orange juice (http://www.investopedia.com/university/commodities/commodities14.asp). The option fee for this call option is \$3000 and the strike price is \$1.1515/lb. Discuss the consequences of this decision if the spot price on the expiry date is:

a \$1.10/lb
b \$1.20/lb
c \$1.1715/lb.

2.10 The Amazing Steel Company is a consumer of electrical energy and Borduria Power is a generating company. In order to insulate themselves from the vagaries of the price on the Bordurian electricity market, these companies have signed a contract for difference for 100 MWh at 10 \$/MWh. Discuss what happens when the price of electrical energy on the Bordurian electricity market is:

a 10 \$/MWh
b 9 \$/MWh
c 12 \$/MWh.

2.11 A speculator has bought 1000 tons of wheat at $100 per ton for delivery in three months time. She hopes to make a profit by reselling this wheat at a higher price. Her analysis suggests that the price of wheat at the time of delivery follows a uniform probability distribution[1] stretching from a minimum price of $80 per ton to a maximum price of $130 per ton. She is concerned about limiting her loss if the price of wheat drops and considers purchasing a put option to limit this loss.

- **a** What should be the strike price of this put option if she wants to limit her potential loss to $10 000? (Do not consider the option fee in this calculation.)
- **b** What is her expected profit if she does not buy this option?
- **c** What is her expected profit if she does buy this option at a fee of $1.50 per ton of wheat?
- **d** Discuss these results.

2.12 Economists estimate that the variable cost of production of electrical energy in the Bordurian electricity market is given by the following expression:

$$C(Q) = 20\,000 + 500Q + 10Q^2 + 0.001e^{(Q-85)}(\$)$$

where Q is in MWh.

They also estimate that the demand curve for electricity is given by the following expressions:

$$\text{For the hour of maximum load}: Q = 110 - 0.0025\pi \text{ (MWh)}$$
$$\text{For the hour of minimum load}: Q = 50 - 0.002\text{(MWh)}$$

where π is the price in \$/MWh.

Determine the following quantities at the market equilibrium for the hours of minimum and maximum load:

- **a** The quantity traded
- **b** The market price
- **c** The revenue collected by the producers
- **d** The total variable cost of production for all the producers
- **e** The economic profit collected by these producers
- **f** The aggregated net consumers' surplus
- **g** The price elasticity of the demand.

2.13 Syldavia is considering the introduction of competition in electricity supply. Government economists estimate that the generation cost for electrical energy fits the following formula:

$$C = 1.5\,Q^2 + 100\,Q$$

where:

- Q is the quantity produced (MWh).
- C is the total cost (\$).

---

1 This is not a very realistic assumption. However, the calculations are too complex if we assume a more realistic probability distribution.

They also estimate that the demand curve obeys the following relation for the hour of peak demand:

$$P = -20\,D + 4000$$

where:

- D is the demand (MWh).
- P is the price (\$/MWh).
  - a Draw the supply curve.
  - b Calculate the price and quantity at which the market would clear for the hour of peak demand.
  - c Calculate the elasticity of demand at equilibrium.
  - d Calculate the corresponding consumers' gross surplus and net surplus.
  - e Calculate the corresponding producers' revenue and economic profit.
  - f The Syldavian government is considering imposing a tax of \$20.00/MWh produced on generators. Calculate how much revenue this tax would raise during the hour of peak demand and the effect that it would have on the global welfare.
  - g The Syldavian government is considering imposing a tax of \$20.00/MWh produced. Assuming that the generators pass on this tax to the consumers in full, calculate how much revenue this tax would raise during the hour of peak demand and the effect that it would have on the global welfare.

Carry out numerical calculations with two decimal places.

2.14 What are the principal characteristics of contracts?

2.15 Describe briefly the characteristics of spot contracts, forward contracts, futures contracts, and option contracts.

2.16 Using a simple numerical example, explain the operation of a two-way contract for difference.

2.17 The demand curve, in terms of quantity Q(kWh) bought by a consumer, in a given period, as a function of price π(¢/kWh) is given by the following expression:

$$\pi = -0.01\,Q + 5$$

- a Calculate the level of consumption at $\pi = 3.5$ ¢/kWh, net consumer surplus, demand charges, and revenue received by suppliers.
- b If the producer decides to increase the price by 10%, determine the reduction in consumption and the new revenue received by the producer.
- c What is the price elasticity of the demand in this case?

Given that the production cost in the period under consideration in (a) is given by

$$C = 0.0075Q^2 + 1.3Q$$

- d Determine the expression for the marginal cost and then find the equilibrium price and demand at which the social welfare is maximized.
- e Calculate the producer revenue, profit, and average cost at the equilibrium point.

2.18 Economists estimate that the variable cost of production of electrical energy in an electricity market is given by the following expression:

$$C(Q) = 20\,000 + 500Q + 10Q^2(\$) \qquad \text{for } Q \leq 90\,\text{MWh}$$
$$C(Q) = 141\,400 + 287.5(Q - 86)^2(\$) \quad \text{for } Q \geq 90\,\text{MWh}$$

They also estimate that the demand curve for electricity is given by the following expressions:

$$\text{For the hour of maximum load} : Q = 100 - 0.00125\pi(\text{MWh})$$
$$\text{For the hour of minimum load} : Q = 55 - 0.001\pi(\text{MWh})$$

where π is the price in \$/MWh.

Sketch the supply and demand curves for this market.

Determine the following quantities at the market equilibrium for the hours of minimum and maximum load:

a The quantity traded
b The market price
c The revenue collected by the producers
d The total variable cost of production for all the producers
e The economic profit collected by these producers
f The price elasticity of the demand
g The price elasticity of the supply.

2.19 The demand for a particular commodity is given by the following demand curve:

$$\pi(Q) = a - bQ$$

Assume that there is a monopoly supplier for this commodity and that its cost for producing this commodity is given by:

$$C(Q) = d + cQ$$

How much of this commodity should this monopolist produce to maximize its profit? How does this amount compare to the amount that would be traded if this market were perfectly competitive and had the same aggregated production cost curve?

2.20 Borduria Power is a generating company and Syldavia Electricity is a retailer. These two companies are negotiating a long-term contract for the delivery of electricity and both forecast that the average spot price for the duration of the contract will be 20.00 \$/MWh. If Borduria Power is less risk averse than Syldavia Electricity, what is the price that is more likely to result from these negotiations:

a 21.00 \$/MWh
b 20.00 \$/MWh
c 19.00 \$/MWh.

Both of these companies must physically trade through the Southern Antarctica Power Pool, but they have signed a two-way contract for difference for 100 MWh at 20.00 \$/MWh. If everything happens as planned, describe how trading is settled if the price on the Southern Antarctica Power Pool is 23.00 \$/MWh.

Further Reading

The reader interested in studying microeconomics in more depth will find an abundance of textbooks on this subject. Most of these textbooks cover the same topics albeit at different levels. Engineers may find introductory texts somewhat frustrating because they assume that the reader does not know calculus. Explanations thus tend to be long and wordy. Intermediate level books, such as Varian (2014), are probably a better choice. An even more rigorous and mathematical treatment of the subject can be found in Gravelle and Rees (2007). Tirole (1988) analyzes in considerable depth the theory of the firm. Hunt and Shuttleworth (1996) provide a very readable introduction to the various types of contracts. A discussion of how the concept of elasticity can be applied in electricity markets can be found in Kirschen et al. (2000). A wealth of information about how various types of markets operate, as well as how to make (or lose) money by trading on these markets, can be found at https://www.investopedia.com/dictionary/.

Gravelle, H. and Rees, R. (2007). *Microeconomics*, 4e. Pearson Education.

Hunt, S. and Shuttleworth, G. (1996). *Competition and Choice in Electricity*. Wiley.

Kirschen, D.S., Strbac, G., Cumperayot, P., and Mendes, D.P. (2000). Factoring the elasticity of demand in electricity prices. *IEEE Trans. Power Syst.* 15 (2): 612–617.

Tirole, J. (1988). *The Theory of Industrial Organization*. MIT Press.

Varian, H.R. (2014). *Intermediate Microeconomics: A Modern Approach*, 9e. W. W. Norton.

3

Markets for Electrical Energy

3.1 What Is the Difference Between a Megawatt-hour and a Barrel of Oil?

The development of electricity markets is based on the premise that electrical energy can be treated as a commodity. However, there are important differences between electrical energy and other commodities such as bushels of wheat, barrels of oil, or even cubic meters of gas. These differences have a profound effect on the organization and the rules of electricity markets.

The most fundamental differences are that the supply of electrical energy is inextricably linked with a physical system and that its delivery occurs on a continuous rather than a batch basis. In this physical power system, generators, consumers, and the wires that connect them are mutually dependent and cannot be treated like fully independent entities. In particular, supply and demand – generation and load – must be balanced on a second-by-second basis. If this balance is not maintained, or if the capacity of the transmission network is exceeded, the system collapses with catastrophic consequences. Such a breakdown is intolerable because it is not only trading that stops. An entire region or country may be without power for many hours because restoring a power system to normal operation following a complete collapse is a very complex process that may take 24 h or longer in large industrialized countries. The social and economic consequences of such a system-wide blackout are so severe that no sensible government would agree to the implementation of a market mechanism that could significantly increase the likelihood of large supply interruptions. Balancing the supply and the demand for electrical energy in the short run is thus a process that simply cannot be left to an unaccountable entity such as a market: a single organization must ultimately be responsible for maintaining this balance at practically any cost.

Another significant, but somewhat less fundamental, difference between electrical energy and other commodities is that the energy produced by one generator cannot be directed to a specific consumer. Conversely, a consumer cannot take energy from only one generator. Instead, the power produced by all generators is pooled on its way to the loads. This pooling is possible because units of electrical energy produced by different generating units are indistinguishable. It is also desirable because pooling produces valuable economies of scale: the maximum generation capacity must be commensurate with the maximum aggregated demand rather than with the sum of the maximum

Fundamentals of Power System Economics, Second Edition. Daniel S. Kirschen and Goran Strbac.

individual demands. On the other hand, a breakdown in a system where the commodity is pooled affects everybody, not just the parties to a particular transaction.

Finally, the demand for electrical energy exhibits predictable daily and weekly cyclical variations. Electricity is by no means the only commodity for which the demand is cyclical. To take a simple example, the consumption of coffee exhibits two or three rather sharp peaks every day, separated by periods of lower demand. However, trading in coffee does not require special mechanisms because consumers can easily store it in solid or liquid form. On the other hand, electrical energy must be produced at the same time as it is consumed. Since its short-run price elasticity of demand is very small, matching supply and demand requires production facilities capable of following the large and rapid changes in consumption that take place over the course of a day. However, not all of these generating units will be producing throughout the day. When the demand is low, only the most efficient units are likely to be competitive and the others will be shut down temporarily. These less efficient units are needed only to supply the peak demand. Since the marginal producer changes as the load increases and decreases, we should expect the marginal cost of producing electrical energy (and hence its spot price) to vary over the course of the day. Such rapid cyclical variations in the cost and price of a commodity are very unusual.

One could argue that trading in gas also takes place over a physical network where the commodity is pooled and the demand is cyclical. However, pipelines store a considerable amount of energy in the form of pressurized gas. An imbalance between production and consumption of gas therefore has to last much longer before it causes a collapse of the pipeline network. It can therefore be corrected through a market mechanism. On the other hand, the only energy that is actually stored and readily available in the power system resides in the rotating masses of synchronized generators. Since this stored kinetic energy is quite small, a sudden deficit in generation would deplete it quite quickly, causing a drop in frequency that could not be corrected sufficiently fast by a market mechanism.

As a first step in our study of electricity markets, we will assume that all generators and loads are connected to the same bus or that they are connected through a lossless network of infinite capacity. We will therefore concentrate first on the trading of electrical energy. Chapter 5 discusses how the transmission network affects this trading and Chapter 6 how the need to maintain stable operation and hence reliability shapes the design of electricity markets.

3.2 Trading Periods

As every engineer knows, electricity is fully characterized by several physical variables: voltage, current, power, and energy. While all of these matter in the operation of the system, the quantity that best defines a tradable commodity is clearly the energy supplied. For consumers, the amount of energy consumed is the variable that is most closely related to the value that they obtain from consuming electricity. For generators, the cost of production is directly linked to the amount of energy generated.

However, the power system got its name from the fact that it is designed to supply energy on a continuous basis, i.e. power. Strictly speaking, electricity is therefore not traded in terms of energy but rather in terms of power over a certain interval of time. This is reflected by the fact that we quantify the amount of electrical energy flowing through the grid using kilowatt-hours rather than joules.

Designing an electricity market therefore involves choosing a time interval that serves as a trading period. Power is then translated into tradable energy by integrating it over each time interval. Looking at this the other way, a certain amount of energy traded over a given time period is thus deemed to be delivered at constant (i.e. average) power over that time interval. Because the physical and market conditions are assumed constant over each trading period, the price of energy is also deemed uniform over each trading period.

If we adopt a short time interval (say 5 min), trading reflects more accurately the instantaneous conditions in the physical system because the actual power generated or consumed will not deviate significantly from its average over each trading period. This is particularly important in systems with a significant amount of wind or solar generation because of the rapid and unpredictable changes in supply that these renewable sources can cause.

However, trading over many short time intervals is not always practical. For example, if a generating company has the ability to produce a large amount of power for a period of hours or a retailer must supply a constant base load, they probably do not want to trade in 5-min intervals. They would rather agree on a quantity and price that remain fixed over a much longer period because it reduces their risks. In particular, before a generating company brings a large thermal unit on-line, it will want some certainty that this unit will remain synchronized long enough to recover its startup cost and respect its technical operating constraints.[1]

To resolve this dilemma, electricity trading is organized as a sequence of forward markets with progressively shorter trading periods and a spot market. Forward markets handle trading in large amounts of energy over long periods of time. They operate slowly and close far in advance of the beginning of the delivery period (also known as "real time"). Markets for smaller amounts of energy to be delivered over shorter time periods operate faster and can therefore operate much closer to real time. The spot market is the market of last resort and thus operates closest to real time. It typically handles only a small fraction of the overall energy needs. This arrangement not only helps market participants manage their risks but also gives the system operator time to identify conditions that might affect the operational reliability of the system. In the next section, we discuss the various types of forward markets for electrical energy. We then discuss the need for a spot market and how it differs from spot markets for other commodities. Subsequent chapters discuss the markets for ancillary services, generation capacity, and transmission capacity that have been put in place to meet the reliability requirements.

3.3 Forward Markets

Forward markets normally involve a substantial number of bids to buy and offers to sell. Over time, repeated interactions between the participants, leading to bilateral trades,

1 The billing of residential consumers provides another good example of a longer trading period. Until recently, electricity meters performed a mechanical integration of the power consumed over 1 month and customers were billed on the basis of the retail price in effect for that month. A one-month trading period kept the transaction cost (mostly the cost of manual meter reading) at a reasonable level and completely insulated residential consumers from the short-term vagaries in the supply and demand. The deployment of "smart" meters that can be read automatically and remotely provides an opportunity for retail trading in electrical energy to take place at prices that more accurately reflect the real-time conditions.

clear the market. While a number of forward markets for electrical energy are based on this decentralized model, others have adopted a centralized approach where bids and offers from all the participants are considered together to determine the market equilibrium in a single pass.

3.3.1 Bilateral or Decentralized Trading

Depending on the amount of time available and the quantities to be traded, buyers and sellers make use of different forms of bilateral trading.

Customized long-term contracts: The terms of such contracts are negotiated privately to meet the needs and objectives of both parties and are thus very flexible. They usually involve the sale of large amounts of power (hundreds or thousands of MW) over long periods of time (several months to several years). Negotiating such contracts carries substantial transaction costs and thus make them worthwhile only when the parties want to trade large amounts of energy. For example, a generating company planning the construction of a new power plant will often have to arrange the sale of a substantial proportion of the expected output of this plant to secure the funding that it needs for this project. Large retailers may be interested in being the counterparty to such a contract if it guarantees them a good price for their base load over several years.

Trading "over the counter": These transactions involve smaller amounts of energy to be delivered according to a standard profile, i.e. a standardized definition of how much energy should be delivered over different periods of the day and week. This form of trading has much lower transaction costs and is used by producers and consumers to refine their position as the time of delivery approaches.

Electronic trading: Participants use electronic trading platforms to advertise offers to sell or bids to buy energy. All participants in such a computerized marketplace can observe the quantities and prices submitted by other parties but do not know the identity of the party that submitted each bid or offer. When a party enters a new bid, the software underpinning the exchange checks if there is a matching offer for the period of delivery of the bid. If it finds an offer whose price is lower than or equal to the price of the bid, a deal is automatically struck and the price and quantity are displayed for all participants to see. If no match is found, the new bid is added to the list of outstanding bids and remains there until a matching offer is made or the bid is withdrawn. A similar procedure is used each time a new offer is entered in the system. A flurry of trading activity often takes place in the minutes and seconds before the close of the market as generators and retailers fine-tune their position ahead of the delivery period. This form of trading is extremely fast and cheap.

The essential characteristic of bilateral trading is that the price of each transaction is set independently by the parties involved. There is thus no "official" price and trading can be described as "decentralized." While the details of negotiated long-term contracts are usually kept private, independent reporting services gather anonymized data about over-the-counter (OTC) trading and publish summary information about prices and quantities in a form that does not reveal the identity of the parties involved. This type of market reporting and the display of the last transaction arranged on the electronic marketplace enhance the efficiency of electricity trading by giving participants a clearer idea of the state and direction of the market.

Example 3.1

Borduria Power trades on the Bordurian electricity market that operates on a bilateral basis. It owns three generating units whose characteristics are given in Table 3.1. To keep things simple, we assume that the marginal cost of these units is constant over their range of operation. Because of their large startup costs, Borduria Power tries to keep unit A synchronized to the system at all times and to produce as much as possible with unit B during the daytime. The startup cost of unit C is negligible.

Table 3.1 Characteristics of the generating units of Example 3.1.

Unit	Type	P^{min} (MW)	P^{max} (MW)	MC ($/MWh)
A	Large coal	100	500	10.0
B	Medium coal	50	200	13.0
C	Gas turbine	0	50	17.0

Let us focus on the contractual position of Borduria Power for the 2:00–3:00 p.m. trading period on June 11. From Table 3.2, we can see that Borduria Power has entered into two long-term contracts and three OTC transactions that cover this trading period.

Table 3.2 Contractual position of Borduria Power.

Type	Contract date	Buyer	Seller	Amount (MWh)	Price ($/MWh)
Long term	January 10	Cheapo Energy	Borduria Power	200	12.5
Long term	February 7	Borduria Steel	Borduria Power	250	12.8
OTC	April 7	Quality Electrons	Borduria Power	100	14.0
OTC	May 31	Borduria Power	Perfect Power	30	13.5
OTC	June 8	Cheapo Energy	Borduria Power	50	13.8

Note that Borduria Power has taken advantage of price fluctuations in the forward market to buy back at a profit some of the energy that it had sold. Toward mid-morning on June 11, Fiona, the trader on duty at Borduria Power, must decide if she wants to adjust this position by trading on the electronic Bordurian Power Exchange (BPeX); see Table 3.3. Borduria Power has so far contracted to deliver 570 MWh, but it has a total production capacity of 750 MW available during that hour. Fiona's BPeX trading screen displays the following bids and offers, ordered by price.

Based on her experience with this market, Fiona believes that it is unlikely that the bid prices will increase. Since she still has 130 MW of spare capacity on unit B, she decides to grab bids B1, B2, and B3 before one of her competitors does. These trades are profitable because their price is higher than the marginal cost of unit B. After completing these transactions, Fiona sends revised production instructions for this hour to the power

Table 3.3 Bids and offers on the BPeX at mid-morning.

June 11 14:00–15:00	Identifier	Amount (MW)	Price ($/MWh)
Offers to sell energy	O5	20	17.50
	O4	25	16.30
	O3	20	14.40
	O2	10	13.90
	O1	25	13.70
Bids to buy energy	B1	20	13.50
	B2	30	13.30
	B3	10	13.25
	B4	30	12.80
	B5	50	12.55

plants. Unit A is to generate at rated power (500 MW), while unit B is to set its output at 130 MW. Unit C is to remain on standby.

Shortly before BPeX closes trading for the period between 14:00 and 15:00 (see Table 3.4), Fiona receives a phone call from the operator of plant B. He informs her that the plant has developed some unexpected mechanical problems. Plant B will be able to remain on-line until the evening but will not be able to produce more than 80 MW. Fiona quickly realizes that this failure leaves Borduria Power exposed and that she has three options:

1) Do nothing, leaving Borduria Power short by 50 MWh that would have to be paid for at the spot market price.
2) Make up this deficit by starting up unit C.
3) Try to buy some replacement power on BPeX.

Table 3.4 Bids and offers on the BPeX in the afternoon.

June 11 14:00–15:00	Identifier	Amount (MW)	Price ($/MWh)
Offers to sell energy	O5	20	17.50
	O4	25	16.30
	O3	20	14.40
	O6	20	14.30
	O8	10	14.10
Bids to buy energy	B4	30	12.80
	B6	25	12.70
	B5	50	12.55

Since spot market prices have been rather erratic lately, Fiona is not very keen on remaining unbalanced. She therefore decides to see if she can buy energy on BPeX for less than the marginal cost of unit C. Since she last traded on BPeX, some bids have disappeared and new ones have been entered.

Fiona immediately selects offers O8, O6, and O3 because they allow her to restore the contractual balance of the company for this trading period at a cost that is less than the cost of covering the deficit with unit C. On balance, when trading closes for this trading period, Borduria Power is committed to producing 580 MWh. Note that Fiona based all her decisions on the incremental cost of producing energy. We will revisit this example below, once we have discussed the operation of the spot market.

3.3.2 Centralized Trading

3.3.2.1 Principles of Centralized Trading

Rather than relying on repeated interactions between suppliers and consumers to reach a market equilibrium, a centralized electricity market provides a mechanism for determining this equilibrium in a systematic way. While there are many possible variations, a centralized market essentially operates as follows:

- Generating companies submit offers to supply a certain amount of electrical power at a certain price for the period under consideration. These offers are ranked in order of increasing price. From this ranking, a curve showing the offer price as a function of the cumulative quantity offered can be built. This curve is deemed to be the supply curve of the market.
- Similarly, the demand curve of the market can be established by asking consumers to submit bids specifying the quantity they need and the price that they would be willing to pay. These offers are then ranked in decreasing order of price to create a demand curve. Since the demand for electricity is highly inelastic, this step is often omitted and the demand is set at a value determined using a forecast of the load. In other words, the demand curve is assumed to be a vertical line at the value of the load forecast.
- The intersection of these constructed supply and demand curves determines the market equilibrium. All the offers submitted at a price lower than or equal to the market clearing price are accepted, and generators are instructed to produce the amount of energy corresponding to their accepted offers. Similarly, all the bids submitted at a price greater than or equal to the market clearing price are accepted and the consumers are informed of the amount of energy that they are allowed to draw from the system.
- The market clearing price represents the price of one additional megawatt-hour of energy and is therefore called the System Marginal Price or SMP. Generators are paid this SMP for every megawatt-hour that they produce and consumers pay the SMP for every megawatt-hour that they consume, irrespective of the offers and bids that they submitted.
- The process of finding the intersection between these constructed supply and demand curves emulates the operation of a market and thus maximizes the global welfare. If the demand is assumed constant and the offers reflect the marginal cost of production, a centralized market performs an economic dispatch, i.e. it determines how much each generator should produce to supply the load at minimum cost.

Paying the SMP for all the accepted generation offers may appear surprising at first glance. Why should generators be paid more than their asking price? Wouldn't paying every generator its asking price reduce the average price of electricity? The main reason why such a pay-as-bid scheme is not adopted is that it would discourage generators from submitting offers that reflect their marginal cost of production. All generators would instead try to guess what the SMP is likely to be and would offer at that level to maximize their revenues. Therefore, at best, the SMP would remain unchanged. However, inevitably some low-cost generators would occasionally overestimate the SMP and offer at too high a price. These generators would then be left out of the schedule and be replaced by generators with a higher marginal cost of production. The SMP would then be somewhat higher than it would be if all generators were paid the same price. Furthermore, this substitution is economically inefficient because optimal use is not made of the available resources. In addition, generators are likely to increase their prices slightly to compensate themselves for the additional risk of losing revenue caused by the uncertainty associated with trying to guess what the SMP might be. An attempt to reduce the price of electricity would therefore result in a price increase!

Example 3.2

The centralized forward market of Syldavia has received the bids and offers shown in Table 3.5 for the 9:00–10:00 a.m. trading period on June 11.

Table 3.5 Bids and offers in the centralized Syldavian market.

	Company	Quantity (MWh)	Price ($/MWh)
Offers	Red	200	12.00
	Red	50	15.00
	Red	50	20.00
	Green	150	16.00
	Green	50	17.00
	Blue	100	13.00
	Blue	50	18.00
Bids	Yellow	50	13.00
	Yellow	100	23.00
	Purple	50	11.00
	Purple	150	22.00
	Orange	50	10.0
	Orange	200	25.00

Figure 3.1 shows how these offers and bids stack up to form the supply and demand curves, respectively. The intersection of these two curves shows that for this trading period the SMP is 16.00 $/MWh and that 450 MWh will be traded. Table 3.6 shows how much energy each generator will be instructed to produce and how much energy each

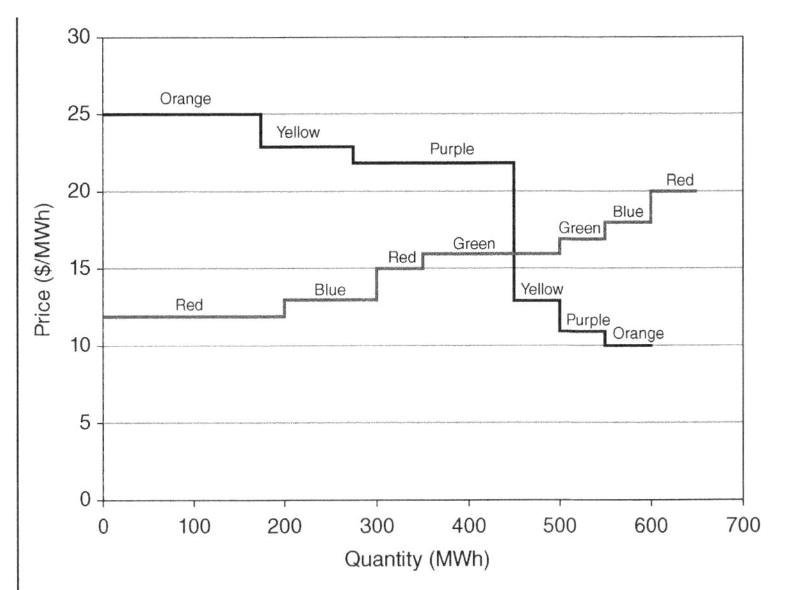

Figure 3.1 Stacks of bids and offers of Example 3.2.

consumer will be allowed to draw. It also shows the revenues and expenses for each company.

If, instead of asking consumers to submit bids, the Syldavian forward market relied on a load forecast to represent the demand side and if the load for this period was forecast to be 450 MWh, we would obtain exactly the same results.

Table 3.6 Generating schedule for the Syldavian market.

Company	Production (MWh)	Consumption (MWh)	Revenue ($)	Expense ($)
Red	250		4000	
Blue	100		1600	
Green	100		1600	
Orange		200		3200
Yellow		100		1600
Purple		150		2400
Total	450	450	7200	7200

3.3.2.2 Day-ahead Centralized Trading

Example 3.2 illustrates how a centralized electricity market clears for a single trading period of 1 h. Treating each hourly trading period separately is not practical for a number of technical and economic reasons:

- Synchronizing a large thermal generating unit to the system can take several hours.

- Once synchronized, such a unit must often continue producing power for a certain number of hours to avoid mechanical damage caused by excessive temperature gradients.
- Similarly, once it has been shut down, it must typically remain idle for a certain number of hours before it can be restarted.
- The output of large generating units usually cannot be reduced to zero. Instead, it must remain above its "minimum stable generation," which can be a significant fraction of its rated output.
- The rate at which the output of a unit can increase or decrease is also limited.

These constraints would often make it impossible for a generating unit to follow dispatch instructions stemming from an hour-by-hour market clearing. Its owner would then be obliged to buy or sell energy on the spot market to compensate for the difference between what the market clearing said it should produce and what it was actually able to do. Furthermore, a large, efficient thermal generating unit usually burns a significant amount of fuel before it can begin injecting electrical energy into the grid. The cost of this fuel represents a fixed "startup cost" that must be amortized over the sale of enough electrical energy to justify starting this unit.

These constraints and this startup cost would make it rather risky for large thermal generators to participate in such an hour-by-hour market. Most centralized markets therefore operate on a day-ahead basis, which means that the market does not clear for each trading period separately but instead simultaneously for all of the trading periods of the next day while respecting the constraints on the operation of the generating units.

3.3.2.3 Formulation as an Optimization Problem

Instead of submitting simple offers consisting of a price and a quantity, each generator participating in a centralized day-ahead market must provide a complex bid. This bid consists of an incremental cost curve, a startup cost, and the parameters of constraints that limit its operation. Note that in these bids, the "costs" do not have to be true values. We will discuss later the conditions under which generators may want to inflate or deflate their costs to maximize their profits or minimize their risks.

The system operator combines these complex bids with the day-ahead load forecast to determine a schedule showing when each generator should be turned on and off and how much power each of them must produce during each trading period. In essence, it carries out a unit commitment calculation similar to the one that a vertically integrated monopoly utility performs, albeit based on bids rather than actual costs. Mathematically, this optimization problem is formulated as follows:

Objective function: Minimize the total cost of operating the system over the next day:

$$\min_{x_i(t),u_i(t)} \sum_{i=1}^{N} \sum_{t=1}^{T} [C_i(P_i(t)) + SC_i(u_i(t), u_i(t-1))] \tag{3.1}$$

where:

$P_i(t)$ is the power produced by unit i during period t.
$u_i(t)$ is the status of unit i during period t ($u_i(t) = 1$ if unit i is on, 0 otherwise).
N is the number of generating units that have submitted bids.
T is the number of trading periods into which the day is divided.

$C_i(P_i(t))$ is the cost of producing a constant $P_i(t)$ with unit i during period t.
$SC_i(u_i(t), u_i(t-1))$ is the cost of starting up unit i at period t. This cost is nonzero only if the unit was off at period $t-1$ and is on at period t.

Load generation balance constraint: The power to be produced by all the generators must be equal to the load:

$$\sum_{i=1}^{N} P_i(t) = L(t), \quad \forall t = 1, \ldots, T \tag{3.2}$$

where $L(t)$ is the load forecast for period t.

Generation limits:

$$P_i^{\min} \leq P_i(t) \leq P_i^{\max}, \quad \forall t = 1, \ldots, T \& \forall i = 1, \ldots, N \tag{3.3}$$

where:

$P_i^{\min}$ is the minimum stable generation of generator i.
$P_i^{\max}$ is the maximum output of generator i.

Ramp rate limits:

$$P_i(t) - P_i(t-1) \leq \Delta P_i^{\text{up}}, \quad \forall t = 1, \ldots, T \& \forall i = 1, \ldots, N \tag{3.4}$$

$$P_i(t-1) - P_i(t) \leq \Delta P_i^{\text{down}}, \quad \forall t = 1, \ldots, T \& \forall i = 1, \ldots, N \tag{3.5}$$

where:

ΔP_i^{up} is the maximum amount by which unit i is able to increase its output between two consecutive trading periods.
ΔP_i^{down} is the maximum amount by which unit i is able to decrease its output between two consecutive trading periods.

Minimum up time constraints: A generating unit is not allowed to shut down unless it has been on for at least its minimum up time $T_{\text{up}}^{\min}$:

$$\text{if } \left\{ u_i(t-1) = 1 \& \exists \tau > t - T_{\text{up}}^{\min} \text{ such that } u_i(\tau) = 0 \right\} \Rightarrow u_i(t) = 1, \quad \forall t = 1, \ldots, T \& \forall i = 1, \ldots, N \tag{3.6}$$

Minimum down time: A generating unit is not allowed to start up unless it has been off for at least its minimum down time $T_{\text{down}}^{\min}$:

$$\text{if } \left\{ u_i(t-1) = 0 \& \exists \tau > t - T_{\text{down}}^{\min} \text{ such that } u_i(\tau) = 1 \right\} \Rightarrow u_i(t) = 0, \quad \forall t = 1, \ldots, T \& \forall i = 1, \ldots, N \tag{3.7}$$

3.3.2.4 Market Clearing Price

Given the schedule produced by this optimization, one can calculate the market clearing price for each trading period. Since in a competitive market this price should reflect the cost of producing one additional megawatt-hour, this price should be set at the marginal cost of the most expensive generating unit scheduled to produce power for each trading period. Mathematically speaking, these marginal costs are given by the Lagrange multipliers associated with constraints (3.2). However, the Lagrange multipliers cannot be directly obtained when the problem involves binary variables. A continuous version of this optimization problem is then solved with the binary decision variables set equal to the optimal values given by the solution of the unit commitment problem. The Lagrange multipliers of this continuous problem constitute the electricity marginal prices at each period and are denoted by π_t^*.

Example 3.3

Let us consider a simple example, involving a market with 3 generators, a scheduling horizon of 3 h, and a trading period of 1 h. Tables 3.7 and 3.8 give the data for the generating units and the demand over the scheduling horizon.

Table 3.7 Generating unit data.

Generating unit i	P_i^{min} (MW) minimum generation	P_i^{max} (MW) maximum generation	β_i (\$/MWh) marginal cost	α_i (\$/h) fixed cost
1	0	500	10	500
2	100	350	30	250
3	0	200	35	100

Table 3.8 Demand data.

Time period t	1	2	3
$L(t)$ (MW)	550	750	1050

For simplicity, we ignore the startup costs, the ramp rate limits as well as the minimum up- and down-time constraints. We also assume that the cost functions of the generators involve only a fixed cost α_i and a constant marginal cost β_i:

$$C_i(P_i(t), u_i(t)) = \alpha_i u_i(t) + \beta_i P_i(t) \tag{3.8}$$

Figure 3.2 illustrates the minimum cost generation schedule for this unit commitment problem. Because of the simplifications that we have made, we can consider each period separately.

Period 1: The 550 MW demand can be met with three combinations of units: 1&2, 1&3, 2&3.

- $P_1 = 450$ MW, $P_2 = 100$ MW, which would result in a total cost for this period of: $500 + 10 \times 450 + 250 + 30 \times 100 = \$8250$
- $P_1 = 500$ MW, $P_3 = 50$ MW, for a total cost: $500 + 10 \times 500 + 100 + 35 \times 50 = \$7350$
- $P_2 = 350$ MW, $P_3 = 200$ MW, for a total cost of: $250 + 30 \times 350 + 100 + 35 \times 200 = \$17\,850$.

Unit 1 has a large fixed cost but by far the lowest variable cost. It should therefore be dispatched as much as possible. Unit 2 has a lower marginal cost than that of unit 3, but a higher fixed cost and a larger minimum stable generation. Dispatching unit 2 during period 1 would force us to reduce the output of unit 1 to 450 MW, resulting in a larger total cost. The optimal dispatch for period 1 is thus $P_1 = 500$ MW, $P_3 = 50$ MW. Unit 3 is the

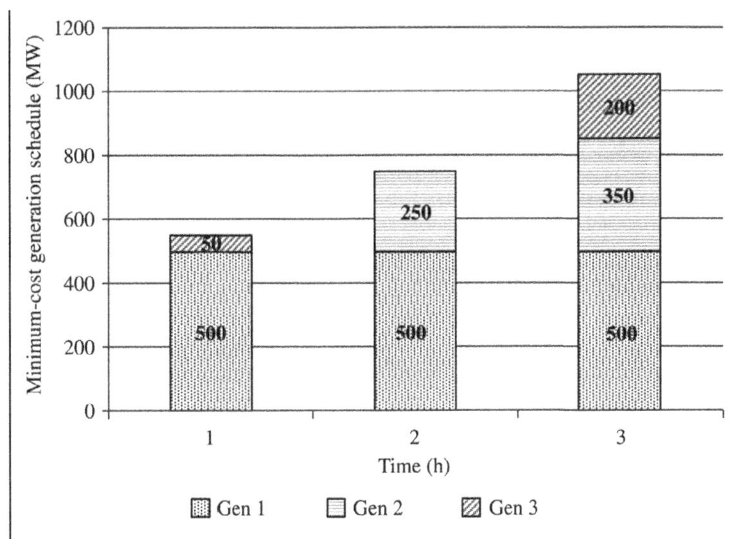

Figure 3.2 Minimum cost generation schedule.

marginal unit, and the market price for this period is therefore set at its marginal cost, i.e. $\pi_1^* = 35\ \$/\text{MWh}$.

Period 2: Unit 1 must be dispatched during this period to be able to meet the larger load. Because a combination of units 1 and 3 would not be able to meet the load, unit 2 must be dispatched. Since unit 2 has a lower marginal cost than unit 3, adding unit 3 is unnecessary because all the power not produced by unit 1 will be produced by unit 2. The dispatch for period 2 is then:

- $P_1 = 500$ MW, $P_2 = 250$ MW, for a total cost: $500 + 10 \times 500 + 250 + 30 \times 250 = \$13\,250$.

Since unit 2 is the marginal unit, the price for period 2 is set at its marginal cost of 30 $/MWh. Note that this price is lower than the price for period 1, even though the demand is higher.

Period 3: All three units are needed to meet the load during this period and the optimal dispatch is:

- $P_1 = 500$ MW, $P_2 = 350$ MW, $P_3 = 200$ MW, for a total cost: $500 + 10 \times 500 + 250 + 30 \times 350 + 100 + 35 \times 200 = \$23\,350$.

Units 1 and 2 being loaded at their maximum output, unit 3 is the marginal unit and the price for this period is $\pi_3^* = 35\ \$/\text{MWh}$.

3.3.2.5 Recovering the Fixed Costs

Because the price for each period is determined by the marginal cost of the marginal unit and does not factor in fixed costs, there is no guarantee that generating units will be able to recover their fixed costs (i.e. their no-load and startup costs).

Example 3.4

Table 3.9 summarizes the output, revenue, cost, and profit for each unit of Example 3.3 during each period, assuming that all generators bid their actual costs.

While unit 1 is profitable during each period, units 2 and 3 are unprofitable during the periods where they are the marginal unit, because they set the price at their marginal cost and therefore do not recover their fixed costs.

Table 3.9 Generation schedule for Example 3.4.

Period	Price ($/MWh)		Unit 1	Unit 2	Unit 3
1	35	Output (MWh)	500	0	50
		Revenue ($)	17 500	0	1 750
		Cost ($)	5 500	0	1 850
		Profit ($)	12 000	0	**−100**
2	30	Output (MWh)	500	250	0
		Revenue ($)	15 000	7 500	0
		Cost ($)	5 500	7 750	0
		Profit ($)	9 500	**−250**	0
3	35	Output (MWh)	500	350	200
		Revenue ($)	17 500	12 250	7 000
		Cost ($)	5 500	10 750	7 100
		Profit ($)	12 000	1 500	**−100**

Negative values are in bold.

Several approaches have been implemented or proposed to ensure that all generators at least recover their costs when bidding in a market where they do not have direct control over the price or the amount of energy that they have to produce.

Approach 1: Uplift Payments Compensating the Unrecovered Costs

In this approach, the market operator calculates the unrecovered cost $\Gamma_i(t)$ incurred by each generator at each period based on its complex bid and the marginal price π_t^*. To be able to reimburse these costs to the generators, the market operator must charge the consumers an uplift payment above the price calculated using marginal costs. This uplift is usually allocated on the basis of each consumer's demand during each period.

As Equation (3.9) shows, the total payment by the consumers for time period t consists of an energy payment at a price π_t^* and an uplift payment.

$$K(t) = \pi_t^* L(t) + \sum_{i=1}^{N} \Gamma_i \times \frac{L(t)}{\sum_{t=1}^{T} L(t)}, \quad \forall t \in T \tag{3.9}$$

Example 3.5

Based on the results of Example 3.4, generators 2 and 3 should receive uplift payments of \$250 and \$200, respectively. Table 3.10 shows how much money the consumers will pay at each hour for energy and as uplift payments calculated using Equation (3.9).

Table 3.10 Demand payments under approach 1.

Time period t	1	2	3
Energy payments at π_t^* (\$)	19 250	22 500	36 750
Uplift payments (\$)	105.3	143.6	201.1

Approach 2: Uplift Payments Compensating for Profit Suboptimality

Another way of looking at this issue is to treat each generating unit as a profit-seeking entity whose ability to maximize its profits is constrained by centralized scheduling. Uplift payments should therefore compensate generators for the difference between the profits that they collect under the minimum cost schedule and the profits that they would achieve if they were allowed to self-schedule given the market prices π_t^* determined by the centralized schedule. For each generating unit, this hypothetical self-schedule is given by the solution of the following optimization problem:

$$\max_{P_i(t),u_i(t)} \sum_{t=1}^{T} \left[\pi_t^* P_i(t) - C_i(P_i(t), u_i(t))\right] \tag{3.10}$$

subject to constraints (3.3)–(3.7)

Since the self-schedule always results in a larger profit than does the centralized schedule, the difference between these two values is called profit suboptimality. Under this approach, every generator that suffers from profit suboptimality is entitled to an uplift payment equal to this loss of profit. These payments are allocated to the demand side on the same basis as in the previous approach, i.e. according to Equation (3.9).

Example 3.6

Figure 3.3 shows what this self-schedule would be for the conditions of Example 3.3. It is clearly not a feasible schedule because it does not respect the load/generation balance constraint. Self-scheduling would make no difference for unit 1 because it produces its maximum output and does not set the marginal price. On the other hand, unit 2, which was not scheduled to produce during period 1 under the centralized schedule, would find it profitable to produce its maximum output at the marginal price $\pi_1^* = 35$ \$/MWh set by unit 3 because this price is high enough to allow it to recover its fixed costs. On the other hand, given the opportunity to self-schedule, unit 2 would not produce during period 2, and unit 3 would not produce during periods 1 and 3 because they set the marginal prices during those periods and do not recover their fixed costs. Table 3.11 summarizes the profits of each generator under centralized scheduling and self-scheduling, while

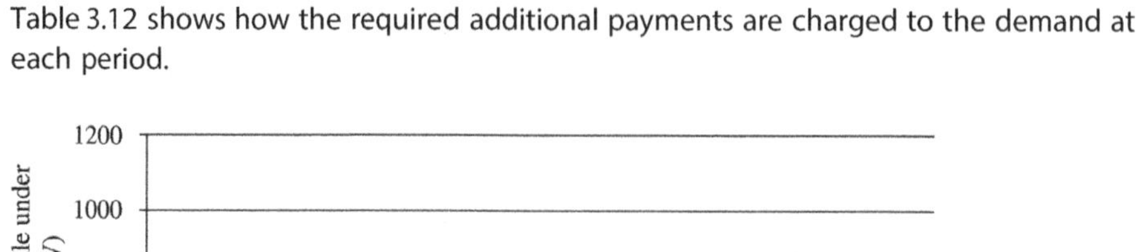

Table 3.12 shows how the required additional payments are charged to the demand at each period.

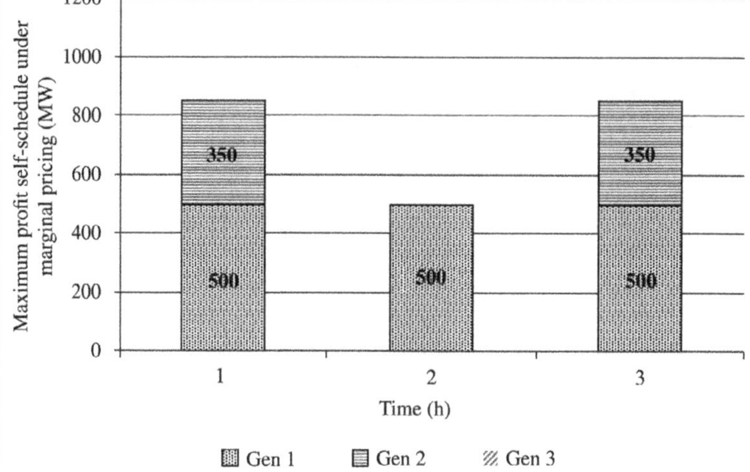

Figure 3.3 Profit-maximizing self-schedule under the marginal prices calculated by the centralized schedule.

Table 3.11 Minimum-cost schedule and self-schedule generation profits under marginal pricing.

Generator	1	2	3
Minimum-cost schedule profit ($)	33 500	1 250	−200
Self-schedule profit at π_t^* ($)	33 500	3 000	0
Profit suboptimality at π_t^* ($)	0	1 750	200

Table 3.12 Demand payment under approach 2.

Time period t	1	2	3
Energy payment at π_t^* ($)	19 250	22 500	36 750
Uplift payment ($)	456.4	622.3	871.3

Approach 3: Convex Hull Pricing

Uplift payments are somewhat controversial because they represent an out-of-market intervention. An alternative approach, called convex hull pricing, aims to minimize total uplift payments by adjusting the market prices. Under this scheme, electric energy prices are set to minimize the profit suboptimality (i.e. the difference between the profits from self-scheduling and centralized scheduling) across all generation units and time periods. These prices are called convex hull prices and are given by the Lagrange multipliers of the dual of the centralized scheduling problem. See Gribik et al. (2007) for a detailed discussion of this and other approaches.

3.3.3 Comparison of Centralized and Decentralized Trading

In the early days of the introduction of competition in electrical energy trading, bilateral trading was seen by many as too big a departure from existing practice. Since electrical energy is pooled as it flows from the generators to the loads, it was felt that trading might as well be done in a centralized manner and involve all producers and consumers. In fact, some of the centralized electricity markets currently in operation (PJM, for example) evolved from collaborative electricity pools that monopoly utility companies with adjacent service territories set up to take advantage of cost saving opportunities. The adoption of a centralized or a decentralized market model was therefore influenced by history and the character of industry organizations.

Day-ahead centralized electricity markets are usually run by the system operator. As we will see in subsequent chapters, this facilitates the integration of forward markets with the needs to maintain the balance of the system and to ensure operational reliability. This advantage is particularly important in systems where network constraints have a significant impact on the market. While this combination of roles avoids the multiplication of organizations, it also blurs the lines between the various functions that need to be performed in an electricity market.

Most small and medium electricity consumers have very little incentive to take an active part in an electricity market. Even when they are aggregated, the retailer that represents them has only a limited ability to adjust consumption in responses to changes in prices. The transaction costs are therefore reduced significantly if demand is deemed to be passive and is represented by a load forecast as is the case in many centralized markets. Many economists are unhappy with this approach because they feel that direct negotiations between consumers and producers are essential if efficient prices are to be reached. Some economists thus dislike centralized markets simply because they are only an administered approximation of a market rather than a true market.

Centralized markets also provide a mechanism for reducing the scheduling risk faced by generators and hence, hopefully, the cost of electrical energy. When a generator sells energy for each market period separately on the basis of simple offers, it runs the risk that for some periods it may not have sold enough energy to keep the plant on-line. At that point it must decide whether to sell energy at a loss to keep the unit running or to shut it down and face the expense of another startup at a later time. Either option increases the cost of producing energy with this unit and forces the generator to raise its average offer price. On the other hand, in a centralized market, the generation schedule produced by the unit commitment calculation avoids unnecessary unit shutdowns and the market rules typically ensure that generators recover their startup and no-load costs. Since these factors reduce the risks faced by the generators, they should, in theory, foster lower average prices. However, this reduction in risk requires the implementation of more complex market rules. Such rules reduce the transparency of the price setting process and increase opportunities for price manipulations.

Generators and retailers who participate in a day-ahead centralized market are typically obliged to sell their entire production or buy their total consumption through this market. However, they usually retain the right to enter into bilateral financial contracts of the type described in Section 2.5.5 to manage the price risks that they are exposed to in this day-ahead centralized market.

Some centralized day-ahead markets also allow bilateral trading. The quantities traded bilaterally are taken into account in the clearing process, but these transactions do not set the market price.

3.4 Spot Markets

Retailers and large consumers cannot predict their needs for electrical energy consumption with perfect accuracy. Similarly, generators cannot guarantee that they will be able to produce the exact quantity that they have sold on forward markets, particularly if all or part of this energy is produced from stochastic renewable sources such as wind or solar. At the time of delivery, any market participant who has more or less energy than it needs or has contracted to buy or sell must be able to trade on the spot market to cover the difference.

While spot markets for other commodities rely on direct interactions between buyers and sellers to cover imbalances, this approach is not considered sufficiently dependable to ensure the operational reliability of the power system. The physical counterpart of an imbalance between demand and supply is an imbalance between load and generation, which must be covered very rapidly if the power system is to remain stable. While electronic markets for some financial instruments have become extremely fast in recent years, this form of automated trading relies on a limited number of price signals. On the other hand, maintaining the stability of the power system requires monitoring and assessing a large number of physical quantities. In the current state of the technology, it is difficult to conceive a mechanism where enough generators and consumers would have the ability and the inclination to process this information sufficiently fast to enter into trades that would restore the balance between load and generation in a reliable manner.

The market of last resort for electrical energy is therefore not a true spot market. It is instead what one might call a "managed spot market" because the system operator is counterparty to all trades. This means that market participants do not buy or sell from each other but deal only with the system operator, who decides how much power it needs to buy or sell to maintain the system's stability. The spot price is then calculated based on the bids and offers submitted by the market participants and selected by the system operator. Each system operator runs its own spot market and all generators, retailers, large consumers, and speculators who trade energy within the area overseen by a system operator are subject to its rules. In particular, any difference between what a market participant was contractually committed to do and what it actually did is settled at the spot market price. Let us assume, for example, that a generator had sold 100 MW to various retailers but produced only 97 MW during a given spot market trading period. To maintain the system in balance, the system operator covered the difference through purchases on the spot market. From a financial perspective, the generator is deemed to have purchased the corresponding amount of energy at the spot price. Similarly, a retailer who contracted for 100 MW but consumed only 96 MW during a trading period is deemed to have sold the difference at the spot market price.

Although the need to manage the spot market stems from technical considerations, this spot market must operate in an economically efficient manner. Being out of balance may be unavoidable for producers and consumers, but it should not be cost-free. To encourage efficient behavior, producers and consumers who are in imbalance must pay the true cost of the electrical energy that is bought or sold on the spot market to restore the balance between load and generation. Note that if a market participant is out of balance but in the opposite direction of the system as a whole, it should be rewarded for that.

An efficient spot market gives market participants the confidence that imbalances will be settled in a fair manner. Once such a mechanism is in place, electrical energy can be traded in forward markets like other commodities. Depending on the jurisdiction, the spot market is called "real-time market," "balancing market," "intraday market," or "balancing mechanism" and the trading period ranges from 1 h to 5 min.

In the next paragraphs, we discuss the functionality of a generic spot market for electricity. Actual implementations can differ substantially from this blueprint.

3.4.1 Obtaining Balancing Resources

If market participants were able to predict with enough lead time and with perfect accuracy the amount of energy that they will consume or produce, the system operator would not have to take balancing actions. The participants themselves could trade to cover their deficits and absorb their surpluses. In practice, there are always small imbalances and the system operator must obtain adjustments in generation or load. Integrated over time, these adjustments translate into purchases and sales of electrical energy that can be settled at a spot price reflecting the market's willingness to provide or absorb extra energy. In keeping with the free market philosophy, any party that is willing to adjust its production or consumption should be allowed to do so on a competitive basis. This approach provides the system operator with the widest possible choice of balancing options and should therefore help reduce the cost of balancing. Balancing resources can be offered either for a specific period or on a long-term basis. Resources for a specific trading period are normally offered by market participants on the spot market after the forward market for that period has closed. Generating units that are not fully loaded can submit bids to increase their output. A generating unit can also offer to pay to reduce its output. This is a profitable proposition if the price that this generator has to pay is less than the incremental cost of producing energy with this unit. A generating unit that submits such an offer is in effect trying to replace its own generation by cheaper power purchased on the spot market.

The demand side can also provide balancing resources. A consumer could offer to reduce its consumption if the price it would receive for this reduction is greater than the value it places on consuming electricity during that period. Such demand reductions have the advantage that they can be implemented very quickly. It is also conceivable that consumers might offer to increase their demand if the price is sufficiently low.

Since these offers of balancing resources are submitted shortly before real time, the system operator may be concerned about the amount or the price of balancing resources that will be offered. To reduce the risk of not having enough balancing resources or of having to pay a very high price for these resources, it can purchase balancing resources on a long-term basis. Under such contracts, the supplier is paid a fixed price (often called the option fee) to keep available some generation capacity. The contract also specifies the price or exercise fee to be paid for each megawatt-hour produced using this capacity. The system operator would call upon this contract only if the exercise fee is lower than what it would have to pay for a similar balancing resource offered on a short-term basis. As the terminology suggests, these contracts are equivalent to the option contracts used in financial and commodity markets. Their purpose is the same: to give the buyer (in this case the system operator) some flexibility and some protection against high prices while guaranteeing some revenues to the supplier.

Imbalances due to forecasting errors by the participants are relatively small, evolve gradually, and have a known probability distribution. On the other hand, imbalances caused by generator failures are often large, unpredictable, and sudden. Many generating units can adjust their output at a rate that is sufficient to cope with the first type of imbalances. Handling the second type of imbalances requires resources that can increase their output rapidly and sustain this increased output for a sufficiently long time. We will explore the issue of reserve generation capacity in more detail when we discuss operational reliability in Chapter 6. In the meantime, it is important to realize that all the units of energy that are traded to keep the system in balance do not have the same value. A megawatt obtained by increasing slightly the output of a large thermal plant costs considerably less than a megawatt of load that must be shed to prevent the system from collapsing. To be able to keep the system in balance at minimum cost, the system operator should therefore have access to a variety of balancing resources. When producers and consumers bid to supply balancing resources, their bids must specify not only a quantity and a price but also what constraints limit their ability to deliver a change in power injection.

3.4.2 Gate Closure

As we argued above, forward markets must close at some point before real time to give the system operator control over what happens in the system. How much time should elapse between this gate closure and real time depends on the perspective that one has on the market. System operators prefer a longer interval because it gives them more time to develop their plans and more flexibility in their selection of balancing resources. For example, if the gate closes half an hour before real time, there is not enough time to bring on-line a large coal-fired plant to make up a deficit in generation. Generators and retailers, on the other hand, usually prefer a shorter interval between gate closure and real time because it reduces their exposure to risk. A load or wind generation forecast calculated 1 h ahead of real time is usually much more accurate than a forecast calculated 4 h ahead. If a generating unit fails after gate closure, there is nothing that its owner can do except hope that the spot market price will not be too high. On the other hand, if it fails before gate closure, its owner can try to make up the deficit in generation by purchasing at the best possible price on the electronic exchange. Market participants would therefore prefer to trade electronically up to the last minute to match their contractual position with their anticipated load or production. This is considered preferable to relying on the spot market where participants are exposed to prices over which they have no control. Furthermore, traders prefer dealing on a true market that is driven solely by market forces rather than on a spot market that is heavily influenced by complex considerations about the operation of the power system.

3.4.3 Operation of the Spot Market

Figure 3.4 is a schematic representation of the operation of a spot market for electrical energy. At gate closure, the producers and the consumers must inform the system

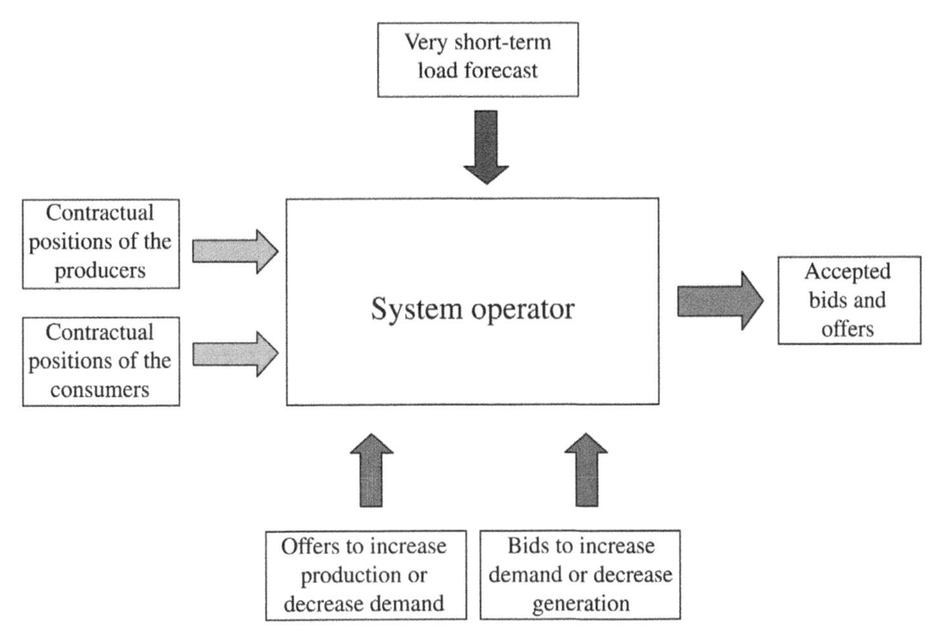

Figure 3.4 Schematic diagram of the operation of a managed spot market for electricity.

operator of their contractual positions, i.e. how much power they intend to produce or consume during the upcoming trading period. The system operator combines that information with its own forecast of the total load to determine by how much the system is likely to be in imbalance. If generation exceeds the load, the system is said to be long. If the opposite holds, the system is short. The system operator must then decide which balancing bids and offers it will use to cover the imbalances.

In a centralized electricity market, the balancing function is often so closely integrated with the energy market function that they are difficult to separate.

Example 3.7

Let us revisit our trader for Borduria Power from Example 3.1 and see how she handles the spot market. When the gate closes for bilateral trading, Fiona has contracted to produce a net amount of 580 MWh for the period under consideration. She informs the system operator that her company intends to produce this amount as follows:

Unit	Scheduled production (MW)
A	500
B	80
C	0

Fiona then has to decide what bids and offers she wants to make in the spot market. To help her in this decision, she considers the scheduled output and the characteristics of Borduria Power's generating units:

Unit	$P^{scheduled}$ (MW)	P^{min} (MW)	P^{max} (MW)	MC ($/MWh)
A	500	100	500	10.0
B	80	50	80	13.0
C	0	0	50	17.0

The only bid to increase generation that she can offer involves unit C because units A and B are scheduled to produce their maximum output. Such a bid would be for a maximum of 50 MWh and would have to be priced at 17.00 $/MWh or above to be profitable if we assume that the startup cost of unit C is negligible.

Fiona also considers the possibility of reducing the production of units A and B. She would be willing to pay up to 10 $/MWh to reduce the output of unit A and up to 13 $/MWh to reduce the output of unit B because that is the incremental cost of producing power with these units. The output of these units can be reduced by 400 and 30 MW, respectively, without affecting the plans for the following periods, if we assume that there are no restrictions on the ramp rate of these units. Further reductions would require their shutdown and might then preclude Borduria Power from meeting its commitments for later trading periods. In addition, the cost of restarting these units would reduce their profitability.

In Chapter 2 we argued that, in a perfectly competitive market, the optimal strategy of each participant is to offer at its incremental cost or to bid at its incremental value. As we will discuss in the next chapter, electricity markets are usually not perfectly competitive. Some participants can increase their profits by offering above their marginal costs or bidding below their marginal value. Based on her experience, Fiona decides that the following bids and offers are likely to maximize Borduria Power's profit:

Type	Identifier	Price ($/MWh)	Amount (MW)
Bid	SMB-1	17.50	50
Offer	SMO-1	12.50	30
Offer	SMO-2	9.50	400

We will revisit this example one more time when discussing the settlement process.

3.4.4 Interactions Between the Spot Market and the Forward Market

Since the spot market is the market of last resort, it has a strong influence on forward markets. If the spot price tends to be low, purchasers of energy will not be unduly concerned about being short because they can make up their deficits in the spot market at a reasonable price. They might therefore buy slightly less than they need in forward markets and will thus push down the price of energy in these markets. On the other hand, if the spot price tends to be high, these same purchasers will push up the price in the forward market as they buy more to make sure that they cover all their needs at the better price. If electricity was a simple commodity, such discrepancies should even themselves out over time and the forward price should reflect the expected value of the spot price. Some centralized electricity markets support "virtual bidding" to facilitate this convergence between forward and spot prices. Virtual bids are submitted by market participants who do not own physical assets and thus have no intention of producing or consuming electricity. Instead, they submit a bid to buy (or sell) in the day-ahead market and an equivalent bid to sell (or buy) in the spot market. Allowing these virtual transactions increases the number of market participants, and there is some evidence that this helps the day-ahead prices converge toward the spot prices. It may also reduce the potential for abuse of market power.

Electricity is certainly not the only commodity whose spot price is highly volatile. A weather forecast predicting frost in the coffee-producing regions of Brazil will send the spot price of coffee soaring. This price may very well come crashing down the next day if the forecast turns out to be inaccurate or the damage to the crop is more limited than was feared. The difference between coffee and electrical energy is that coffee traded on the spot market is produced exactly the same way as coffee sold under a long-term contract. On the other hand, a megawatt-hour sold on the spot market often will have been produced by a plant that is much more flexible than the plants that generate the bulk of the energy consumed at that time. Flexible plants may not be competitive in the forward market, but their flexibility makes it possible for them to occupy this market niche. However, because the amount of electrical energy that they deliver is often relatively small, they may have to charge a high price per megawatt-hour to collect enough revenue to stay in business. Including the cost of this flexible energy in the calculation of the spot price will often result in sharp price spikes. These price spikes do not reflect a sudden deficit in electrical energy in the market. They are a consequence of an acute but temporary lack of liquidity. Price spikes occur because for a short time the number of participants able to provide this energy is very small and because consumers are not able or not willing to reduce their demand at short notice. Price spikes represent a risk for companies that are forced to buy from the spot market and will encourage them to purchase more in the forward market and hence drive up forward prices. These forward prices are thus artificially inflated by the need to generate a small fraction of the total energy demand at short notice.

3.5 The Settlement Process

Commercial transactions are normally settled directly between the two parties involved: following the delivery of the goods by the seller to the buyer, the buyer pays the seller the agreed price. If the amount delivered is less than the amount contracted, the buyer is entitled to withhold part of the payment. Similarly, if the buyer consumes more than the agreed amount, the seller is entitled to an additional payment. This process is more complex for electricity markets because the energy is pooled during its transmission from the producers to the consumers. This is the reason why a centralized settlement system is needed.

For bilateral transactions in electrical energy, the buyer pays the seller the agreed price as if the agreed quantity had been delivered exactly. Similarly, the anonymous transactions arranged through screen-based trading are settled through the intermediary of the power exchange as if they had been executed perfectly. However, there will always be inaccuracies in the completion of the contracts. If a generator fails to produce the amount of energy that it has contracted to sell, the deficit cannot simply be withheld from this generator's customers. Instead, to maintain the stability of the system, the system operator buys replacement energy on the spot market on behalf of the generator. Similarly, if a large user or retailer consumes less than it has bought, the system operator sells the excess on the spot market. These balancing activities make all bilateral contracts look as if they have been fulfilled perfectly. They also carry a cost. In most cases, the amount of money paid by the system operator to purchase replacement energy is not equal to the amount of money earned when selling excess energy. The parties that are responsible for the imbalances should pay the cost of these balancing activities.

The first step in the settlement process consists therefore in determining the net position of every market participant. To this end, each generator must report to the settlement system the net amount of energy that it had contracted to sell for each period. This amount is subtracted from the amount of energy that it actually produced. If the result is positive, the generator is deemed to have sold this excess energy to the system. On the other hand, if the result is negative, the generator is treated as if it had bought the difference from the system.

Similarly, all large consumers and retailers must report the net amount of energy that they had contracted to buy for each period. This amount is subtracted from the amount of energy actually consumed. Depending on the sign of the result, the consumer or retailer is deemed to have sold energy to the system or bought energy from the system.

These imbalances are charged at the spot market price. If this market is suitably competitive, this price should reflect the incremental cost of balancing energy.

Settlement in a centralized electricity market is more straightforward because all physical transactions are handled by the system operator. These markets typically implement what is called a "two-settlement system." In such a system, quantities scheduled on the day-ahead market are settled at the day-ahead price, while deviations between scheduled and actual quantities for each market period are settled at the spot market (real-time) price for that market period.

Example 3.8

In Examples 3.1 and 3.7 we looked at the trading activities of Borduria Power for the trading period from 2:00 to 3:00 p.m. on June 11 in the bilateral market and the spot market. Let us assume that the following events took place after gate closure:

- Faced with a deficit in generation, the system operator called 40 MWh of Borduria Power's spot market bid (SMB-1).
- The troubles with Borduria Power's unit B turned out to be worse than anticipated, forcing its complete shutdown soon after the beginning of the period. It was only able to produce 10 of the 80 MWh that it was scheduled to produce, leaving Borduria Power with a deficit of 70 MWh.
- The spot price of electrical energy was 18.25 $/MWh for this trading period.

Table 3.13 shows the detail of the flows of money in and out of Borduria Power's trading account.

Table 3.13 Borduria Power trading account.

Market	Identifier	Amount (MWh)	Price ($/MWh)	Income ($)	Expense ($)
Futures and forwards	Cheapo Energy	200	12.50	2500.00	
	Borduria Steel	250	12.80	3200.00	
	Quality Electrons	100	14.00	1400.00	
	Perfect Power	−30	13.50		405.00
	Cheapo Energy	50	13.80	690.00	
Power exchange	B1	20	13.50	270.00	
	B2	30	13.30	399.00	
	B3	10	13.25	132.50	
	O3	−20	14.40		288.00
	O6	−20	14.30		286.00
	O8	−10	14.10		141.00
Spot market	SMB-1	40	18.25	730.00	
	Imbalance	−70	18.25		1277.50
Total		550		9321.50	2397.50

Bilateral trades are settled directly between Borduria Power and its counterparties. Since trades on the power exchange are anonymous, they are settled through BPeX (the company running the power exchange). Finally, activity on the spot market (both voluntary and compulsory) is settled through the system operator or its settlement agent. The bottom line of this table indicates that the net trading revenue of Borduria Power for this period amounts to $6924.00. To determine whether this trading period was profitable, we would have to compute the cost of producing the energy that Borduria Power delivered. However, carrying out this computation for a single trading period would not be meaningful because there is no unambiguous way to allocate the startup costs of the generating units.

3.6 Problems

3.1 Choose an electricity market about which you have access to sufficient information, preferably the same market that you studied for the problems of Chapter 1. Describe the implementation of this market. In particular, determine the aspects that are based on bilateral trading and those that are centrally operated. Discuss the mechanism used to set prices in the spot market.

3.2 The rules of the Syldavian electricity market stipulate that all participants must trade energy exclusively through the Power Pool. However, the Syldavia Aluminum Company (SALCo) and the Northern Syldavia Power Company (NSPCo) have signed a contract for difference for the delivery of 200 MW on a continuous basis at a strike price of 16 $/MWh.

a Trace the flow of power and money between these companies when the pool price takes the following values: 16 \$/MWh, 18 \$/MWh, and 13 \$/MWh.

b What happens if during 1 h the Northern Syldavia Power Company is only able to deliver 50 MWh and the pool price is 18 \$/MWh?

c What happens if during 1 h the Syldavia Aluminum Company consumes only 100 MWh and the pool price is 13 \$/MWh?

3.3 The following six companies participate, along with others, in the Southern Antarctica electrical energy market:

- *Red*: A generating company owning a portfolio of plants with a maximum capacity of 1000 MW.
- *Green*: Another generating company with a portfolio of plants with a maximum capacity of 800 MW.
- *Blue*: A retailer of electrical energy.
- *Yellow*: Another retailer of electrical energy.
- *Magenta*: A trading company with no generating assets and no demand.
- *Purple*: Another trading company with no physical assets.

The following information pertains to the operation of this market for Monday, February 29, 2016 between 1:00 and 2:00 p.m.

Load Forecasts

Blue and Yellow forecast that their customers will consume, respectively, 1200 and 900 MW during that hour.

Long-term contracts

June 2015: Red signs a contract for the supply of 600 MW at 15 \$/MWh for all hours between January 1, 2015 and December 31, 2020.

July 2015: Blue signs a contract for the purchase of 700 MW for all hours between February 1, 2016 and December 31, 2016. The price is set at 12 \$/MWh for off-peak hours and at 15.50 \$/MWh for peak hours.

August 2015: Green signs a contract for the supply of 500 MW at 16 \$/MWh for peak hours in February 2016.

September 2015: Yellow signs a contract for the purchase of electric energy. The contract specifies a profile of daily and weekly volumes and a profile for daily and weekly prices. In particular, on weekdays between 1:00 and 2:00 p.m., the volume purchased is 550 MW at 16.25 \$/MWh.

Futures contracts: All contracts are for delivery on February 29, 2016 between 1:00 and 2:00 p.m.

Date	Company	Type	Amount	Price
10/9/15	Magenta	Buy	50	14.50
20/9/15	Purple	Sell	100	14.75
30/9/15	Yellow	Buy	200	15.00
10/10/15	Magenta	Buy	100	15.00
20/10/15	Red	Sell	200	14.75
30/10/15	Green	Sell	250	15.75

(*Continued*)

Date	Company	Type	Amount	Price
30/10/15	Blue	Buy	250	15.75
10/11/15	Purple	Buy	50	15.00
15/11/15	Magenta	Sell	100	15.25
20/11/15	Yellow	Buy	200	14.75
30/11/15	Blue	Buy	300	15.00
10/12/15	Red	Sell	200	16.00
15/12/15	Red	Sell	200	15.50
20/12/15	Blue	Sell	50	15.50
15/1/16	Purple	Sell	200	14.50
20/1/16	Magenta	Buy	50	14.25
10/2/16	Yellow	Buy	50	14.50
20/2/16	Red	Buy	200	16.00
25/2/16	Magenta	Sell	100	17.00
28/2/16	Purple	Buy	250	14.00
28/2/16	Yellow	Sell	100	14.00

Option contracts

In November 2015, Red bought a put option for 200 MWh at 14.75 $/MWh. The option fee was $50.

In December 2015, Yellow bought a call option for 100 MWh at 15.50 $/MWh. The option fee was $25.

Outcome

- The spot price on the Southern Antarctica electricity market was set at 15.75 $/MWh for February 29, 2016 between 1:00 and 2:00 p.m.
- Due to difficulties at one of its major plants, Red was only able to generate 800 MW. Its average cost of production was 14.00 $/MWh.
- Green generated 770 MW at an average cost of 14.25 $/MWh.
- Blue's demand turned out to be 1250 MW. Its average retail price was 16.50 $/MWh.
- Yellow demand turned out to be 850 MW. Its average retail price was 16.40 $/MWh.

Assuming that all imbalances are settled at the spot market price, calculate the profit or loss made by each of these participants.

3.4 The operator of a centralized market for electrical energy has received the bids shown in Table 3.14 for the supply of electrical energy during a given period:

a Build the supply curve.

b Assume that this market operates unilaterally, i.e. that the demand does not bid and is represented by a forecast. Calculate the market price, the quantity produced by each company, and the revenue of each company for each of the following loads: 400 MW, 600 MW, 875 MW.

Table 3.14 Bids in the centralized market of Problem 3.4.

Company	Amount (MWh)	Price ($/MWh)
Red	100	12.5
Red	100	14.0
Red	50	18.0
Blue	200	10.5
Blue	200	13.0
Blue	100	15.0
Green	50	13.5
Green	50	14.5
Green	50	15.5

c Suppose that instead of being treated as constant, the load is represented by its inverse demand curve, which is assumed to have the following form:

$$D = L = 4.0 \cdot \pi$$

where D is the demand, L is the forecasted load, and π is the price. Calculate the effect that this price sensitivity of demand has on the market price and the quantity traded.

3.5 The Syldavian Power and Light Company owns one generating plant and serves some load. It has been actively trading in the electricity market and has established the following positions for June 11 between 10:00 and 11:00 a.m.:

- Long-term contract for the purchase of 600 MW during peak hours at a price of 20.00 $/MWh.
- Long-term contract for the purchase of 400 MW during off-peak hours at a price of 16.00 $/MWh.
- Long-term contract with a major industrial user for the sale of 50 MW at a flat rate of 19.00 $/MWh.
- The remaining customers purchase their electricity at a tariff of 21.75 $/MWh.
- Futures contract for the sale of 200 MWh at 21.00 $/MWh.
- Futures contract for the purchase of 100 MWh at 22.00 $/MWh.
- Call option for 150 MWh at an exercise price of 20.50 $/MWh.
- Put option for 200 MWh at an exercise price of 23.50 $/MWh.
- Call option for 300 MWh at an exercise price of 24.00 $/MWh.

The option fee for all the options is 1.00 $/MWh. The peak hours are defined as being the hours between 8:00 a.m. and 8:00 p.m.

The outcome for June 11 between 10:00 and 11:00 is as follows:

- The spot price is set at 21.50 $/MWh.
- The total load of the Syldavian Power and Light Company is 1200 MW, including the large industrial customer.
- The power plant produces 300 MWh at an average cost of 21.25 $/MWh.

a Assuming that all imbalances are settled at the spot market price, calculate the profit or loss made by the company during that hour.

b What value of the spot market would reduce the profit or loss of the company to zero? Would this change in the spot price affect any of the option contracts?

3.6 Borduria Energy is involved in several commercial activities related to electrical energy: generation, bulk sales to large consumers, retail sales to small consumers, and energy trades with other market participants. The following transactions are in effect for the trading period between 10:00 and 11:00 am on July 21:

- Long-term contract with a generation company for the purchase of 500 MW during peak hours at a price of 23.00 $/MWh.
- Long-term contract for the purchase of 300 MW during off-peak hours at a price of 14.00 $/MWh.
- Long-term contract with another electricity retailer for the sale of electrical energy. Figure P3.6a shows the quantities sold at each hour and Figure P3.6b the prices.
- Futures contract for the sale of 200 MWh at 22.00 $/MWh.
- Futures contract for the purchase of 100 MWh at 24.00 $/MWh.
- Call option for 250 MWh at an exercise price of 23.50 $/MWh.
- Put option for 200 MWh at an exercise price of 22.50 $/MWh.
- Call option for 300 MWh at an exercise price of 21.75 $/MWh.
- The tariff paid by small residential and commercial customers is 26.00 $/MWh.

When the market closes for that hour, the spot price is set at 22.25 $/MWh. The total load of the small consumers served by Borduria Energy turns out to be 800 MWh during that hour, while power plants owned by the company produced 300 MWh at an average cost of 21.25 $/MWh. Note that the option fee for all the options is 2.00 $/MWh and that the peak hours are defined as being the hours between 8:00 a.m. and 8:00 p.m.

Assuming that all imbalances are settled at the spot market price, calculate the profit or loss made by the company during that hour.

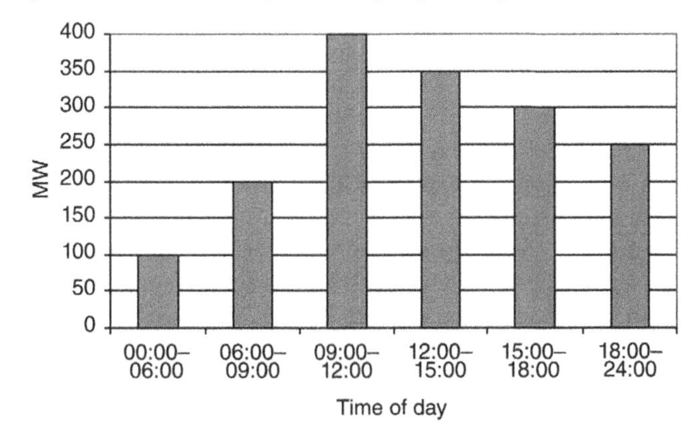

Figure P3.6a Quantities sold under long-term contract.

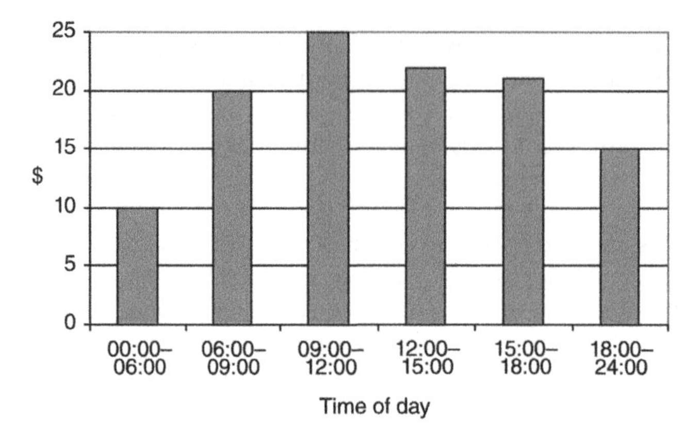

Figure P3.6b Prices for long-term contract ($/MWh).

3.7 A company called Borduria Energy owns a nuclear power plant and a gas-fired power plant. Its trading division has entered into the following contracts for January 25:

T-1. A forward contract for the sale of 50 MW at a price of 21.00 $/MWh. This contract applies to all hours.

T-2. A long-term contract for the sale of 300 MW during off-peak hours at a price of 14.00 $/MWh.

T-3. A long-term contract for the sale of 350 MW at 20 $/MWh during peak hours.

In addition, for the trading period from 2:00 to 3:00 p.m. on that day, it has entered into the following transactions:

T-4. A futures contract for the purchase of 600 MWh at 20.00 $/MWh.

T-5. A futures contract for the sale of 100 MWh at 22.00 $/MWh.

T-6. A put option for 250 MWh at an exercise price of 23.50 $/MWh.

T-7. A call option for 200 MWh at an exercise price of 22.50 $/MWh.

T-8. A put option for 100 MWh at an exercise price of 18.75 $/MWh.

T-9. A bid in the spot market to produce 50 MW using its gas-fired plant at 19.00 $/MWh.

T-10. A bid in the spot market to produce 100 MW using its gas-fired plant at 22.00 $/MWh.

The option fee for all call and put options is $2.00/MWh. The peak hours are defined as being the hours between 8:00 a.m. and 8:00 p.m.

Borduria Energy also sells electrical energy directly to small consumers through its retail division. Residential customers pay a tariff of 25.50 $/MWh and commercial consumers pay a tariff of 25.00 $/MWh. Borduria Energy does not sell electricity to industrial consumers.

The graph in Figure P3.7 shows the stack of bids that the spot market operator has received for the trading period from 2:00 to 3:00 p.m. on January 25. In order to balance the load and generation, it accepted bids for 225 MW in increasing order of price for that hour. The spot price was set at the price of the last accepted bid.

During that hour, the residential customers served by Borduria Energy consumed 300 MW while its commercial customers consumed 200 MW. The nuclear power plant produced 400 MWh at an average cost of 16.00 $/MWh. Its gas-fired plant produced 200 MWh at an average cost of 18.00 $/MWh. All imbalances were settled at the spot market price.

a Calculate the profit or loss made by Borduria Energy during that hour.

b Calculate the effect that the sudden outage of the nuclear generating plant at 2:00 p.m. on January 25 would have on the profit (or loss) of Borduria Energy for that hour.

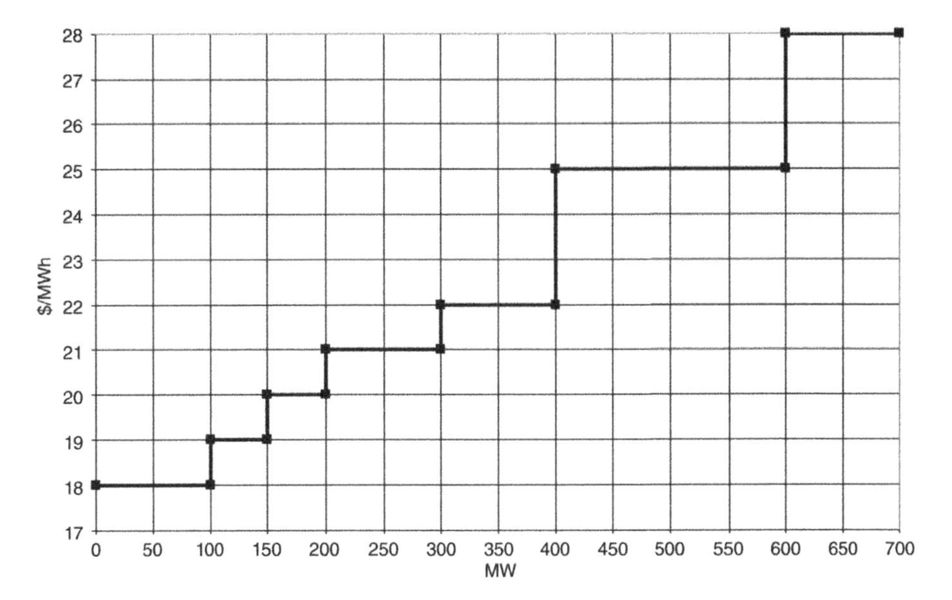

Figure P3.7 Stack of bids.

3.8 A company called "Dragon Power" owns a single generating unit whose cost function and operating limits are shown in Figure P3.8. This company participates in the Syldavian electricity market, which consists of bilateral trading followed by a managed spot market.

Consider the trading period from 9:00 to 10:00 a.m. on June 11. Prior to gate closure for that trading period, Dragon Power has made the following relevant bilateral trades:

Reference	Type	Quantity (MWh)	Price ($/MWh)
A1	Sold	200	0.16
A2	Sold	100	0.22
A3	Sold	75	0.30
A4	Bought	125	0.28
A5	Sold	25	0.25

Dragon Power can submit bids to increase or decrease its energy production in the managed spot market. Considering the bilateral trades that Dragon Power has concluded, determine the quantity and price of competitive bids that could be submitted. Assume that the managed spot market is perfectly competitive and that the generating unit cannot be shut down.

During the settlement process for this trading period, it turns out that:

- None of the bids submitted by Dragon Power in the managed spot market was accepted.
- The generating unit owned by Dragon Power produced 225 MWh during that hour.
- The spot price on the managed spot market was 0.35 $/MWh. All imbalances are traded at this spot price.

Calculate the overall operating profit or loss made by Dragon Power during that hour.

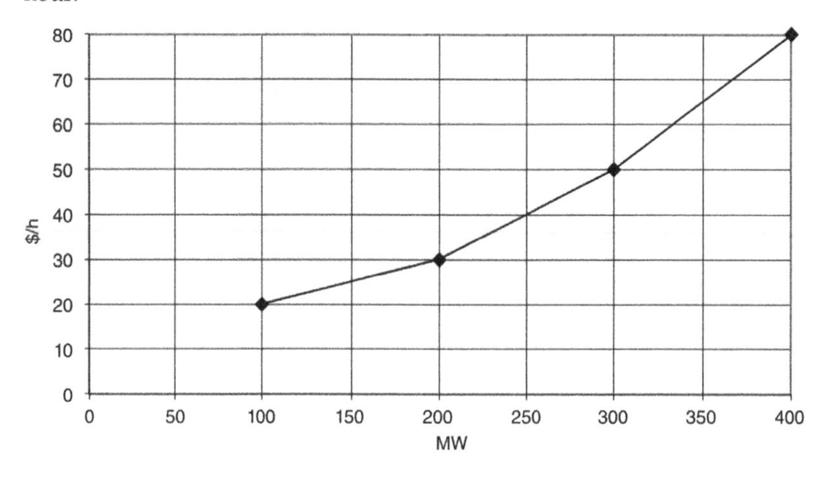

Figure P3.8 Cost function of the generating unit.

3.9 A spot market and a futures and option market have been established for trading electrical energy in Syldavia. There are a number of participants in this market, but we consider only the following three players:

Syldavian Genco: A generating company owning a portfolio of plants with a maximum capacity of 800 MW.

Syldavian Power and Light: A company that owns one generating plant with a maximum capacity of 500 MW and serves load for one industrial user and residential consumers.

Borduria Investment: A trading company with no generating assets and no demand.

We focus on contracts for delivery on May 14 between 1:00 and 2:00 p.m. These companies are parties to the following trades during this period.

Long-term contracts

June 11, 2011: Syldavian Genco signed a contract for the supply of 600 MW during peak hours at a price of 20.00 $/MWh.

July 04, 2012: Syldavian Genco signed a contract for the supply of 400 MW during off-peak hours at a price of 16.00 $/MWh.

August 04, 2013: Syldavian Power and Light signed a long-term contract with the industrial user for the sale of 50 MW at a flat rate of 19.00 $/MWh.

January 01, 2015: The regulatory agency required Syldavian Power and Light to sell electricity to residential customers at a tariff of 21.75 $/MWh.

June 11, 2009: Syldavian Power and Light signed a contract for the purchase of 800 MW for all hours from January 1, 2015 to December 31, 2015. The price is 18 $/MWh for off-peak hours and 21 $/MWh for peak hours.

Futures contracts

Company	Type	Amount	Price
Borduria Investment	Buy	50	17.50
Borduria Investment	Buy	100	19.00
Syldavian Genco	Sell	200	22.75
Syldavian Power and Light	Buy	100	22.00
Borduria Investment	Sell	100	20.25
Syldavian Power and Light	Sell	150	24.00
Syldavian Genco	Buy	200	19.25
Syldavian Power and Light	Sell	50	19.25
Borduria Investment	Buy	50	18.25
Syldavian Genco	Buy	200	19.00
Borduria Investment	Sell	100	20.00
Borduria Investment	Sell	100	22.00

Option Contracts

- On March 1, 2015, Syldavian Genco bought a put option for 200 MWh at $20.75/MWh.
- On May 1, 2015, Syldavian Power and Light bought a call option for 150 MWh at an exercise price of 20.50 $/MWh.
- On May 7, 2015, Syldavian Power and Light bought a put option for 200 MWh at an exercise price of 23.50 $/MWh.
- On May 10, 2015, Syldavian Power and Light bought a call option for 300 MWh at an exercise price of 24.00 $/MWh.

Spot Market

Outcome for May 14, 2015 between 1:00 and 2:00 p.m.

- The spot price was 21.50 $/MWh.
- The total load of the Syldavian Power and Light Company was 1400 MW, including the large industrial customer.
- The power plant of Syldavian Power and Light produced 300 MWh at an average cost of 21.25 $/MWh.
- Syldavian Genco generated 730 MW at an average cost of 21.20 $/MWh.

The option fee for all the options is 1.00 $/MWh.

The peak hours are defined as being the hours between 8:00 a.m. and 8:00 p.m.

a Assuming that all imbalances are settled at the spot market price, calculate the profit or loss made by each company during that hour.

b What value of the spot price would reduce the profit or loss of Syldavian Genco to zero?

c Assume that the power plant Syldavian Power and Light owns has a cost function of $C = 0.015P^2 + 9P + 2325$ and that this company has full control over the output of this plant at that hour. Was it optimal for this plant to produce 300 MWh during this period? If not, what should it have produced?

3.10 The load profile shown in Table 3.15 must be supplied using the units whose characteristics are summarized in Table 3.16. Table 3.17 summarizes three solutions to this problem. Check the feasibility of each of these solutions. If a solution is

Table 3.15 Load profile for Problem 3.10.

Hour	1	2	3	4
Load (MW)	400	500	600	400

Table 3.16 Characteristics of the units of Problem 3.10.

Unit	Minimum MW	Maximum MW	Min up time	Min down time	Startup cost ($)	Initial status
A	25	100	1	1	5	Down for 6 h
B	50	150	3	3	200	Down for 1 h
C	150	250	3	3	600	Up for 6 h
D	200	400	6	6	800	Up for 12 h

Table 3.17 Potential solutions to the unit commitment of Problem 3.10.

Solution	Units	Hour 1	Hour 2	Hour 3	Hour 4
S1	A	OFF	OFF	OFF	OFF
	B	OFF	100	OFF	OFF
	C	100	250	200	200
	D	250	150	400	200
S2	A	OFF	OFF	100	OFF
	B	OFF	OFF	OFF	OFF
	C	150	250	250	200
	D	250	250	250	200
S3	A	OFF	50	OFF	OFF
	B	OFF	OFF	40	100
	C	150	OFF	110	150
	D	250	450	450	150

infeasible, indicate **all** the constraints that are not satisfied. Assume that 60 MW of reserve must be carried at all times. Using the cost curves of these units shown in Figure P3.10, calculate the total cost of all the feasible solutions.

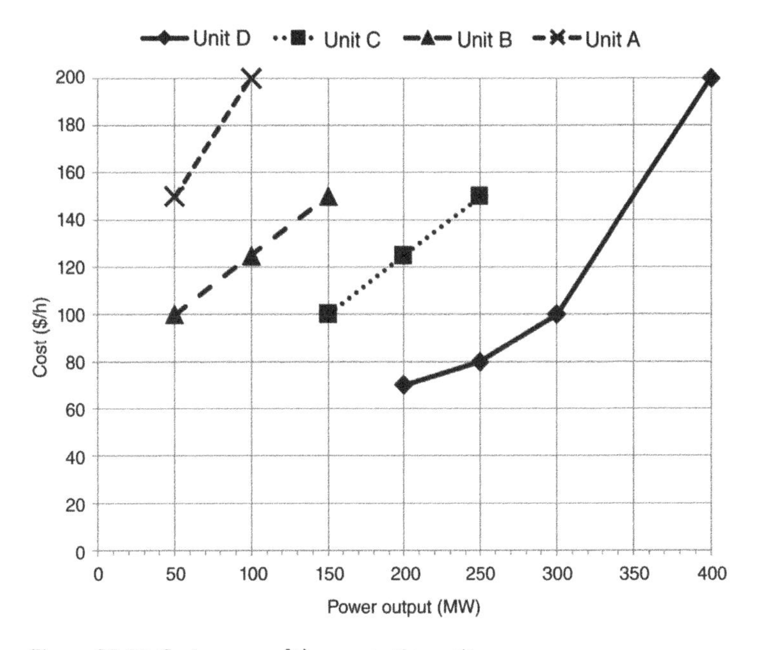

Figure P3.10 Cost curves of the generating units.

3.11 A small power system is supplied by three generators. The technical and cost characteristics of these generators are shown in Table 3.18.

These generators must supply the load profile shown in Table 3.19. Assume that no reserve is required (a bad idea but it makes the problem much simpler!).

Table 3.18 Characteristics of the generating units of Problem 3.11.

Unit	P_{min} (MW)	P_{max} (MW)	Min up (h)	Min down (h)	No-load cost ($)	Marginal cost ($/MWh)	Startup cost ($)	Initial status
A	180	250	3	3	0	10	1000	OFF for 5 h
B	70	100	2	2	0	12	600	ON for 3 h
C	10	50	1	1	0	20	150	ON for 3 h

Table 3.19 Load profile for Problem 3.11.

Hour	1	2	3
Load (MW)	320	250	260

a Identify the feasible unit combinations at each hour.
b Taking into account the initial state of the system and the minimum up- and down-time constraints, identify the feasible transitions between feasible states.
c Calculate the running cost for each feasible state.
d Identify the startup costs associated with each feasible transition.
e Calculate the accumulated cost for each feasible state. In accordance with Bellman's optimality principle, consider only the cheapest way of reaching each state.
f Identify the lowest cost solution to the problem.

3.12 The Western Antarctica generating company must supply a load of 1000 MW using its three generating units. The cost characteristics of these generating units are given by the following expressions (where the powers are expressed in MW):

$$C_1 = 200 + 8P_1 + 0.07\,P_1^2(\$/\text{h})$$
$$C_2 = 300 + 9P_2 + 0.10\,P_2^2(\$/\text{h})$$
$$C_3 = 350 + 5P_3 + 0.09\,P_3^2(\$/\text{h})$$

a Calculate the optimal economic dispatch, neglecting any limits on the production of the generating units.
b How is this dispatch affected if the following limits are imposed on the production of the generating units? What is the total cost of producing this amount of power?

$$200 \leq P_1 \leq 600\ \text{MW}$$
$$100 \leq P_2 \leq 200\ \text{MW}$$
$$200 \leq P_3 \leq 600\ \text{MW}$$

c The Eastern Antarctica power company would like to buy 200 MW from the Western Antarctica generating company and is willing to pay $12 000 for this amount of power. Considering the limits on the output of its generating units, should the Western Antarctica generating company agree to this transaction?

References

Gribik, P.R., Hogan, W.W., and Pope, S.L. (2007). Market-clearing electricity prices and energy uplift. www.hks.harvard.edu/fs/whogan/Gribik_Hogan_Pope_Price_Uplift_123107.pdf.

Further Reading

The theory of spot pricing was first applied to electrical energy by Schweppe et al. (1988). Schweppe's book is often credited with providing the theoretical underpinning for the introduction of competition in power systems. Baldick et al. (2005) and O'Neill et al. discuss issues in market design in significant details. A considerable amount of material on the organization of electricity markets is available from regulatory bodies such as the Federal Energy Regulatory Commission (FERC) in the United States, the Office of Gas and Electricity

Markets (OFGEM) in the United Kingdom, the Council of European Energy Regulators (CEER) in the European Union or from market operators such as PJM. Gribik et al. (2007) and Galiana et al. provide a detailed discussion of make-whole payments. Isemonger (2006) discusses the pros and cons of virtual bidding. Additional information on this topic can also be found in the PJM report (2015).

Baldick, R., Helman, U., Hobbs, B.F., and O'Neill, R.P. (2005). Design of efficient generation markets. *Proc. IEEE* 93 (11): 1998–2012.

CEER. http://www.ceer.eu (accessed 14 March 2018).

FERC. http://www.ferc.gov (accessed 14 March 2018).

Galiana, F.D., Motto, A.L., and Bouffard, F. (2003). Reconciling social welfare, agent profits and consumer payments in electricity pools. *IEEE Trans. Power Syst.* 18 (2): 452–459.

Isemonger, A.G. (2006). The benefits and risks of virtual bidding in multi-settlement markets. *Electr. J.* 19 (9): 26–36. doi: 10.1016/j.tej.2006.09.010.

OFGEM. http://www.ofgem.gov.uk/public/adownloads.htm#retabm (accessed 14 March 2018).

O'Neill, R.P., Helman, U., Hobbs, B.F., and Baldick, R. (2006). Independent system operators in the United States: history, lessons learned, and prospects (Chapter 14). In: *Electricity Market Reform: An International Perspective, Global Energy Policy and Economics Series* (ed. F. Sioshansi and W. Pfaffenberger), 479–528. Elsevier.

PJM. http://www.pjm.com (accessed 14 March 2018).

PJM Interconnection (2015). Virtual transactions in the PJM energy markets. https://www.pjm.com/~/media/documents/reports/20151012-virtual-bid-report.ashx (accessed 12 October 2015).

Schweppe, F.C., Caramanis, M.C., Tabors, R.D., and Bohn, R.E. (1988). *Spot Pricing of Electricity*. Kluwer Academic Publishers.

Wood, A.J. and Wollenberg, B.F. (1996). *Power Generation, Operation and Control*, 2e. Wiley.

4

Participating in Markets for Electrical Energy

4.1 Introduction

In the previous chapter, we discussed the basic principles of markets for electrical energy and we illustrated through some examples how market participants interact with these markets. In this chapter, we discuss in more detail the decisions that generators, consumers, and others take to optimize the benefits that they derive from selling or buying electrical energy.

We will first discuss why consumers have a much more passive role than producers do in electricity markets and how retailers serve as their intermediaries in these markets. We will then adopt the perspective of a generating company and consider the case where this company faces a perfectly competitive market. Since the company's actions do not affect the prices in such a market, it can optimize its activities independently of what other producers or consumers might do. This assumption is somewhat unrealistic in the context of electricity markets because the short-term elasticity of the demand for electricity is very low and because in most markets, the bulk of the electrical energy is produced by a small number of producers. We will therefore discuss some of the techniques that have been proposed to analyze the operation of imperfectly competitive markets and to curb the exercise of market power.

Finally, we will explore how nonconventional resources, such as renewable generation, storage, and demand response, affect electricity markets.

4.2 The Consumer's Perspective

Microeconomic theory suggests that consumers of electricity, like consumers of all other commodities, increase their demand up to the point where the marginal benefit they derive from the electricity is equal to the price they have to pay. For example, a manufacturer will not produce widgets if the cost of the electrical energy required to build them makes their sale unprofitable. Similarly, the owner of a fashion boutique will increase the lighting level only up to the point where it attracts more customers. Finally, at home during a cold winter evening, there comes a point at which most of us will put on some extra clothes rather than turning up the thermostat and face a very large electricity bill. Since this chapter deals only with the short-term behavior of consumers, we do not

Fundamentals of Power System Economics, Second Edition. Daniel S. Kirschen and Goran Strbac.

consider the option of purchasing new appliances, machinery, or other facilities that would change the source of energy or the pattern of consumption.

If these industrial, commercial, and residential customers pay a flat rate for each kilowatt-hour that they consume, they are insulated from the spot price of electricity and their demand is affected only by their activities. Averaged over a few weeks or months, their demand reflects only their willingness to pay this flat rate. But what happens when the price of electrical energy fluctuates more rapidly? Empirical evidence suggests that demand does decrease in response to a short-term price increase, but this effect is relatively small. In other words, the price elasticity of the demand for electricity is small. On a price vs quantity diagram, the slope of the demand curve is therefore very steep. Determining the shape of the demand curve with any kind of accuracy is practically impossible for a commodity such as electrical energy. It is interesting, however, to compare the average wholesale price for electrical energy sold on a competitive market with a measure of the value that consumers place on the availability of electrical energy. One such measure is the value of lost load (VoLL), which is obtained through surveys of consumers and represents the average price per megawatt-hour that consumers would be willing to pay to avoid being disconnected without notice. For example, from 2007 to 2013, the average day-ahead energy price at the MISO trading hubs was 35.85 $/MWh, while MISO estimates VoLL to be 3500 $/MWh. We will revisit the concept of VoLL when we discuss operational reliability in Chapter 6.

Two economic and social factors explain this weak elasticity. First, the cost of electrical energy makes up only a small portion of the total cost of producing most industrial goods and represents only a small fraction of the cost of living for most households. At the same time, electricity is indispensable in manufacturing and most individuals in the industrialized world regard it as essential to their quality of life. Industrial consumers are therefore unlikely to reduce their production drastically to avoid a short-term increase in electricity prices because the savings might be more than offset by the loss of profit. Similarly, most residential consumers will probably not reduce their comfort and convenience to cut their electricity bill by a small percentage. The second factor explaining this weak elasticity is historical. Since the early days of commercial electricity generation over a century ago, electricity has been marketed as a commodity that is easy to use and always available.[1] This convenience has become so ingrained that it is fair to say that very few people carry out a cost/benefit analysis each time they turn on the light!

Rather than simply reducing their demand in response to a sudden increase in the price of electrical energy, consumers instead may decide to delay this demand until a time when prices are lower. For example, a manufacturer may decide to delay the completion of a particularly energy-intensive step of a production process until the night shift if the price of electrical energy is expected to be lower at that time. Similarly, residential consumers in some countries take advantage of lower nighttime tariffs by waiting until early morning hours to wash and dry clothes or heat hot water. Shifting demand is possible only if the consumer is able to store intermediate products, heat, electrical energy, or dirty clothes. Unless the spread between period of low and high prices becomes very large, the actual savings to the consumers may not be very significant because only a fraction of the domestic and small commercial loads can be shifted in time without causing a significant loss of comfort or revenue. Most small residential and commercial consumers are

1 One could indeed argue that the wall-mounted light switch constitutes the first ever "killer app."

therefore unlikely to be very interested in being charged on the basis of prices that change every hour or faster. If they are, their electrical loads would have to be controlled automatically by a "Consumer Energy Management System" that would receive price information and be programmed to reflect each consumer's preferences. We will discuss in more detail the perspective of consumers with flexible demand later in this chapter.

For the foreseeable future, a large majority of these consumers will probably continue purchasing electrical energy on the basis of a tariff, i.e. at a constant price per kilowatt-hour that is adjusted at most a few times per year. Such a tariff insulates them from the fluctuations in the wholesale prices and therefore reduces to zero their contribution to the short-term price elasticity of the demand. A very low elasticity has undesirable effects on the operation of markets for electrical energy. In particular, we will see later in this chapter that it facilitates the exercise of market power by the producers.

4.3 The Retailer's Perspective

Consumers whose peak demand is at least a few hundred kilowatts may be able to save significant amounts of money by employing specialized personnel to forecast their demand and trade in the electricity markets to obtain lower prices. These consumers can be expected to participate directly and actively in the markets. On the other hand, such active trading is not worthwhile for smaller consumers who usually prefer purchasing on a tariff. Electricity retailers are in business to bridge the gap between the wholesale market and these smaller consumers.

The challenge for them is that they have to buy energy at a variable price on the wholesale market and sell it at a fixed price at the retail level. A retailer will typically lose money during periods of high prices because the price it has to pay for energy is higher than the price at which it resells this energy. On the other hand, it makes a profit during periods of low prices because its retail sale price is higher than its purchase price. To stay in business, the quantity-weighted average price at which a retailer purchases electrical energy must therefore be lower than the rate it charges its customers. This is not always easy to achieve because the retailer does not have direct control over the amount of energy that its customers consume. Each retailer is deemed to have sold to its customers the amount of energy that went through their meters. If for any period the aggregate amount over all its customers exceeds the amount that it has contracted to buy, the retailer has to purchase the difference on the spot market at whatever value the spot price reached for that period. Similarly, if the amount contracted exceeds the amount consumed by its customers, the retailer is deemed to have sold the difference on the spot market.

To reduce its exposure to the risk associated with the unpredictability of the spot market prices, a retailer therefore tries to forecast as accurately as possible the demand of its customers. It then purchases energy on the forward markets to match this forecast. A retailer thus has a strong incentive to understand the consumption patterns of its customers. It will often encourage its customers to install meters that record the energy consumed during each period so it can offer them more attractive tariffs if they reduce their energy consumption during peak price hours. By taking into account all the meteorological, astronomical, economic, cultural, and special factors that influence the consumption of electricity and using the most sophisticated forecasting techniques available, it is possible to predict the value of the demand at any hour with an average accuracy of about 1.5–2%. However, such accuracy is possible only with large groups of consumers where the aggregation effects reduce the relative importance of random fluctuations. A retailer that does not have a monopoly on the supply of electricity in a given region is not able to forecast the demand of its customers with the same accuracy as a monopoly utility could achieve. This problem is exacerbated if, as one would expect, customers have the opportunity to change retailer to get a better tariff. An unstable customer base makes it much harder for the retailer to gather the reliable statistical data it needs to refine its demand forecast.

Example 4.1

Pretty Smart Energy is a retailer who forecasts the demand of its consumers, purchases energy on the forward markets (long-term bilateral, forwards, futures, screen-based transactions) to cover this demand and resells this energy to the consumers on a retail tariff. Let us assume first that this retail tariff is flat, i.e. that consumers are charged the same rate for the energy that they consume at every hour. The bars on Figure 4.1 show Pretty Smart Energy's demand forecast for a 12-h period, while the line gives the average price it had to pay on the forward markets to purchase this energy. This average price is higher than the retail rate of 37.00 $/MWh during the periods of high demand and lower than this rate during other periods. Figure 4.2 shows that during periods of low wholesale prices, Pretty Smart Energy makes a profit but it loses money during period of high wholesale prices. Table 4.1 details the operations and shows that the total profit is actually a loss of $2846 over this 12-h period. Our retailer has to hope that this is not a typical period and that the average cost of purchasing electricity will be lower on other days. If this turns out to be a typical day, the retail rate will have to be raised to above the demand-weighted average cost of purchasing energy, which is 37.08 $/MWh in this case.

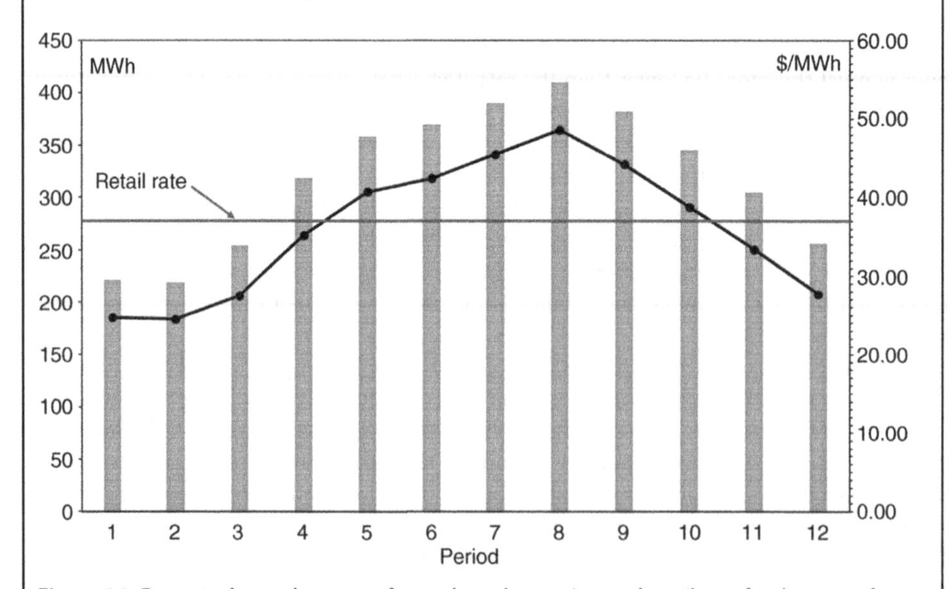

Figure 4.1 Forecast demand, average forward purchase price, and retail rate for the case of a flat retail tariff.

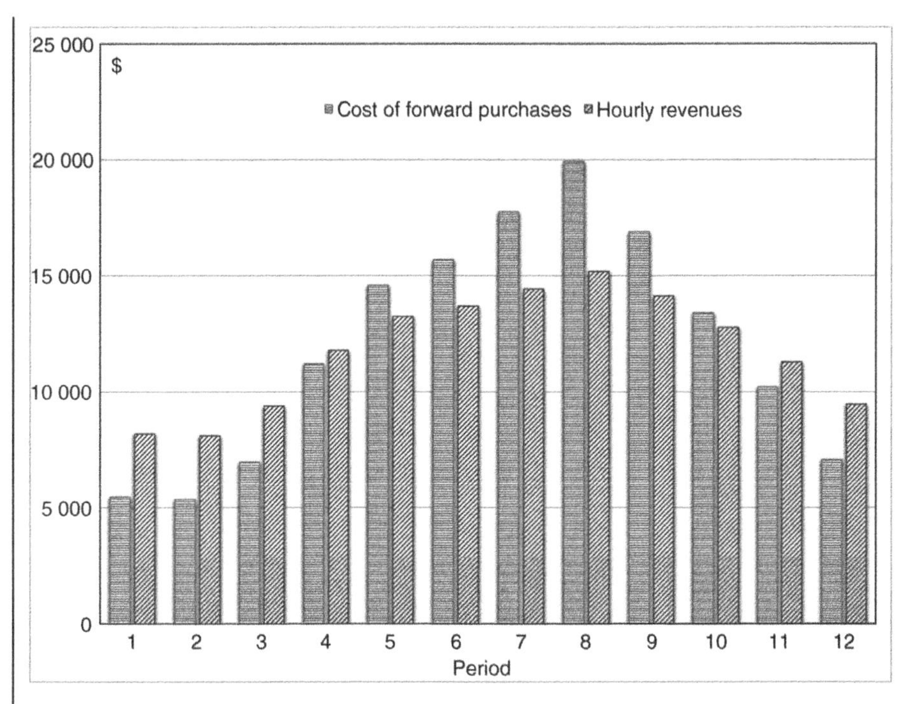

Figure 4.2 Cost of forward purchases and hourly retail revenues for the case of a flat tariff.

Alternatively, Pretty Smart Energy can try to modify the consumption pattern of its customers by offering them an on-peak/off-peak tariff. Figure 4.3 illustrates what might happen if the retail rate is set at 36 $/MWh for hours 1, 2, 3, and 12 (the off-peak hours) and at 38 $/MWh for the on-peak hours. This type of tariff tends to flatten the consumers' demand profile and hence to reduce the amount of energy that the retailer has to purchase during hours of high prices. Figure 4.4 shows the corresponding hourly forward purchase costs and revenues. Table 4.2 gives the details and reveals that this more sophisticated tariff yields a profit of $1399. Note that we have kept the total amount of energy consumed over the 12-h period unchanged.

So far, we have assumed that Pretty Smart Energy purchases in the forward markets the exact amount of energy that its customers consume at each period. In practice, there are always forecasting errors and retailers have to purchase or sell on the spot market the difference between what they purchased on the forward markets and what their customers actually consumed. Figure 4.5 illustrates these imbalances and the resulting cost of the implied trades on the spot market, given the spot market prices shown in Figure 4.6. Table 4.3 gives the details based on the case of a flat tariff of 37 $/MWh and shows that these imbalances significantly increase the retailer's loss.

Table 4.1 Retail operations over a 12-h period for the case of flat retail tariff of 37 \$/MWh.

Hour	1	2	3	4	5	6	7	8	9	10	11	12	Totals
Load forecast (MWh)	221	219	254	318	358	370	390	410	382	345	305	256	3 828
Forward purchases (MWh)	221	219	254	318	358	370	390	410	382	345	305	256	3 828
Average forward prices (\$/MWh)	24.70	24.50	27.50	35.20	40.70	42.40	45.50	48.60	44.20	38.80	33.40	27.70	
Cost of forward purchases (\$)	5 459	5 366	6 985	11 194	14 571	15 688	17 745	19 926	16 884	13 386	10 187	7 091	144 482
Revenues (\$)	8 177	8 103	9 398	11 766	13 246	13 690	14 430	15 170	14 134	12 765	11 285	9 472	141 636
Profits (\$)	2 718	2 737	2 413	572	−1 325	−1 998	−3 315	−4 756	−2 750	−621	1 098	2 381	**−2 846**

Table 4.2 Retail operations over a 12-h period for an on-peak retail rate of 38 \$/MWh and an off-peak retail rate of 36 \$/MWh.

Hour	1	2	3	4	5	6	7	8	9	10	11	12	Totals
Load forecast (MWh)	264	262	297	299	337	348	367	385	359	324	287	299	3 828
Forward purchases (MWh)	264	262	297	299	337	348	367	385	359	324	287	299	3 828
Average forward prices (\$/MWh)	24.70	24.50	27.50	35.20	40.70	42.40	45.50	48.60	44.20	38.80	33.40	27.70	
Cost of forward purchases (\$)	6 521	6 419	8 168	10 525	13 716	14 755	16 699	18 711	15 868	12 571	9 586	8 282	141 821
Revenues (\$)	9 504	9 432	10 692	11 362	12 806	13 224	13 946	14 630	13 642	12 312	10 906	10 764	143 220
Profits (\$)	2 983	3 013	2 524	837	−910	−1 531	−2 753	−4 081	−2 226	−259	1 320	2 482	**1 399**

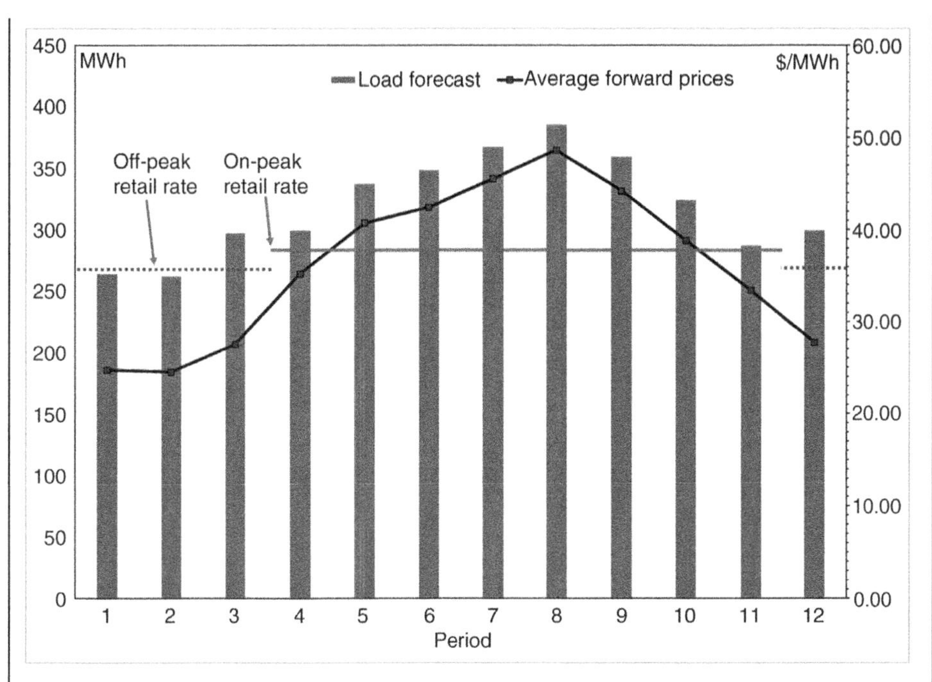

Figure 4.3 Forecast demand, average forward purchase prices, and retail rate for the case of an on-peak/off-peak tariff.

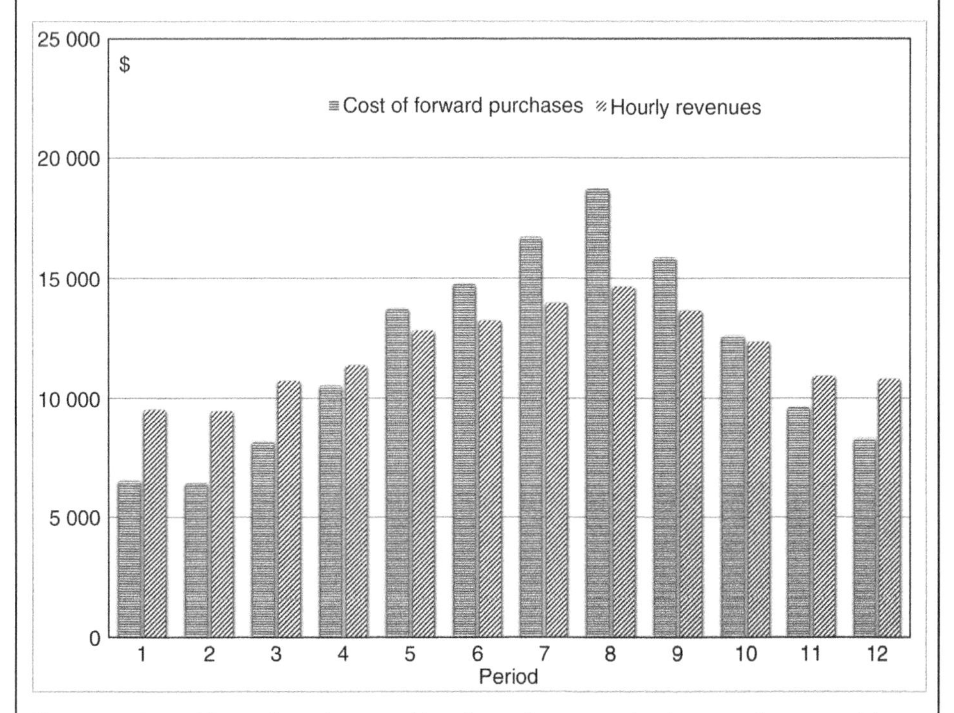

Figure 4.4 Cost of forward purchases and hourly retail revenues for the case of an on-peak/off-peak tariff.

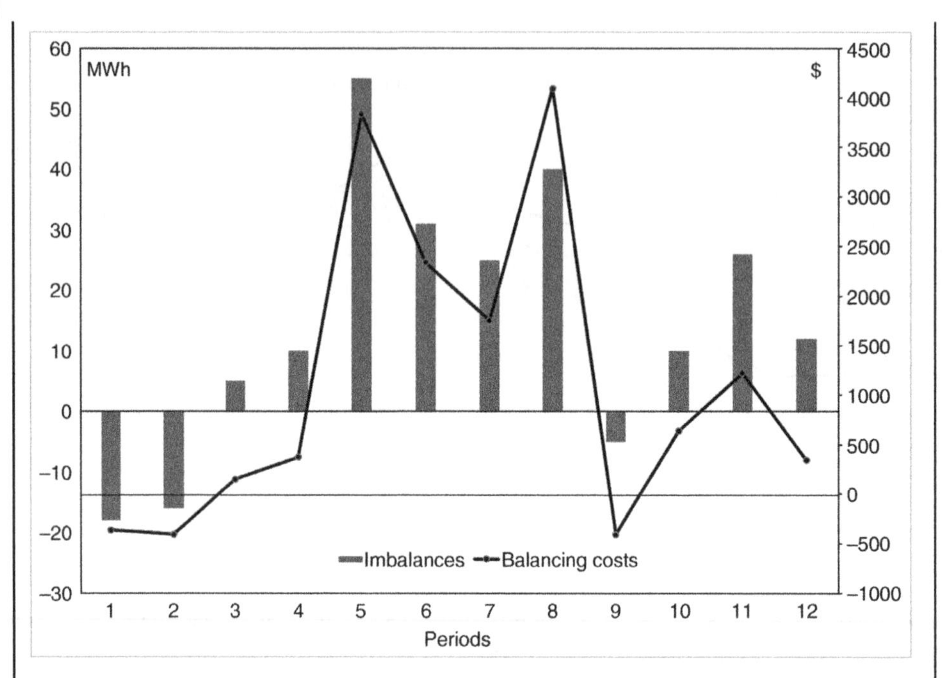

Figure 4.5 Imbalances between forward purchases and actual energy consumed and corresponding balancing costs.

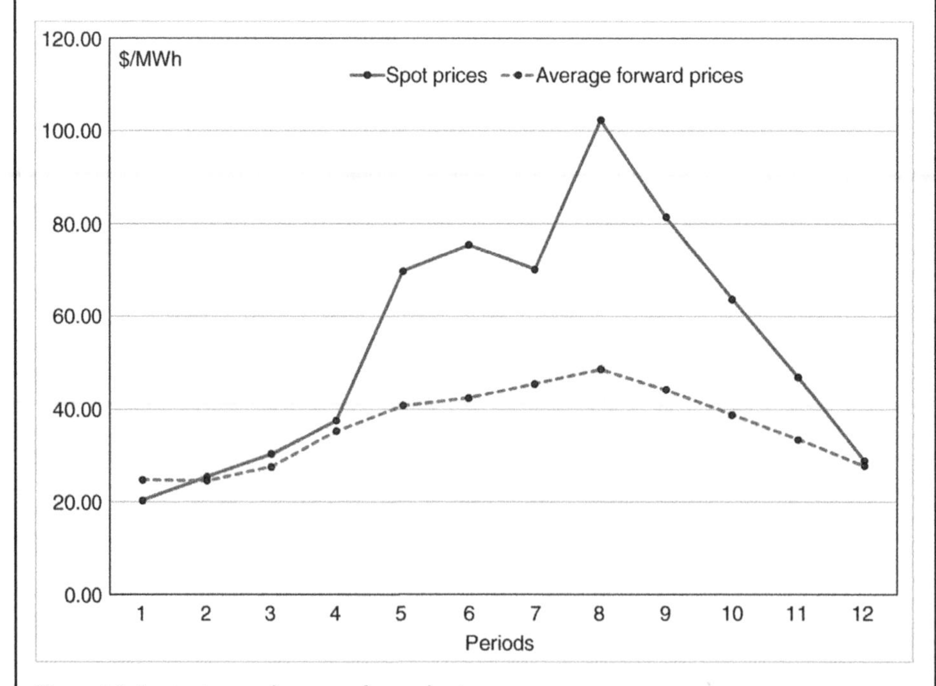

Figure 4.6 Spot prices and average forward prices.

Table 4.3 Retail operations over a 12-h period for a flat retail rate of 37 $/MWh considering forecasting errors and implied spot market trades.

Hour	1	2	3	4	5	6	7	8	9	10	11	12	Totals
Forecast load	221	219	254	318	358	370	390	410	382	345	305	256	3 828
Forward energy purchases	221	219	254	318	358	370	390	410	382	345	305	256	3 828
Average forward price	24.70	24.50	27.50	35.20	40.70	42.40	45.50	48.60	44.20	38.80	33.40	27.70	
Forward purchase costs	5 459	5 366	6 985	11 194	14 571	15 688	17 745	19 926	16 884	13 386	10 187	7 091	144 482
Actual loads	203	203	259	328	413	401	415	450	377	355	331	268	4 003
Imbalances	−18	−16	5	10	55	31	25	40	−5	10	26	12	175
Spot prices	20.30	25.40	30.30	37.50	69.70	75.40	70.10	102.30	81.40	63.70	46.90	28.90	
Balancing costs	−365	−406	152	375	3 834	2 337	1 753	4 092	−407	637	1 219	347	13 568
Total hourly cost	5 094	4 960	7 137	11 569	18 405	18 025	19 498	24 018	16 477	14 023	11 406	7 438	158 050
Revenues	7 511	7 511	9 583	12 136	15 281	14 837	15 355	16 650	13 949	13 135	12 247	9 916	148 111
Profits	2 417	2 551	2 446	567	−3 124	−3 188	−4 143	−7 368	−2 528	−888	841	2 478	**−9 939**

4.4 The Producer's Perspective

In this section, we take the perspective of a generating company that tries to maximize the profits it derives from the sale of electrical energy produced by a single generating unit. We first consider this decision in the context of a market with perfect competition. We then discuss how the bidding behavior of generators can be modeled when they are able to exert market power.

4.4.1 Perfect Competition

4.4.1.1 Basic Dispatch

Let us first consider how to maximize the profit of generating unit i over a period of 1 h, assuming that all quantities remain constant during that period. This profit is the difference between the revenue resulting from the sale of the energy it produces and the cost of producing this energy:

$$\max\Omega_i = \max[\pi \cdot P_i - C_i(P_i)] \tag{4.1}$$

where P_i is the power produced by unit i during that hour, π is the price at which this energy is sold, and $C_i(P_i)$ is the cost of producing energy. If we assume that the only variable over which the company has direct control is the power produced, the necessary condition for optimality is:

$$\frac{\mathrm{d}\Omega_i}{\mathrm{d}P_i} = \frac{\mathrm{d}(\pi \cdot P_i)}{\mathrm{d}P_i} - \frac{\mathrm{d}C_i(P_i)}{\mathrm{d}P_i} = 0 \tag{4.2}$$

The first term in this expression represents the marginal revenue of unit i, i.e. the revenue the company would get for producing an extra megawatt during this hour. The second term represents the cost of producing this extra megawatt, i.e. its marginal cost. To maximize profits, the production of unit i must therefore be adjusted up to the level where its marginal revenue is equal to its marginal cost:

$$\mathrm{MR}_i = \mathrm{MC}_i \tag{4.3}$$

If competition is perfect (or if the potential output of the unit is very small compared to the size of the market), the market price π is not affected by changes in P_i. The marginal revenue of unit i is thus:

$$\mathrm{MR}_i = \frac{\mathrm{d}(\pi \cdot P_i)}{\mathrm{d}P_i} = \pi \tag{4.4}$$

which simply expresses the fact that a price-taking generator collects the market price for each megawatt-hour that it sells. Under these conditions, if the marginal cost is a monotonically increasing function of the power produced, the generating unit should increase its output up to the point where the marginal cost of production is equal to the market price:

$$\frac{\mathrm{d}C_i(P_i)}{\mathrm{d}P_i} = \pi \tag{4.5}$$

The marginal cost includes the costs of fuel, maintenance, and all other items that vary with the power produced by the unit. Costs that are not a function of the amount of power produced (for example, the amortized cost of building the plant or the fixed operating costs) are not factored in the marginal cost and are thus irrelevant when making short-term generation dispatch decisions.

As long as competition is perfect, the output of each generating unit should be determined using Equation (4.5). Since the market price is given, all generating units can be dispatched independently, even if a generating company owns more than one unit. In a later section, we discuss the much more complicated case where the total capacity of the generating units owned by a single company is large enough to influence the price of energy.

Example 4.2

Fossil-fuel-fired generating units are characterized by input–output curves that specify the amount of fuel (usually expressed in MJ/h or MBTU/h) required to produce a given and constant electrical power output for 1 h.

Consider a coal-fired steam unit whose minimum stable generation (i.e. the minimum amount of power that it can produce continuously) is 100 MW and whose maximum output is 500 MW. Based on measurements taken at the plant, the input–output curve of this unit is estimated as:

$$H_1(P_1) = 110 + 8.2P_1 + 0.002P_1^2\,(\mathrm{MJ/h})$$

The hourly cost of operating this unit is obtained by multiplying the input–output curve by the cost of fuel F in \$/MJ:

$$C_1(P_1) = 110F + 8.2FP_1 + 0.002F\ P_1^2\,(\$/\mathrm{h})$$

If we assume that the cost of coal is 1.3 \$/MJ, the cost curve of this unit is:

$$C_1(P_1) = 143 + 10.66P_1 + 0.0026P_1^2\,(\$/\mathrm{h})$$

If the price at which electrical energy can be sold is 12 \$/MWh, the output that this unit should produce is given by:

$$\frac{\mathrm{d}C_1(P_1)}{\mathrm{d}P_1} = 10.66 + 0.0052P_1 = 12\,(\$/\mathrm{MWh}) \quad \text{or} \quad P_1 = 257.7\ \mathrm{MW}$$

In practice, optimally dispatching even a single generating unit is more complex than Equation (4.5) suggests. In the following subsections, we examine how the cost and technical characteristics of the generating units affect the basic dispatch.

4.4.1.2 Unit Limits

Suppose that the maximum power $P_i^{\max}$ that can be produced by generating unit i is such that:

$$\left.\frac{\mathrm{d}C_i(P_i)}{\mathrm{d}P_i}\right|_{P_i^{\max}} \le \pi \tag{4.6}$$

This generating unit should therefore produce $P_i^{\max}$. On the other hand, if the minimum stable generation of unit i is such that:

$$\left.\frac{\mathrm{d}C_i(P_i)}{\mathrm{d}P_i}\right|_{P_i^{\min}} > \pi \tag{4.7}$$

This unit cannot generate profitably at that price and the only way to avoid losing money on its operation is to shut it down.

Example 4.3

The generating unit of Example 4.2 should operate at its maximum output whenever the price of electrical energy is greater than or equal to:

$$\left.\frac{\mathrm{d}C_i(P_i)}{\mathrm{d}P_i}\right|_{500\mathrm{MW}} = 10.66 + 0.0052 \cdot 500 = 13.26\,\$/\mathrm{MWh}$$

On the other hand, this unit cannot operate profitably if the price drops below:

$$\left.\frac{\mathrm{d}C_i(P_i)}{\mathrm{d}P_i}\right|_{100\mathrm{MW}} = 10.66 + 0.0052 \cdot 100 = 11.18\,\$/\mathrm{MWh}$$

4.4.1.3 Piecewise Linear Cost Curves

Input–output curves are drawn on the basis of measurements taken while the generating unit is operating at various levels of output. Even if every effort is made to make these measurements as accurate as possible, the data points usually do not line up along a smooth curve. A piecewise linear interpolation of this data is therefore just as acceptable as a quadratic function. Figure 4.7 shows a piecewise linear cost curve and its associated marginal cost curve.

Since each segment of the cost curve is linear, each segment of the marginal cost curve is constant. This makes the process of dispatching the unit in response to electrical energy

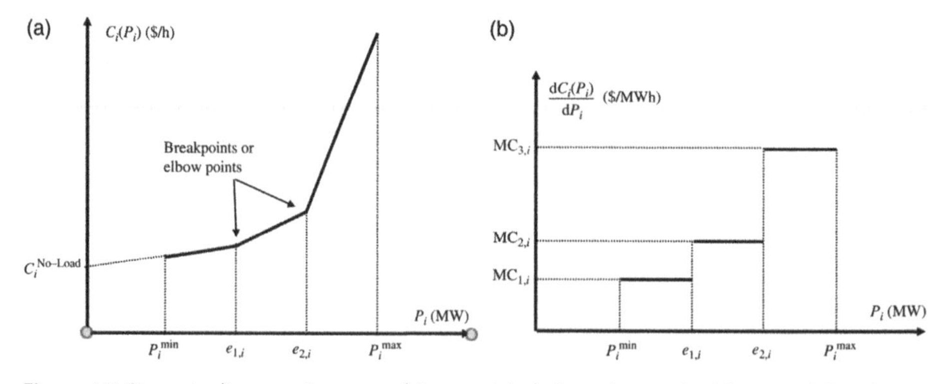

Figure 4.7 Piecewise linear cost curve and its associated piecewise constant incremental cost curve.

prices very simple:

$$\begin{array}{l} \pi < \mathrm{MC}_{1,i} \Rightarrow P_i = P_i^{\min} \\ \mathrm{MC}_{1,i} < \pi < \mathrm{MC}_{2,i} \Rightarrow P_i = \mathrm{e}_{1,i} \\ \mathrm{MC}_{2,i} < \pi < \mathrm{MC}_{3,i} \Rightarrow P_i = \mathrm{e}_{2,i} \\ \mathrm{MC}_{3,i} < \pi \Rightarrow P_i = P_i^{\max} \end{array} \tag{4.8}$$

If the price is exactly equal to the marginal cost of one of the segments of the curve, the generation can take any value within that segment. The marginal cost at a breakpoint is equal to the slope of the next segment because the marginal cost is traditionally defined as the cost of the next megawatt, not the cost of the previous megawatt.

Example 4.4

The quadratic cost curve of Example 4.2 can be approximated by the following three-segment piecewise linear cost curve:

$$\begin{array}{l} 100 \le P_1 \le 250 : C_1(P_1) = 11.57P_1 + 78.0 \\ 250 \le P_1 \le 400 : C_1(P_1) = 12.35P_1 - 117.0 \\ 400 \le P_1 \le 500 : C_1(P_1) = 13.00P_1 - 377.0 \end{array}$$

Figure 4.8 shows how this unit should be dispatched as the price paid for electrical energy varies.

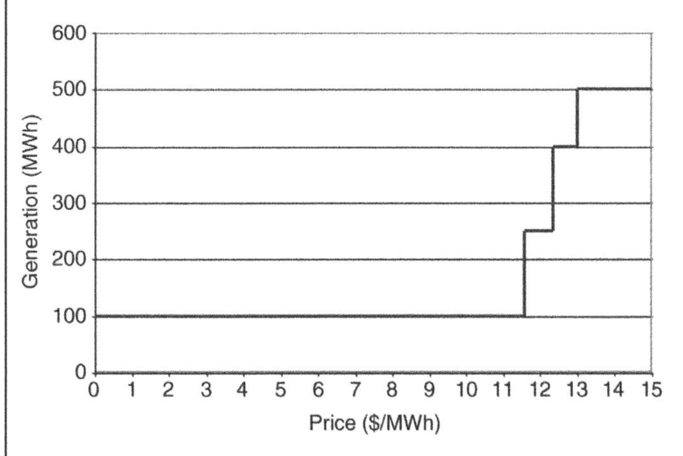

Figure 4.8 Dispatch of the generating unit as a function of the price of electrical energy.

4.4.1.4 No-load Cost

While the optimality condition derived above guarantees that the profit will be maximum, it does not ensure that the generating unit will not operate at a loss. Producers must also consider the pseudo-fixed costs associated with the operation of a generating unit, i.e. the costs that are incurred only if the unit is generating but are independent of the amount of power generated. The first type of pseudo-fixed cost is the no-load cost. If it is possible for the unit to remain connected to the system while supplying no electrical power, the no-load cost represents the cost of the fuel required to keep the unit running. Such a mode of operation is not possible for most thermal generating units. The no-load cost is then simply the constant term in the cost function and does not have a physical meaning.

As we discussed in Chapter 2, selling at the marginal cost is profitable only if this marginal cost is greater than the *average* cost of production.

Example 4.5

Let us assume that the unit of Example 4.4 is always dispatched optimally as the market price for electrical energy varies. This means that it is dispatched according to Figure 4.8. Figure 4.9 shows that its profits increase in a piecewise linear fashion with the price of electrical energy. Because of the no-load cost, the unit becomes profitable only when the price reaches 11.882 $/MWh.

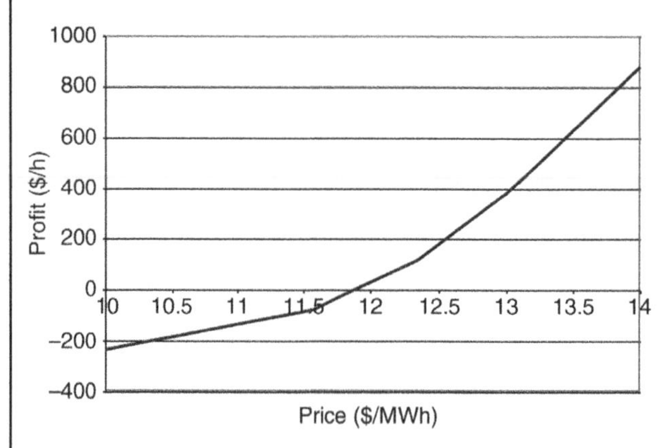

Figure 4.9 Profit accrued by the generating unit if it is dispatched optimally as the price of electrical energy varies.

4.4.1.5 Scheduling

Since the demand for electrical energy changes over time, the price that a generator gets for its production varies. As we saw in the previous chapter, the price for electrical energy is usually constant for a period of time whose duration ranges from a few minutes to an hour depending on the market. Given a profile of prices extending over a day or more, the optimal dispatch could be calculated as described above for each market period taken separately. However, the resulting production schedule would not be optimal because it neglects the cost of starting up generating units. It would also often be technically infeasible because this approach ignores the constraints on the transitions that generating units are able to make between operating states. In addition, other economic opportunities and environmental constraints may also affect the optimization of the sale of electrical energy. These different types of constraints are discussed below.

Generating units that have large startup costs or are subject to restrictive operating constraints therefore do not maximize their profits if their output is optimized over each period individually. Instead their operation must be scheduled over a horizon ranging from one day to a week or more. This problem has some similarities to the unit commitment problem that monopoly utilities solve to determine how to meet a given load schedule at minimum cost with a given set of generating units. In essence, both problems balance the pseudo-fixed and variable elements of the cost while satisfying the constraints. In the unit commitment problem, the production of all units is optimized together because their total output must be equal to the total load. On the other hand, if we assume that a generator is a price taker, its production can be optimized independently of the production of other generators. Even when this price-taking approximation holds, scheduling generation to maximize profits is computationally complex. The on/off nature of some of the decision variables makes the problem nonconvex and a rigorous treatment of the constraints significantly increases the dimensionality of the problem. Techniques such as dynamic or mixed-integer programming have been successfully used to solve this problem.

4.4.1.6 Startup Cost

The startup cost of a generating unit represents the cost of getting this unit running and ready to produce from a shutdown state. It is thus another type of pseudo-fixed cost. Diesel generators and open-cycle gas turbines have low startup costs because these types of units start quickly. On the other hand, large thermal units must burn a considerable amount of fuel before the steam is at a sufficient temperature and pressure to sustain the generation of electric power. These units therefore have a large startup cost. To maximize the profitability of a thermal unit, this startup cost must be amortized over a long period. This may even involve operating the unit at a loss for a few hours rather than shutting it down and having to reincur the startup cost when prices increase again.

Example 4.6

Let us examine how the coal-fired plant of Example 4.2 should be scheduled over a period of several hours. We will assume that the price at which electrical energy can be sold is set on an hourly basis and that the prices for the next few hours are shown in Figure 4.10. Suppose also that the generating unit is started up at hour 1 and that the cost of bringing it on-line is $600. Table 4.4 summarizes the results.

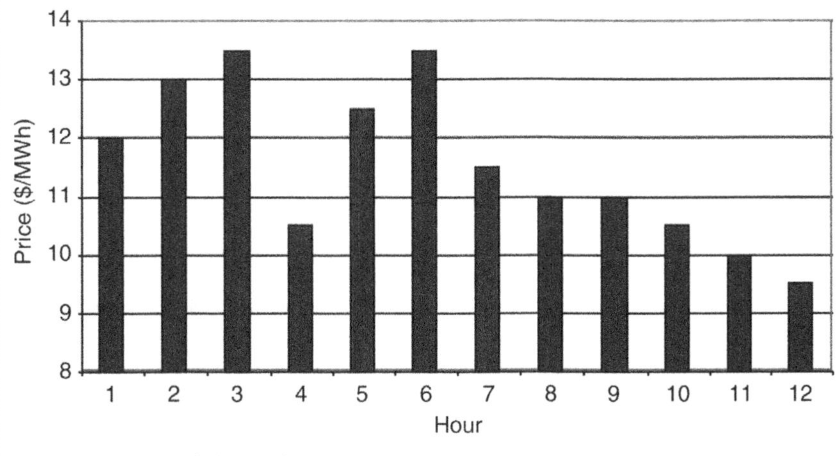

Figure 4.10 Price of electrical energy.

Table 4.4 Operation and profitability of the coal-fired plant of Example 4.6.

Hour	1	2	3	4	5	6	7
Price ($/MWh)	12.0	13.0	13.5	10.5	12.5	13.5	11.5
Generation (MW)	257.7	450.0	500.0	100.0	353.8	500.0	161.5
Revenue ($)	3092	5850	6750	1050	4423	6750	1858
Running cost ($)	3063	5467	6123	1235	4240	6123	1933
Startup cost ($)	600	0	0	0	0	0	0
Total cost ($)	3663	5467	6123	1235	4240	6123	1933
Profit ($)	−571	383	627	−185	183	627	−75
Cumulative profit ($)	−571	−188	439	254	437	1064	989

The first thing to notice is that the optimal generation varies substantially as the price of electrical energy fluctuates. The unit generates at maximum capacity at hours 3 and 6 and at minimum capacity at hour 4. The operation of this unit records a deficit at hour 1 because of the cost of starting up the unit. By hour 3, this startup cost has been recovered and the unit begins making a profit. The price at hour 4 is so low that the unit shows a loss even though it operates at its minimum capacity. Not shutting down the unit at this hour is, however, the best decision because it avoids incurring the startup cost again at hour 5. At hour 7, the unit records a deficit even though it is not running at its minimum capacity. This is because the unit does not generate enough to recover its no-load cost. If the price continues to decline over the next few hours, the best strategy would be to shut the unit down at the end of hour 6 and to wait until prices are higher before starting it up again.

4.4.1.7 Operating Constraints

Starting up or shutting down a thermal generating unit or even increasing or decreasing its output by more than a small amount causes a considerable amount of mechanical stress in the prime mover. Excessive stress damages the plant and shortens its life. Limits are therefore often placed on such changes to avoid damaging these expensive assets. These safeguards have long-term benefits but short-term costs. In particular, placing a limit on the rate at which a unit can increase or decrease its output may prevent it from achieving its economically optimal output in successive periods. Minimizing the cost of these ramp-rate limits requires that the operation of the unit be optimized over at least several hours.

To limit the damage caused by frequent startups and shutdowns, a minimum is often placed on the number of hours that a thermal unit must remain connected to the system once it has been started. A similar limit is usually placed on the number of hours that a unit must remain idle once it has been shut down. These limits ensure that there is enough time for the temperature gradients in the turbine to subside. These minimum up-time and minimum down-time constraints reduce the opportunities to change the status of the unit and can have a significant impact on the optimal schedule. For example, the minimum down-time constraint could force a unit to continue generating at a loss during a period of low prices because shutting it down would prevent it from reaping larger profits later on.

4.4.1.8 Environmental Constraints

Generating plants must abide by environmental regulations that may affect their ability to operate at their economic optimum. Emissions of certain pollutants by fossil-fuel-fired power plants are increasingly regulated. In some cases, the rate at which a certain pollutant is released in the atmosphere is limited, thereby reducing the maximum power output of the plant. In other cases, it is the total amount of a pollutant released over a year that is capped, putting a complex integral constraint on the operation of the plant.

While hydroelectric plants do not emit pollutants and are more flexible than thermal plants, there may be constraints on their use of water. These constraints may be designed to ensure the availability of water for recreation or to help preserve endangered species of fish. Water must also be made available for irrigation and other hydroelectric plants. Optimizing the operation of hydroelectric plants is a very complex problem, particularly in river basins with multiple interconnected plants.

4.4.1.9 Other Economic Opportunities

The amount of electric power produced by cogeneration or combined heat and power plants is often determined by the needs of the associated industrial process. The ability of such plants to take advantage of opportunities to sell energy on the electricity market therefore may be limited.

Besides electrical energy, generators can provide other services such as operating reserves, load following, frequency regulation, and voltage regulation. These other services, which are usually called ancillary or system services, constitute a source of revenue that is distinct from the sale of electrical energy. We will discuss the issues related to the provision of these services in Chapter 6. At this point, we simply need to note that a producer's ability to trade electrical energy may be affected by contracts that it has entered into for the provision of ancillary services. Conversely, the production of electrical energy may hamper a generator's ability to provide ancillary services.

4.4.1.10 Forecasting Errors

Optimally scheduling the production of a generating unit over a time horizon requires a forecast of the price of electrical energy at each period. Such forecasts are never perfectly accurate, and forecasting errors result in scheduling and dispatch decisions that turn out to be less than optimal. Forecasting prices accurately is difficult because of the number of factors involved and the lack of information on some of these factors. Since the price of electrical energy depends on the market equilibrium, it is influenced by both the load and the generation. On the load side, all the temporal, meteorological, economic, and special factors that are used in load forecasting must also be taken into account when forecasting prices. The generation side is even more troublesome because some events occur at random (e.g. failures of generating units) and others are not always publicized in advance (e.g. planned outages for maintenance).

4.4.2 The Produce Vs Purchase Decision

Consider the case of a generating company that has signed a contract for the supply of a given load L during a single hour. Let us first assume that this company decides to meet its contractual obligation to supply this load using its portfolio of N generating plants. It will obviously try to produce the energy required at minimum cost to itself. Mathematically, if we ignore the constraints on the operation of the generating units, this can be formulated as the following optimization problem:

$$\text{Minimize} \sum_{i=1}^{N} C_i(P_i) \quad \text{subject to} \sum_{i=1}^{N} P_i = L \tag{4.9}$$

where P_i represents the production of unit i of the portfolio and $C_i(P_i)$ the cost of producing this amount of power with this unit. From calculus, we know that forming a Lagrangian function ℓ that combines the objective function and the constraint is the easiest way to solve such an optimization problem:

$$\ell(P_1, P_2, \ldots, P_N, \lambda) = \sum_{i=1}^{N} C_i(P_i) + \lambda\left(L - \sum_{i=1}^{N} P_i\right) \tag{4.10}$$

where λ is a new variable called a *Lagrange multiplier.*

Setting the partial derivatives of this Lagrangian function to zero gives the necessary conditions for optimality and solving these equations gives the optimal solution:

$$\begin{aligned} \frac{\partial \ell}{\partial P_i} &\equiv \frac{\mathrm{d}C_i}{\mathrm{d}P_i} - \lambda = 0, \quad \forall i = 1, \ldots, N \\ \frac{\partial \ell}{\partial \lambda} &\equiv \left(L - \sum_{i=1}^{N} P_i\right) = 0 \end{aligned} \tag{4.11}$$

From these optimality conditions, we conclude that all the generating units in the portfolio should be operated at the same marginal cost and that this marginal cost is equal to the value of the Lagrange multiplier λ:

$$\frac{dC_1}{dP_1} = \frac{dC_2}{dP_2} = \ldots \frac{dC_N}{dP_N} = \lambda \tag{4.12}$$

The value of the Lagrange multiplier is thus equal to the cost of producing one additional megawatt-hour with any of the generating units. This Lagrange multiplier is therefore often called the *shadow price* of electrical energy.

Let us now suppose that this generating company can participate in a spot market for electricity where the price of energy is π. If the market price is lower than the shadow price λ at which it can produce energy, our generating company should purchase energy on the market and reduce its own production up to the point where:

$$\frac{dC_1}{dP_1} = \frac{dC_2}{dP_2} = \ldots \frac{dC_N}{dP_N} = \pi \tag{4.13}$$

If the amount of energy involved is significant, the market may not have enough liquidity to handle this transaction without an increase in the price π. This issue will be discussed in more detail later in this chapter.

An increasing number of industrial consumers operate processes that cannot be shut down because of interruptions in the electricity supply without causing significant financial losses. Such consumers often install emergency generators capable of supplying at least part of their load during period of outages. When the power system operates normally but prices are high, these consumers may find that, even though the marginal cost of operating these emergency generators may be high, it is lower than the spot price of electrical energy. Under these conditions, they might want to start up their emergency generators to reduce their demand and possibly sell the surplus on the market.

Example 4.7

The 300 MW load of a small power system must be supplied at minimum cost by two thermal generating units and a small run-of-the-river hydro plant. The hydro plant generates a constant 40 MW and the cost functions of the thermal plants are given by the following expressions:

$$\text{Unit A}: \quad C_{\text{A}} = 20 + 1.7P_{\text{A}} + 0.04P_{\text{A}}^2 \quad \$/\text{h}$$

$$\text{Unit B}: \quad C_{\text{B}} = 16 + 1.8P_{\text{B}} + 0.03P_{\text{B}}^2 \quad \$/\text{h}$$

Since the variable operating cost of the hydro unit is negligible, the Lagrangian function of this optimization problem can be written as follows:

$$\ell = C_{\text{A}}(P_{\text{A}}) + C_{\text{B}}(P_{\text{B}}) + \lambda(L - P_{\text{A}} - P_{\text{B}})$$

where L represents the 260 MW load that the thermal units must provide.

Setting the partial derivatives of the Lagrangian equal to zero, we obtain the necessary conditions for optimality:

$$\frac{\partial \ell}{\partial P_{\text{A}}} \equiv 1.7 + 0.08P_{\text{A}} - \lambda = 0$$

$$\frac{\partial \ell}{\partial P_{\text{B}}} \equiv 1.8 + 0.06P_{\text{B}} - \lambda = 0$$

$$\frac{\partial \ell}{\partial \lambda} \equiv L - P_{\text{A}} - P_{\text{B}} = 0$$

Solving this system of equations for λ, we get the marginal cost of electrical energy in this system for this loading condition:

$$\lambda = 10.67\,\$/\text{MWh}$$

We can then calculate the optimal outputs of the thermal units:

$$P_{\text{A}} = 112.13\,\text{MW}$$

$$P_{\text{B}} = 147.87\,\text{MW}$$

Replacing these values in the cost functions, we find the total cost of supplying this load:

$$C = C_{\text{A}}(P_{\text{A}}) + C_{\text{B}}(P_{\text{B}}) = 1651.63\,\$/\text{h}$$

4.4.3 Imperfect Competition

When competition is less than perfect, some firms (the strategic players) are able to influence the market price through their actions. It is quite common for an electricity market to consist of a few strategic players and a number of price takers. A company that owns more than one generating unit is likely to have a greater influence on the market price if it optimizes the combined output of its entire portfolio of units. The total profit of a firm that owns multiple generating units is then:

$$\Omega_f = \pi \cdot P_f - C_f(P_f) \tag{4.14}$$

where P_f represents the combined output of all the units controlled by that firm while $C_f(P_f)$ represents the minimum cost at which this firm is able to produce this power. In this section, we no longer assume that the market price π is beyond the control of any single market participant. Because this market price is no longer set, the power sold by firm f depends not only on its own decisions but also on those of its competitors. We therefore rewrite Equation (4.14) as follows to summarize these dependencies:

$$\Omega_f = \Omega_f(X_f, X_{-f}) \tag{4.15}$$

where X_f represents the actions of firm f and X_{-f} those of its competitors.

Equation (4.15) shows that firm f cannot optimize its profits in isolation. It must consider what the other firms will do. At first sight, this may seem very difficult because these firms are competitors and exchanging information would be illegal. However, it is reasonable to assume that all firms are behaving in a rational manner, i.e. that they are all trying to maximize their profits. In other words, we have to find for each firm f the actions X_f^* such that:

$$\Omega_f\left(X_f^*, X_{-f}^*\right) \geq \Omega_f\left(X_f, X_{-f}^*\right), \qquad \forall f \tag{4.16}$$

where X_{-f}^* represents the optimal action of the other firms.

Such interacting optimization problems form what is called in game theory a *non-cooperative game*. The solution of such a game, if it exists, is called a *Nash equilibrium* and represents a market equilibrium under imperfect competition.

While representing the possible actions or decisions of a firm by the generic variable X_f allowed us to formulate the problem elegantly, it hides the fact that the solution of Equation (4.16) depends on how we model the strategic interactions between the firms. In the following subsections, we discuss four approaches that are used to model imperfect competition.

Imperfect competition can be modeled using a Cournot model, where firms are assumed to decide how much they produce, or a Bertrand model, where firms are assumed to decide at what price they sell their production.

4.4.3.1 Bertrand Model

In the Bertrand model of competition, we assume that the price at which each firm offers its electrical energy is its only decision variable:

$$X_f = \pi_f, \quad \forall f \tag{4.17}$$

The amount of energy sold by firm f is thus a function of its own offer price and the offer prices of its competitors. Firm f's revenue is given by:

$$\pi \cdot P_f = \pi \cdot P_f\left(\pi_f, \pi_{-f}^*\right) \tag{4.18}$$

According to the model, for an undifferentiated product such as electrical energy, firm f can sell as much as it wants as long as its price is lower than the prices of its competitors:

$$P_f\left(\pi_f, \pi^*_{-f}\right) = P_f, \quad \text{if} \quad \pi_f \le \pi^*_{-f} \\ = 0, \quad \text{otherwise} \tag{4.19}$$

Example 4.8

Let us consider the case of an electricity market with only two generating companies. This is known as a *duopoly*. We first assume that these two companies have the following constant marginal costs of production:

$$\mathrm{MC_A} = 35\ \$/\mathrm{MWh}$$
$$\mathrm{MC_B} = 45\ \$/\mathrm{MWh}$$

According to the Bertrand model, these firms compete by setting their prices and letting the market decide how much each firm sells. In this case, by setting its price just below 45 \$/MWh, Generator A would capture the whole market because Generator B would lose money on every MWh sold at that price. The market price in this example is thus $(45 - \varepsilon)$ \$/MWh.

Example 4.9

Let us now assume that these two generating companies have the same constant marginal costs:

$$\mathrm{MC_A} = 35\ \$/\mathrm{MWh}$$
$$\mathrm{MC_B} = 35\ \$/\mathrm{MWh}$$

Neither generating company can set its price above its marginal cost because it would then be undercut by the other company who would then capture the whole market. A sustainable equilibrium is reached only when the price offered by both firms (and thus the market clearing price) is 35 \$/MWh, which is equal to the marginal cost of production.

The results of these two examples are counterintuitive because in the first case one firm captures the entire market while in the other the market price is the same as the price that clears a perfectly competitive market. One would expect duopoly competitors to be able to obtain a higher price than in a perfectly competitive market.

4.4.3.2 Cournot Model

In a Cournot model of competition, each firm decides the quantity that it wants to produce. Let us first consider again the case of a duopoly. If both firms must decide simultaneously how much to produce, each of them will estimate the expected production of the other. If firm 1 estimates that firm 2 will produce quantity y_2^e, it will set its production at a level y_1 that maximizes its expected profit:

$$\max_{y_1} \pi\left(y_1 + y_2^e\right)y_1 - c\left(y_1\right) \tag{4.20}$$

where $\pi(y_1 + y_2^e)$ represents the market price that would result from the expected total output $y_1 + y_2^e$. The optimal production of firm 1 thus depends on its estimate of the production of firm 2. We can express this relation directly in the form of a *reaction function*:

$$y_1 = f_1(y_2^e) \tag{4.21}$$

Since firm 2 follows a similar process to optimize its production, we also have:

$$y_2 = f_2(y_1^e) \tag{4.22}$$

At first, the estimates that each firm makes of the production of their competitor may be incorrect or inaccurate. However, as they gather more information during subsequent market clearings, they revise their estimates and adjust their production accordingly. Ultimately, their productions reach the *Cournot equilibrium*:

$$\begin{aligned} y_1^* &= f_1(y_2^*) \\ y_2^* &= f_2(y_1^*) \end{aligned} \tag{4.23}$$

Once this equilibrium is reached, neither firm would find it profitable to change its output.

Let us now consider the case where there are n firms competing in the market. The total industry output is:

$$Y = y_1 + \cdots + y_n \tag{4.24}$$

Firm i, like all the other firms, seeks to maximize its profit:

$$\max_{y_i} \{y_i \cdot \pi(Y) - c(y_i)\} \tag{4.25}$$

where the market price $\pi(Y)$ is a function of the total industry output. This maximum is achieved when:

$$\frac{\mathrm{d}}{\mathrm{d}y_i}\{y_i \cdot \pi(Y) - c(y_i)\} = 0 \tag{4.26}$$

or

$$\pi(Y) + y_i \frac{\mathrm{d}\pi(Y)}{\mathrm{d}y_i} = \frac{\mathrm{d}c(y_i)}{\mathrm{d}y_i} \tag{4.27}$$

Factoring out $\pi(Y)$ on the left-hand side and multiplying the second term by Y/Y, we get:

$$\pi(Y)\left\{1 + \frac{y_i}{Y}\frac{Y}{\mathrm{d}y_i}\frac{\mathrm{d}\pi(Y)}{\pi(Y)}\right\} = \frac{\mathrm{d}c(y_i)}{\mathrm{d}y_i} \tag{4.28}$$

The right-hand side of this equation is equal to the marginal cost of production of firm i. If we define the *market share* of firm i as $s_i = y_i/Y$ and use the definition of elasticity given in Equation (2.3), we can write Equation (4.28) in the following form:

$$\pi(Y)\left\{1 - \frac{s_i}{|\varepsilon(Y)|}\right\} = \frac{\mathrm{d}c(y_i)}{\mathrm{d}y_i} \tag{4.29}$$

This expression shows that when the market share of a firm is not negligible, it maximizes its profit by setting its production at a level where its marginal cost is less than the market price. It is thus exerting market power through physical withholding. Equation (4.29) suggests that a low elasticity and a high degree of market concentration facilitate the exercise of market power. It is interesting to note that one firm's ability to exert market power benefits all the firms in the market because it raises the price at which price-taking firms sell their products. Actions aimed at reducing market power therefore have to be initiated by regulatory authorities representing the interests of the customers. Such actions usually do not receive support from any of the producers.

Equation (4.29) is applicable to the extreme cases where the firm has a monopoly ($s_i = 1$) and where its market share is negligible ($s_i \approx 0$). The biggest difference between price and marginal cost occurs in the case of a monopoly, where the monopolist's ability to raise prices is limited only by the elasticity of the demand. In the case of a firm with a very small market share, Equation (4.29) reduces to the same form as Equation (4.5) and the firm acts as a price taker.

The Cournot model suggests that firms should be able to sustain prices that are higher than the marginal cost of production, with the difference being determined by the price elasticity of the demand. Numerical results obtained with Cournot models are very sensitive to this elasticity. In particular, for a commodity such as electrical energy that has a very low elasticity, the equilibrium price calculated using a Cournot model tends to be higher than the prices that are observed in the actual market.

Example 4.10

Let us consider the case of a market where two firms (A and B) compete for the supply of electrical energy. Empirical studies have shown that the inverse demand curve at a particular hour is given by:

$$\pi = 100 - D \quad (\$/\text{MWh}) \tag{4.30}$$

where D is the demand for electrical energy at this hour. Let us also suppose that firm A can produce energy more cheaply than firm B:

$$\begin{aligned} C_{\text{A}} &= 36 \cdot P_{\text{A}} \, (\$/\text{h}) \\ C_{\text{B}} &= 46 \cdot P_{\text{B}} \, (\$/\text{h}) \end{aligned} \tag{4.31}$$

If we assume a Bertrand model of competition in this market, firm A would set its price at slightly less than the marginal cost of production of firm B (i.e. 46 $/MWh) and would capture the whole market. At that price, the demand would be 54 MWh and firm A would achieve a profit of $540. Firm B would lose money on any megawatt-hour that it sold and would therefore decide not to produce anything. It would then obviously not make any profit.

On the other hand, if we assume a Cournot model of competition, the state of the market is determined by the production decisions made by each firm. Let us suppose that firms A and B have both decided to produce 5 MWh. According to the Cournot model, the market price must be such that the demand equals the total production. Since the total production is 10 MWh, the total demand is also 10 MWh and, according to Equation (4.30), the market price must be 90 $/MWh. Given the market price and their respective productions, we can easily calculate that firm A makes a profit of \$270 and firm B a profit of \$220. The following cell summarizes this state of the market:

10	270
220	90

Similar cells can be generated for other combinations of production by the two firms and arranged as shown in Table 4.5. This table illustrates the interactions of the two firms under a Cournot model of competition. Toward the top-left corner of the table, generators are driving the price up by limiting production. As production increases (i.e. as we move

Table 4.5 Illustration of Cournot competition in the two-firm market of Example 4.10.

	Production of firm A													
Production of firm B	5		10		15		20		25		30		35	
5	10	270	15	490	20	660	25	780	30	850	35	870	40	840
	220	90	195	85	170	80	145	75	120	70	95	65	70	60
10	15	245	20	440	25	585	30	680	35	725	40	720	45	665
	390	85	340	80	290	75	240	70	190	65	140	60	90	55
15	20	220	25	390	30	510	35	580	40	600	45	570	50	490
	510	80	435	75	360	70	285	65	210	60	135	55	60	50
20	25	195	30	340	35	435	40	480	45	475	50	420	55	315
	580	75	480	70	380	65	280	60	180	55	80	50	−20	45
25	30	170	35	290	40	360	45	380	50	350	55	270	60	140
	600	70	475	65	350	60	225	55	100	50	−25	45	−150	40
30	35	145	40	240	45	285	50	280	55	225	60	120	65	−35
	570	65	420	60	270	55	120	50	−30	45	−180	40	−330	35
35	40	120	45	190	50	210	55	180	60	100	65	−30	70	−210
	490	60	315	55	140	50	−35	45	−210	40	−385	35	−560	30

The numbers in each cell represent the following quantities:

Demand	Profit of A
Profit of B	Price

right or down through the table), the price decreases and the demand increases. Toward the bottom-right corner of the table, the market is flooded and the price drops below firm B's marginal cost of production, causing it to lose money. Among the possibilities shown in Table 4.5, firm A would prefer the situation where it produces 30 MWh and B produces 5 MWh because this would maximize its profit. Similarly, firm B would like A to produce only 5 MWh so that it could produce 25 MWh and maximize its own profit. Game theory suggests that the market will not settle in either of these situations because they are not in the best interests of the other firm. Instead, the market will settle at a point where neither firm can increase its profit through its own actions. The highlighted cell in Table 4.5 corresponds to this equilibrium. The profit of firm A ($600) is the largest that it can achieve in that row, i.e. by adjusting its own production. Similarly, the profit of firm B ($210) is the largest in this column. Therefore, neither firm has an incentive to produce any other amount. While firm A captures a larger share of the market because its marginal cost of production is lower, it does not freeze firm B completely out of the market. Together, these firms manage to maintain a price that is much higher than the marginal cost of production. This price is also higher than the value predicted by the Bertrand model.

Rather than constructing a table for every possible pair of productions, we can formulate and solve this problem mathematically as follows. Since each firm uses the quantity it produces as its decision variable, the profits earned by each firm are given by the following expressions:

$$\Omega_A(P_A, P_B) = \pi(D) \cdot P_A - C_A(P_A) \tag{4.32}$$

$$\Omega_B(P_A, P_B) = \pi(D) \cdot P_B - C_B(P_B) \tag{4.33}$$

where $\pi(D)$ represents the inverse demand curve. If each firm tries to maximize its profit, we have two separate optimization problems. These two optimization problems cannot be solved independently because both firms compete in the same market and the supply must be equal to the demand. Therefore, we must also have:

$$D = P_A + P_B \tag{4.34}$$

For each of these problems, we can write a condition for optimality:

$$\frac{\partial \Omega_A}{\partial P_A} = \pi(D) - \frac{dC_A}{dP_A} + P_A \cdot \frac{d\pi}{dD} \cdot \frac{dD}{dP_A} = 0 \tag{4.35}$$

$$\frac{\partial \Omega_B}{\partial P_B} = \pi(D) - \frac{dC_B}{dP_B} + P_B \cdot \frac{d\pi}{dD} \cdot \frac{dD}{dP_B} = 0 \tag{4.36}$$

Inserting the values given by Equations (4.30) and (4.31) in Equations (4.34)–(4.36), we get the following reaction curves:

$$P_A = \frac{1}{2}(64 - P_B) \tag{4.37}$$

$$P_B = \frac{1}{2}(54 - P_A) \tag{4.38}$$

Solving these two equations gives:

$$P_A = 24.7\,\text{MWh};\, P_B = 14.7\,\text{MWh};\, D = 40\,\text{MWh};\, \pi = 60.7\,\$/\text{MWh}$$

which is close to the equilibrium that we found using discrete values of P_A and P_B to build Table 4.5.

Example 4.11

We can also use the data from the previous example to explore what happens when the number of firms competing in a market increases. For the sake of simplicity, we consider the case where firm A competes against an increasing number of firms identical to firm B. An optimality condition similar to Equation (4.35) or Equation (4.36) can be written for each of these firms, and this system of equations can be solved together with the inverse demand relation (4.30) and the equation expressing that all these firms compete in the same market:

$$D = P_A + P_B + \cdots + P_N \tag{4.39}$$

where N represents the number of firms competing in this market. In this particular case, these equations are easy to solve for an arbitrary number of firms because firms B to N are identical and thus produce the same amount of energy. Since firm A produces electrical energy at a lower cost than the other firms, it has a competitive advantage in this market. Figure 4.11 shows that it always produces more than any other firm does. While its share of the market decreases as the number of competing firms increases, it does not tend to zero like the individual share of the other firms. Figure 4.12 shows that an increase in the number of competing firms depresses the market price, even if the new firms have the same marginal cost of production as the existing ones. In this case, however, the price asymptotically tends toward 46 $/MWh, which is the marginal cost of production of firms B to N. This heightened competition induces an increase in demand and therefore benefits the consumers. Finally, as Figure 4.13 shows, this increased competition also

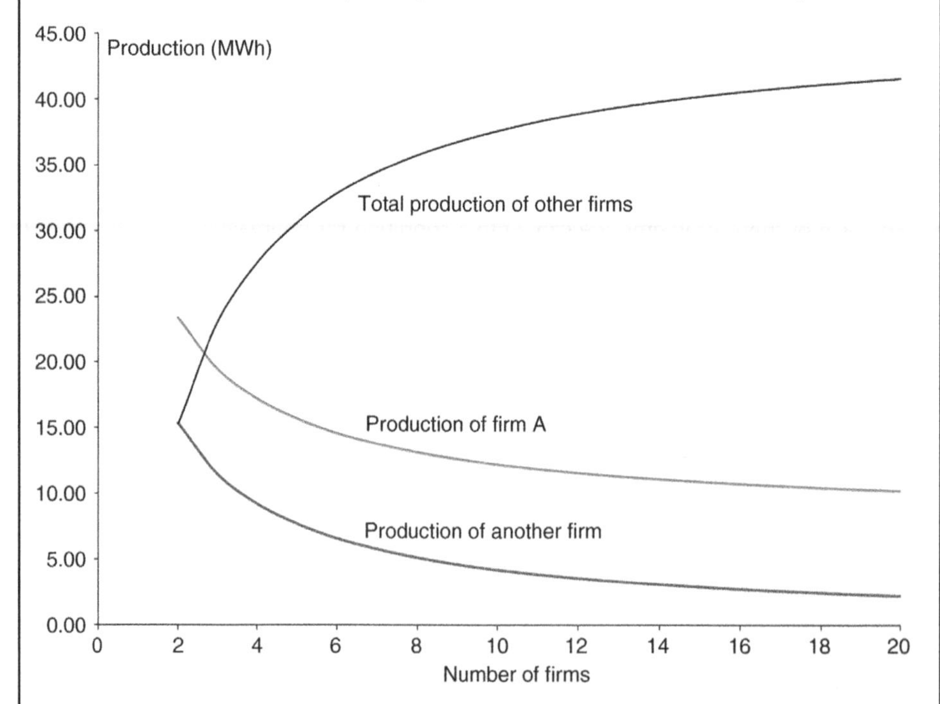

Figure 4.11 Evolution of the production of each firm as the number of competitors increases in the Cournot model.

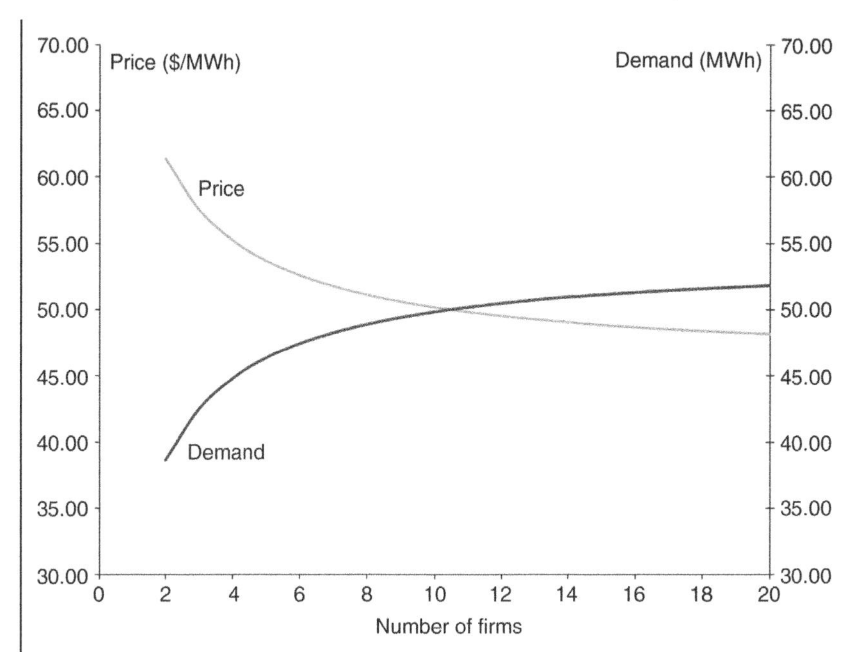

Figure 4.12 Evolution of the price and demand as the number of competitors increases in the Cournot model.

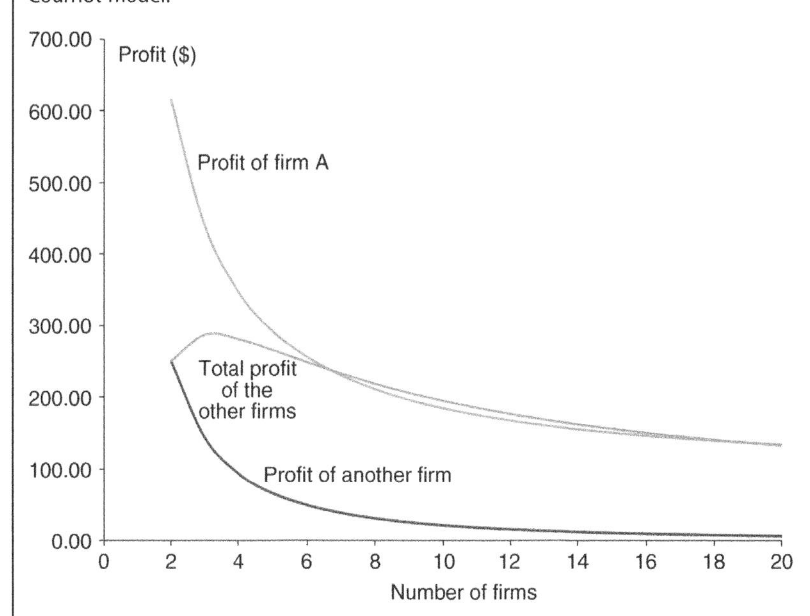

Figure 4.13 Evolution of the profits of each firm as the number of competitors increases in the Cournot model.

reduces the profits made by each firm. Because of its cost advantage, the profits of firm A are larger than the combined profits of all the other firms, and, unlike the profits of the firms in the competitive fringe, it does not tend to zero as the number of competitors increases.

4.4.3.3 Supply Functions Equilibria

While the Cournot model provides interesting insights into the operation of a market with imperfect competition, its application to electricity markets produces unreasonably high forecasts for the market price. More complex representations of the strategic behavior of generating companies have therefore been developed to obtain more realistic market models. In these models, it is assumed that the amount of energy that a firm is willing to deliver is related to the market price through a *supply function*:

$$P_f = P_f(\pi), \quad \forall f \tag{4.40}$$

In this case, the decision variables of each firm are thus neither the price nor the quantity but the parameters of its supply function.

At equilibrium, the total demand is equal to the sum of the quantities produced by all the firms:

$$D(\pi) = \sum_f P_f(\pi) \tag{4.41}$$

The profit of each firm can be expressed as follows:

$$\begin{aligned} \Omega_f &= \pi \cdot P_f - C_f(P_f) \\ &= \pi \cdot \left[D(\pi) - \sum_{-f} P_{-f}(\pi) \right] - C_f\left(D(\pi) - \sum_{-f} P_{-f}(\pi) \right), \quad \forall f \end{aligned} \tag{4.42}$$

These profit functions can be differentiated with respect to the price to get the necessary conditions for optimality, which after some manipulations can be expressed in the following form:

$$P_f(\pi) = \left(\pi - \frac{\mathrm{d}C_f(P_f)}{\mathrm{d}P_f} \right) \cdot \left(-\frac{\mathrm{d}D}{\mathrm{d}\pi} + \sum_{-f} \frac{\mathrm{d}P_{-f}(\pi)}{\mathrm{d}\pi} \right), \quad \forall f \tag{4.43}$$

The solution of this system of equations is an equilibrium point where all firms simultaneously maximize their profits. These optimality conditions are differential equations because the parameters of the supply functions are unknown. In order to find a unique solution to this set of differential equations, the supply and cost functions are usually assumed to have, respectively, linear and quadratic forms:

$$P_f(\pi) = \beta_f(\pi - \alpha_f), \quad \forall f \tag{4.44}$$

$$C_f(P_f) = \frac{1}{2} a_f P_f^2 + b_f P_f, \quad \forall f \tag{4.45}$$

The decision variables are thus:

$$X_f = \{\alpha_f, \beta_f\}, \quad \forall f \tag{4.46}$$

The optimal values of these variables can be computed by inserting Equations (4.44) and (4.45) as well as the inverse demand function into Equation (4.43). Once these optimal values have been computed using an iterative process, it is then possible to calculate the market price, the demand, and the production of each firm. It is interesting to note that, if the inverse demand function is affine (i.e. it includes a linear term plus an offset), the supply functions do not depend on the actual level of the demand.

4.4.3.4 Agent-Based Modeling

The modeling techniques described in the previous sections rely on relatively simple assumptions about the competitive behavior of market participants. While they provide some useful insights about the effects of market power, they typically do not model the market rules in sufficient details to be able to identify the various ways in which market participants are likely to behave.

Instead of assuming a template for the behavior of all the participants, agent-based modeling represents each of them by an independent software entity that strives to maximize its own objective and is able to learn from previous experience. These agents are made to interact repeatedly in an environment that replicates the rules of the market. Through these repeated interactions, each agent discovers the bidding strategy that best suits the characteristics of the market participant that it represents.

The main advantage of this approach is that it can incorporate a detailed representation of the various aspects of the market (e.g. interactions between forward and spot markets, transmission constraints). However, more detailed market models require more repeated interactions for the agents to settle on their optimal strategy and thus more computing time. There is also no guarantee that the optimal (or otherwise interesting) behaviors will emerge from these simulations. Finally, the agents' ability to learn is limited by the algorithm used and the set of parameters that they can adjust.

4.4.3.5 Experimental Economics

Instead of representing each market participant by a software agent, market simulations can also be performed using human subjects (typically students). Each of them is given a financial incentive to do as well as possible in the market using the assets of the company he or she represents. While having humans in the loop slows down the process and reduces the number of iterations that can be performed in the time available, human reasoning abilities and creativity still surpass what machines can achieve.

4.4.3.6 Limitations of These Models

Published applications to electricity markets of the models described in the previous sections have dealt so far mostly with predictions of market shares over a period of years. These models work on the aggregated capacity of each generating company and are probably not yet sophisticated enough to be useful in the daily optimization of individual generating units. In particular, they do not take into account nonlinearities such as no-load and startup costs and dynamic constraints on the output of each unit.

Furthermore, formulating the problem as a short-run profit maximization is probably an oversimplification. In some cases, a generating company that has market power may decide to limit or even drive down the market price. Such a course of action could be justified by a desire to increase or maintain market share, by a strategy to discourage entry in the market by new participants or by a fear of attracting regulatory intervention.

4.5 Perspective of Plants That Do Not Burn Fossil Fuels

So far, we have focused on market participation by plants that convert chemical energy into electrical energy by burning fossil fuels such as coal, oil, or natural gas. The marginal cost of these plants is therefore significant because of the cost of extracting these fuels from the earth. Plants that do not burn fossil fuels have a much lower (e.g. nuclear) or negligible (e.g. hydroelectric, wind, solar) marginal cost. However, these plants tend to have a much higher investment cost per MW of installed capacity. The challenge for their owners is therefore to collect enough revenue to recover their investments.

4.5.1 Nuclear Power Plants

Nuclear units tend to be operated at an almost constant generation level because adjusting their output is technically difficult. Ideally, these plants should be shut down only every 12–18 months for refueling because restarting them is a slow and costly process. In a centralized market, nuclear power plants often bid at 0.00 $/MWh to ensure that they are scheduled to produce their full output. They thus act as price takers, i.e. they let other participants set the market clearing price. In a bilateral market, the owners of these plants enter into long-term contracts for base load power.

Unplanned shutdowns of nuclear power plants can be very costly to their owners because their large capacity and the long duration of such outages require the purchase of a large amount of replacement energy on the spot and short-term forward markets. Such large purchases can significantly drive up prices on these markets.

4.5.2 Hydroelectric Power Plants

The very robust technology used in hydroelectric plants gives them the ability to easily adjust their power output. Their production can be ramped up or down very quickly over a wide range, and they can be shut down frequently without a significant impact on their expected life. As we will see in Chapter 6, this flexibility is very valuable for power system operation. However, while hydroelectric plants are often loosely constrained in terms of power, they can be significantly constrained in terms of the energy that they can or must produce over a given amount of time. The maximum amount of energy that a hydroelectric plant can produce is determined by the amount of rain or snow that falls in its river basin. The minimum amount of energy that it has to produce is dictated by the need to let some water through the dam either to avoid overfilling its reservoir or for environmental reasons (e.g. providing the right amount of water for fish preservation) or for other uses (e.g. irrigation, transportation, recreation). While one can always let water go over the dam without going through turbines, such water spillage is wasteful because the corresponding energy could have been produced at zero marginal cost.

In river basins with cascading dams, these constraints become even more complex because the energy that a particular plant can or must produce is a function of the energy that has been produced by plants located upstream and downstream. The operation of all the hydroelectric plants of a given river basin must therefore be optimized together to maximize the value of the electrical energy that they produce. Given a forecast of prices for electrical energy, this optimization determines how much each plant should produce to maximize the revenue collected while respecting the constraints. If the hydro generation capacity constitutes only a small fraction of the total installed capacity, it can be treated as a price taker. Otherwise, the effect of the hydro production on the market clearing price should be taken into account.

Since the amount of water available for energy production depends on precipitation and since precipitation depends on the season, this optimization should be carried out over a horizon of one year.[2] However, to keep the computational burden manageable, a

2 Or more if the reservoirs can store more than one-year worth of rain and snow.

two-stage approach is usually adopted. The first stage optimizes water usage over one year with a resolution of one month or one week. The second stage uses the results of the first stage as targets and refines the optimization with a much finer time resolution to determine how much energy can be sold into the market at each period.

Because of the long periods of time involved, the forecasts of market prices and water availability are quite inaccurate. Scheduling of hydro generation is therefore usually done using stochastic rather than deterministic optimization techniques.

4.5.3 Wind and Solar Generation

4.5.3.1 Intermittency and Stochasticity

When competing with other forms of generation, owners of wind and solar farms have the big advantage that their primary energy sources are free. On the other hand, they have to deal with two significant problems. First, these sources are intermittent: in most places the wind does not always blow and the sun does not always shine. This means that these renewable generators do not get to choose when to produce electrical energy. Depending on the region and the season, the pattern of availability of wind and sunshine may or may not line up with the periods of peak demand. The best they can do is forecast when and how much energy they expect to produce and sell this energy on forward markets. Which leads us to the second problem: since these sources of energy are stochastic, it is impossible to predict with perfect accuracy when the wind will start or stop blowing and how hard it will blow, or when and how long a large cloud will cast a shadow over a solar farm. Wind and solar generators therefore often face an imbalance between the amount of energy that they have sold and what they have actually produced. Since the cost of covering these imbalances on the spot market can be quite significant, operators of renewable generation use several mitigation techniques. First, they strive to improve the accuracy of their generation forecasts using numerical weather forecasting (for wind generation) and satellite images of cloud covers (for solar generation).[3] Second, they can actively trade in the short-term forward markets to cover their expected imbalances as improved forecasts become available. Third, they can partner with a flexible conventional generator or an energy storage facility. This partner then increases, decreases, or reverses its energy output to compensate for any deficit or surplus in the renewable generation. Together, these partners can enter into "firm" contracts for the delivery of energy. While energy produced by this flexible partner may not always be cheaper than purchases on the spot market, such an arrangement reduces the price risk to which the renewable generator is exposed.

4.5.3.2 Government Policies and Subsidies

These problems are often tempered by government policies aimed at reducing carbon dioxide emissions by encouraging the generation of electrical energy from renewable sources. These policies aim to help renewable generators by either mandating the purchase of renewable energy or by subsidizing investments in renewable generation capacity or the energy produced by these facilities. In other words, sticks and carrots.

Renewable portfolio standards or *renewable energy standards* oblige retailers to produce or buy a certain fraction of the energy that they sell from certain types of renewable sources. This fraction often increases over time. For example, in the State of

---

3 While significant progress has been made in the accuracy of these predictions, an experienced forecaster told us recently that "in this business, you have to accept that sometimes your forecast will be completely wrong!"

California, these percentages are 33% by 2020, 40% by 2024, 45% by 2027, and 50% by 2030. These standards sometimes also specify fractions for different renewable technologies.

Investment tax credits help defray the high cost of investments in renewables by giving the investors a rebate on their taxes for each kilowatt of installed renewable energy generation capacity.

Production subsidies take different forms:

A *production tax credit* is a rebate that the owner of a renewable generating plant receives on its taxes for each kilowatt-hour produced by this plant.

Feed-in tariffs guarantee that all the electrical energy produced from renewable sources will be bought at a favorable per kilowatt-hour rate.

Contracts for difference are set-up between renewable generators and the government. When the average price of electrical energy on the wholesale market is below an agreed strike price, the government pays the generators the difference. When the average market price exceeds the strike price, the generators pay the difference back to the government. With this type of contracts, renewable generators must still participate actively in the wholesale market but are guaranteed a more stable revenue stream.

Renewable energy certificates are given to renewable energy producers for each megawatt-hour that they generate. These certificates can then be sold either on a voluntary or a compliance market. Buyers on the voluntary markets are companies or individuals who want to make sure that an amount of energy equal to what they consume has been produced from renewable source. Buyers on the compliance markets are retailers who must meet their renewable portfolio standard.

Tax credits, feed-in tariffs, and the strike price of contracts for difference usually depend on the renewable generation technology and typically decrease over time to reflect the expectation that the cost of deploying these technologies will decrease to the point where renewable generation achieves "grid parity," i.e. that it no longer requires subsidies to be competitive with conventional generation on the electricity markets. The cost of these subsidies is socialized, which means that it is borne by either taxpayers or electricity consumers.

4.5.3.3 Effect on the Markets

As the proportion of generation capacity from wind and solar grows, its effect on the electricity markets becomes significant. The main outcome is that the average price decreases because renewable generators are willing to sell at a low price because their marginal cost of production is very low and their main challenge is to recover their large investment costs. They thus displace other forms of generation and often force them to retire. On the other hand, when there is no wind or sunshine, prices can rise significantly.

While justified on the basis of environmental policy, subsidies distort the market. For example, renewable generators who receive a production tax credit essentially get paid a fixed amount on top of the market price for every megawatt-hour that they produce. When demand is low and renewable resources are abundant, this can lead to negative market prices (i.e. generators have to pay to produce). Renewable generators can tolerate this better than other because they continue to collect revenues as long as the absolute value of the negative market price does not exceed the production tax credit.

The amount of photovoltaic generation capacity that residential and commercial consumers have installed on their roofs has become significant in some locations. While these consumers do not usually participate directly in the wholesale electricity market, they have an indirect effect on these markets because the aggregation of this distributed production can cause a significant drop in demand during the middle of the day when solar irradiance is strongest.

4.6 The Storage Owner's Perspective

Vertically integrated utilities have used pumped hydro plants for several decades to facilitate the integration of nuclear power plants and reduce the system operating cost by flattening the load profile. These plants consume energy by pumping water uphill during periods of light load and produce energy by releasing this water through turbines during periods of high load. Cycling consumption and production in this manner reduces the difference between the peaks and the troughs in the demand curve. This allows nuclear power plants to operate at a constant power output, reduces the need to cycle conventional power plants on and off or to operate them at less than their optimal efficiency, and thus decreases the system operating cost.

Since dealing with the intermittency and stochasticity of renewable energy sources would be a lot easier if more storage capacity were available, a considerable amount of effort has been devoted in recent years to the development of electrochemical battery energy storage systems. These devices hold the promise of being cheaper, more efficient, and more environmentally friendly compared to pumped storage plants.

In this section, we assume that batteries or other energy storage devices perform only temporal arbitrage, i.e. they buy and store energy when the price is low and release and sell this energy when the price is high. In later chapters, we will discuss the additional benefits that the flexibility of energy storage provides to the system operators.

In a competitive environment, temporal arbitrage can be profitable if the revenue generated by selling energy during periods of high prices is larger than the cost of the energy consumed during periods of low prices. This calculation must take into account the fact that, because of the losses, not all of the energy bought and stored can be sold back.

4.6.1 Self-scheduling

Let us first consider the case of a storage operator who decides ahead of time for the next T periods when to charge and when to discharge a storage device on the basis of a forecast of prices. This operator seeks to maximize its operating profit, which is given by the following expression:

$$\Omega = \sum_{t=1}^{T} \pi(t)(P_{\mathrm{D}}(t) - P_{\mathrm{C}}(t))\Delta t \tag{4.47}$$

where:

$\pi(t)$ is the forecast market price during period t (\$/MWh)
$P_{\mathrm{D}}(t)$ is the rate of discharge of the storage device during period t (MW)
$P_{\mathrm{C}}(t)$ is the rate of charge of the storage device during period t (MW)
Δt is the duration of each period (h).

During each hour, the battery is charging, discharging, or idle, which means that $P_{\mathrm{C}}(t)$ and $P_{\mathrm{D}}(t)$ cannot be nonzero simultaneously. The amount of energy stored (i.e. the state of charge of the storage device) is given by the following expression:

$$E(t) = E(t-1) + [\eta P_C(t) - P_D(t)]\Delta t \tag{4.48}$$

where:

$E(t)$ is the state of charge at the end of period t (MWh)
η is the round-trip efficiency of the storage device.

This optimization is subject to the following constraints:

$$0 \leq E(t) \leq E^{\max}, \quad \forall t = 1, \ldots, T \tag{4.49}$$

$$0 \leq P_D(t) \leq P^{\max}, \quad \forall t = 1, \ldots, T \tag{4.50}$$

$$0 \leq P_C(t) \leq P^{\max}, \quad \forall t = 1, \ldots, T \tag{4.51}$$

$$E(T) = E(0) = E_0 \tag{4.52}$$

where:

$E^{\max}$ is the energy rating of the storage device
$P^{\max}$ is its power rating
E_0 is the initial state of charge.

4.6.2 Centralized Operation

Storage devices can also be treated as another resource that the system operator can use to meet the load at minimum cost. The optimization problem that the system operator must solve to clear the market is similar to the problem discussed in Section 3.3.2.4, with the exception of the load generation balance constraint that becomes:

$$\sum_{i=1}^{N} P_i(t) + P_D(t) - P_C(t) = L(t), \quad \forall t = 1, \ldots, T \tag{4.53}$$

where:

$P_i(t)$ is the power produced by generating unit i during period t
N is the number of generating units
$L(t)$ is the load forecast for period t.

The rates of charge or discharge of the storage device during each period are subject to the constraints on the operation of the storage described by Equations (4.48)–(4.52).

From a mathematical perspective, the availability of storage provides additional decision variables and thus helps reduce the overall cost of operating the system. However, this does not ensure that the storage device will make an operating profit as measured by Equation (4.47).

Example 4.12

Let us consider a simple example, involving a market with three generators, a scheduling horizon of 3 h, and a trading period of 1 h. Tables 4.6 and 4.7 give the characteristics of the generating units and the demand over the scheduling horizon.

For simplicity, we ignore the startup costs, the ramp rate limits as well as the minimum up- and down-time constraints. These simplifications allow us to consider each period

Table 4.6 Generating unit data.

Generating unit i	P_i^{min} (MW) minimum generation	P_i^{max} (MW) maximum generation	α_i (\$/MWh) marginal cost	β_i (\$/h) fixed cost
1	0	500	10	500
2	100	350	25	250
3	0	200	50	100

Table 4.7 Demand data.

Time period t	1	2	3
$L(t)$ (MW)	495	750	505

separately. We also assume that the cost functions of the generators involve only a fixed cost α_i and a constant marginal cost β_i:

$$C_i(P_i(t), u_i(t)) = \alpha_i u_i(t) + \beta_i\, P_i(t) \tag{4.54}$$

Table 4.8 shows how a centralized market operator would schedule these generating units to satisfy this demand at minimum cost. It also gives the market price for each trading hour, as well as the revenue, the operating cost, and the profit for each generating unit. The total cost of operating the system over this horizon is $23 300. Unit 1 makes a substantial profit because it is only marginal during hour 1.

Table 4.8 Market settlement without storage.

Hour	Price ($/MWh)		Unit 1	Unit 2	Unit 3
1	10	Output (MWh)	495	0	0
		Revenue ($)	4 950	0	0
		Cost ($)	5 450	0	0
		Profit ($)	−500	0	0
2	25	Output (MWh)	500	250	0
		Revenue ($)	12 500	6 250	0
		Cost ($)	5 500	6 500	0
		Profit ($)	7 000	−250	0
3	50	Output (MWh)	500	0	5
		Revenue ($)	25 000	0	250
		Cost ($)	5 500	0	350
		Profit ($)	19 500	0	−100
		Total profit ($)	26 000	−250	−100
		Total cost ($)		23 300	

Table 4.9 Market settlement with a 1 MW/10 MWh battery.

Hour	Price ($/MWh)		Unit 1	Unit 2	Unit 3	Battery
1	10	Output (MWh)	496	0	0	−1
		Revenue ($)	4 960	0	0	−10
		Cost ($)	5 460	0	0	0
		Profit ($)	−500	0	0	−10
2	25	Output (MWh)	500	251	0	−0.2
		Revenue ($)	12 500	6 275	0	−5
		Cost ($)	5 500	6 525	0	0
		Profit ($)	7 000	−250	0	−5
3	50	Output (MWh)	500	0	4	1
		Revenue ($)	25 000	0	200	50
		Cost ($)	5 500	0	300	0
		Profit ($)	19 500	0	−100	50
		Total profit ($)	26 000	−250	−100	35
		Total cost ($)		23 265		

Let us introduce in this market a 1 MW/10 MWh battery with a round-trip efficiency of 0.83. This battery is initially completely discharged and it self-schedules based on the published prices to perform temporal arbitrage. We assume that its capacity is small enough compared to the rest of the system that it has no impact on the prices. Table 4.9 summarizes the market settlement under these conditions. The battery takes advantage of the low prices during hour 1 to charge at its maximum 1 MW rate. During period 2 it charges at a rate of 0.2 MW to compensate for its losses and ensure that it can discharge a full 1.0 MWh at the high price of period 3. This arbitrage cycle yields a profit of $35 for the battery and reduces the total generation cost by $50 over the case without storage.

If instead of having a power rating of 1 MW this battery was rated at 10 MW, it would be fully charged during hour 1 and fully discharged during hour 3 to take advantage of the biggest price difference. However, in this case our assumption that this battery would have no effect on the prices and the power balance would be questionable. Table 4.10 shows how this market clears if this 10 MW/10 MWh battery is optimally scheduled along with the generating units to meet the load at minimum cost. In this case, the battery is charged at a rate of 5 MW during hour 1 and at a rate of 1 MW during hour 2. Because Unit 1 is operating at its maximum during hour 1, it does not set the price. Instead, this price is set at 25 $/MWh because a marginal increase in load would result in a marginal shift in the charging of the battery from hour 1 to hour 2. Since Unit 2 is marginal during hour 2, it therefore sets the price not only for hour 2 but also for hour 1. Because the battery is able to discharge 5 MWh during hour 3, Unit 3 does not need to be committed. A marginal increase in load at hour 3 would require the battery to charge more at hour 2. The price at hour 3 is therefore 30.12 $/MWh, which is 25 $/MWh divided by the 0.83 round-trip efficiency. While the battery makes no profit over this scheduling horizon, it flattens the

Table 4.10 Market settlement with a 10 MW/10 MWh battery.

Hour	Price ($/MWh)		Unit 1	Unit 2	Unit 3	Battery
1	25	Output (MWh)	500	0	0	−5
		Revenue ($)	12 500	0	0	−125
		Cost ($)	5 500	0	0	0
		Profit ($)	7 000	0	0	−125
2	25	Output (MWh)	500	251	0	−1
		Revenue ($)	12 500	6 276	0	−26
		Cost ($)	5 500	6 526	0	0
		Profit ($)	7 000	−250	0	−26
3	30.12	Output (MWh)	500	0	0	5
		Revenue ($)	15 060	0	0	151
		Cost ($)	5 500	0	0	0
		Profit ($)	9 560	0	0	151
		Total profit ($)	23 560	−250	0	0
		Total cost ($)		23 026		

load profile and reduces the system operating cost by $274 compared to the case without storage.

4.7 The Flexible Consumer's Perspective

4.7.1 Flexible Demand Vs Storage

Consumers who are able to shift their demand in time, either through self-scheduling or by offering this ability to the system operator, provide a resource that, in some ways, is similar to energy storage. Instead of storing energy in chemical or gravitational form, these consumers store heat, intermediate products in a manufacturing process, or dirty dishes. A significant advantage of flexible demand over storage is that the storage facility has often already been built and therefore does not require the large investments needed for batteries or pumped hydro plants. The downside is that providing services to the power system is not the primary purpose of these facilities. Constraints on the manufacturing process or the comfort of residential users therefore limit their usage as a system resource. Some large industrial consumers are able to shift loads that are sufficiently large to be significant at the system level. On the other hand, flexible demand from smaller consumers must be aggregated to have a measurable effect. Since at any given time, a particular consumer may or may not be willing or able to shift its demand, this aggregation also makes possible a probabilistic quantification of the available capacity that bypasses the need for polling each consumer individually.

4.7.2 Remunerating Flexible Demand

If consumers with a flexible demand are exposed to time-varying prices, they may find it financially worthwhile to schedule their consumption in a way that minimizes their cost while meeting their needs for production efficiency or comfort. This optimization problem is very similar to the storage self-scheduling problem discussed in Section 4.6.1. In this case, the remuneration for having a flexible demand is *price-based*.

Consumers who are not exposed to time-varying prices can also agree to reduce or shift their load in response to a signal from their utility, their retailer, or their aggregator. In exchange for being flexible in their consumption, these consumers are entitled to a favorable tariff. Their remuneration is then said to be *incentive-based*. In addition to the incentive, the contract between the flexible consumer and the entity sending the signal must also specify the following:

- How often the consumer is expected to respond
- How large should the load reduction be
- How much time must elapse before the consumer can return to its normal consumption pattern and start recovering the energy not consumed
- How the consumer's response will be measured.

Measuring the consumer's response is not a trivial problem because the entity paying the incentive wants to make sure that it is getting an actual demand shift, not a random fluctuation in the consumer's demand. This is typically done by comparing the consumer's actual load profile against its profile for similar days.

4.7.3 Implementation Issues

Flexible demand can be incorporated into an energy market in two ways. It can either be considered along with the supply side in a centralized optimal scheduling process or consumers can self-schedule their flexible demand based on prices broadcast by the market operator.

Under the centralized paradigm, consumers submit the technical and economic characteristics of their flexibility to the market operator, who then schedules their operation simultaneously with the generating units through a global optimization problem. Dispatch signals are then sent to each individual flexible load and generating unit. While this centralized optimal scheduling could theoretically achieve the optimal economic outcome, it is doubtful that it could scale up to handle thousands or even millions of flexible appliances. The communication and computational requirements would indeed get too severe and the implementation cost too high. Such a centralized mechanism would also raise privacy concerns because it requires that the consumers reveal a substantial amount of information about how and when they need electricity.

The alternative approach to integrating flexible demand in electricity markets avoids the scalability and privacy concerns of the centralized architecture by giving consumers the opportunity to self-schedule their demand in response to posted prices. Consumers could then take advantage of price differences to reduce their electricity bills. Because prices would be higher during peak demand periods and lower during off-peak demand periods, consumers would have an incentive to shift their demand from peak to off-peak periods. However, a naive application of such a pricing scheme, combined with an automatic response of the appliances to these prices, could concentrate the demand at the periods with the lowest prices, potentially creating new demand peaks during originally periods of low demand, leading to inefficient system operation. Similarly, flexible demand has been shown to "rebound" after a period of high prices.

The following highly simplified examples aim to illustrate these issues and how they might be overcome.

Example 4.13 Centralized scheduling of flexible appliances

Let us consider an electricity market that is centrally scheduled over two, one-hour market periods. The participants in this market are as follows:

- A generator, producing P_t (MW) at hour t with a cost function $C(P_t) = 100^* P_t^2$ (\$) and a maximum output $P^{max} = 8$ MW.
- An inflexible demand, consuming $D_1 = 1$ MW during period 1 and $D_2 = 2$ MW during period 2.
- One thousand identical flexible appliances with continuously adjustable demands d_t. Each of these appliances must consume $E = 6$ kWh over the two market periods but cannot consume more than $d^{max} = 5$ kWh during each period.

The objective of the centralized schedule is to determine the demand of the flexible appliances during each of the two periods that minimize the total generation cost. Because the cost function of the generator is quadratic, minimizing this total cost is equivalent to minimizing the absolute value of the difference between the power produced by the generator during these two periods:

$$\min|P_1 - P_2| \tag{4.55}$$

Since the total generation must be equal to the total demand at each period, we have:

$$P_1 = D_1 + 1000^* d_1 \tag{4.56}$$

$$P_2 = D_2 + 1000^* d_2 \tag{4.57}$$

Inserting (4.56) and (4.57) into the objective function (4.55), we get:

$$\min|(D_1 + 1000^* d_1) - (D_2 + 1000^* d_2)| \tag{4.58}$$

We must also take into account the operating constraints on the generation:

$$D_1 + 1000^* d_1 \le P^{max} \tag{4.59}$$

$$D_2 + 1000^* d_2 \le P^{max} \tag{4.60}$$

And on the flexible appliances:

$$d_1 \le d^{max} \tag{4.61}$$

$$d_2 \le d^{max} \tag{4.62}$$

$$d_1 + d_2 = E \tag{4.63}$$

Table 4.11 shows a set of possible options for scheduling the flexible demand over the two market periods, as well as the corresponding total generation cost over the two

Table 4.11 Feasible flexible scheduling options.

d_1	d_2	P_1	P_2	$C(P_1) + C(P_2)$
5.0	1.0	6.0	3.0	4500
4.5	1.5	5.5	3.5	4250
4.0	2.0	5.0	4.0	4100
3.5	**2.5**	**4.5**	**4.5**	**4050**
3.0	3.0	4.0	5.0	4100
2.5	3.5	3.5	5.5	4250
2.0	4.0	3.0	6.0	4500
1.5	4.5	2.5	6.5	4850
1.0	5.0	2.0	7.0	5300

Optimal values are in bold.

periods. Flattening the demand profile by scheduling $d_1 = 3.5$ and $d_2 = 2.5$ kWh minimizes the total generation cost while meeting the constraints.

Example 4.14 Unlimited self-scheduling of flexible appliances

Suppose that each flexible appliance is informed ahead of time that the price will be π_1 during hour 1 and π_2 during hour 2. Because the inflexible demand is lower during hour 1, the market operator would set $\pi_1 < \pi_2$ to encourage a shift in demand from hour 2 to hour 1. If, as we assume, the appliances respond entirely to prices, as much of the flexible demand as possible would be reallocated to the period of low price, i.e. hour 1. We would then have:

$$d_1 = d^{\max} = 5\,\text{kW} \tag{4.64}$$

$$d_2 = E - d_1 = 1\ \text{kW} \tag{4.65}$$

To balance generation and load, the generation schedule would then have to be:

$$P_1 = D_1 + 1000^* d_1 = 6\ \text{MW} \tag{4.66}$$

$$P_2 = D_2 + 1000^* d_2 = 3\ \text{MW} \tag{4.67}$$

Hour 1 would then no longer be the off-peak period and the total generation cost would be:

$$100{P_1}^2 + 100{P_2}^2 = \$4500 \tag{4.68}$$

which is significantly higher than the cost under optimal centralized scheduling because the prices π_1 and π_2 are not consistent with the actual generation schedule.

The previous example shows that a naive approach to sending prices to the flexible demand can have counterproductive consequences if this demand is allowed to respond in an unconstrained manner. The following examples illustrate two possible approaches to steering the flexible demand toward an optimal solution.

Example 4.15 Externally limited self-scheduling of flexible appliances

In this approach, the market operator sends to the appliances not only a set of prices but also a relative flexibility restriction signal ω. This signal ranges from 0 to 1 and represents the fraction of the maximum demand that the appliance is allowed to consume during each period. It thus prevents the appliances from shifting too much of their total energy requirements to the lowest price periods. A value of ω close to 1 may not sufficiently limit the appliances' ability to concentrate their demand, while a value of ω close to 0 may limit excessively their flexibility and preclude a sufficient flattening of the demand profile. ω must therefore be properly tuned to achieve the desired result.

Let us consider the same system as in the previous example, but assume that, in addition to the time-differentiated prices $\pi_1 < \pi_2$, the system operator also sends to the appliances a flexibility restriction signal ω. The amount of energy that each flexible appliance would consume during hour 1 is now given by:

$$d_1 = \omega^* d^{\max} \tag{4.69}$$

The flexible demand during hour 2 is then:

$$d_2 = E - d_1 \tag{4.70}$$

However, we must also ensure that:

$$d_2 \leq \omega^* d^{\max} \tag{4.71}$$

The generator is then dispatched to meet the load as follows:

$$s_1 = D_1 + 1000^* d_1 \tag{4.72}$$

$$s_2 = D_2 + 1000^* d_2 \tag{4.73}$$

Table 4.12 shows the flexible demand at each hour, the generation dispatch, and the total generation costs for different values of the flexibility restriction signal ω.

Table 4.12 Externally constrained self-scheduling of flexible appliances.

ω	d_1	d_2	P_1	P_2	$C(P_1)+C(P_2)$
1	5.0	1.0	6.0	3.0	4500
0.9	4.5	1.5	5.5	3.5	4250
0.8	4.0	2.0	5.0	4.0	4100
0.7	**3.5**	**2.5**	**4.5**	**4.5**	**4050**
0.6	3.0	3.0	4.0	5.0	4100
0.5	2.5	2.5	Infeasible: Constraint $d_1+d_2=E$ is not satisfied		
0.4	2.0	2.0			
0.3	1.5	1.5			
0.2	1.0	1.0			
0.1	0.5	0.5			

A value of $\omega = 1$ imposes no limit on demand flexibility and leads to the inefficient solution of the previous example. Large values of ω (0.8–0.9) do not sufficiently limit the appliances' flexibility to concentrate their demand during hour 1. On the other hand, $\omega = 0.6$ limits the appliances' flexibility more than is required, and thus fails to balance the overall demand between the 2 hours. Lower values of ω (0.1–0.5) produce infeasible solutions because the power that the appliances are allowed to draw is so low that the total energy constraint ($E = 6$ kWh) cannot be satisfied. The optimal value of ω is 0.7, because it flattens the overall demand profile and results in the same total generation cost as produced by the centralized scheduling.

Example 4.16 Self-scheduling of flexible appliances with quadratic price penalty

Imposing external restrictions on the operation of the appliances, as illustrated in Example 4.15, is unlikely to be acceptable by many consumers. Instead, let us add a nonlinear price penalty proportional to the square of the demand. Such a penalty makes the consumers' cost nonlinear and thus prevents a concentration of the demand on a small number of market periods. This penalty can be modulated using a factor α to achieve the desired result.

Let us consider once again the system of Example 4.13, assuming that $\pi_1 = 1$ and $\pi_2 = 2$ \$/kWh and let us determine the value of the penalty factor α that minimizes the total generation cost. With this penalty, the total amount paid by the consumers is given by:

$$\pi_1{}^* d_1 + \pi_2{}^* d_2 + \alpha^* (d_1)^2 + \alpha^* (d_2)^2 \tag{4.74}$$

Given the constraint $d_1 + d_2 = E$, Equation (4.74) can be rewritten as a function of a single decision variable d_1:

$$\pi_1{}^* d_1 + \pi_2{}^* (E - d_1) + \alpha^* (d_1)^2 + \alpha^* (E - d_1)^2 \tag{4.75}$$

Since all the consumers try to minimize the total amount of money that they pay, the optimal value of d_1 can be found by setting the derivative of this amount with respect to d_1 equal to zero:

$$\frac{\mathrm{d}\left[\pi_1{}^* d_1 + \pi_2{}^* (E - d_1) + \alpha^* (d_1)^2 + \alpha^* (E - d_1)^2\right]}{\mathrm{d}[d_1]} = 0 \tag{4.76}$$

After some manipulation, we get:

$$d_1 = \frac{2^* \alpha^* E + \pi_2 - \pi_1}{4^* \alpha} \tag{4.77}$$

This is the optimal value of d_1 as long as the operating constraint $d_1 \le d^{\max}$ is satisfied. If (4.77) yields a value $d_1 > d^{\max}$, then d_1 is set equal to $d^{\max}$. d_2 is always equal to $E - d_1$.

Table 4.13 shows how the dispatch of the flexible appliances and the total generation cost change as a function of the penalty factor α. When $\alpha = 0$, no penalty is imposed and the results are identical to the inefficient solution of Example 4.14. A value of $\alpha = 0.1$, yields $d_1 = 5.5$ kW, which is infeasible, and d_1 is therefore set equal to $d^{\max} = 5$ kW.

Small values of α (0.1–0.4) do not sufficiently discourage the concentration of the demand at the lowest-priced hour 1, resulting in a shift in the peak demand from hour 2 to hour 1. Larger values of α (0.6–0.9) limit the appliances' flexibility more than required. The

Table 4.13 Self-scheduling of flexible appliances with quadratic price penalty.

α	d_1	d_2	P_1	P_2	$C(P_1) + C(P_2)$
0	5.0	1.0	6.0	3.0	4500.0
0.1	5.0	1.0	6.0	3.0	4500.0
0.2	4.25	1.75	5.25	3.75	4162.5
0.3	3.833	2.167	4.833	4.167	4072.2
0.4	3.625	2.375	4.625	4.375	4053.1
0.5	**3.5**	**2.5**	**4.5**	**4.5**	**4050.0**
0.6	3.417	2.583	4.417	4.583	4051.4
0.7	3.357	2.643	4.457	4.643	4054.0
0.8	3.313	2.687	4.313	4.687	4057.0
0.9	3.278	2.722	4.278	4.722	4059.9

optimal value of α is 0.5, because it flattens the overall demand profile and yields the same optimal total generation cost as the centralized scheduling.

4.8 The Neighbor's Perspective

Some power systems where a competitive electricity market has been introduced are interconnected with neighboring systems that are operated by vertically integrated utilities. These utilities often take part in the competitive market. If the price paid for electrical energy is higher than their marginal cost of production, they will behave like producers in this market. On the other hand, if the price is lower than their marginal cost of production, it is in their best interest to reduce the output of their own generators and purchase power on the competitive market.

4.9 An Overall Market Perspective

4.9.1 Clearing the Market

Figure 4.14 shows an offer curve derived from data collected on the ISO-New England website. This curve was built by stacking 556 price/quantity offers submitted by generators in increasing order of price. Since this data is anonymized, we can only surmise which groups of generator submitted a certain type of offer. We can discern four distinct parts on this curve:

1) About 750 MW of capacity is offered at zero or negative prices. Some of these bids are submitted by nuclear, run-of-the-river hydro, trash burning, and other generators that have to run and thus want to make sure that they are included in the production schedule, no matter the price. Other offers might come from renewable generators who receive production subsidies and can thus remain profitable even if the price is negative.

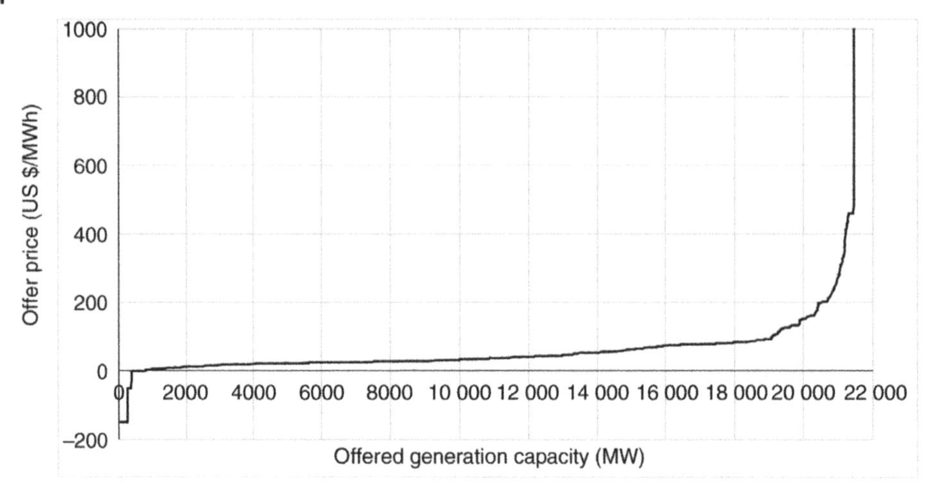

Figure 4.14 Offer curve for the ISO-New England day-ahead market of March 30, 2016.

2) From about 750 MW to about 19 000 MW, the offer price increases gradually and is likely to reflect each generator's marginal production cost. Note that some generating units submit a single price/quantity pair while others divide their offer into 10 segments, which is the maximum allowed by the market rules.
3) From 19 000 to 21 400 MW, the offer price increases much more steeply. These offers are submitted either by generators with a much higher marginal cost or by generators that run infrequently and thus need a much higher price to recover their fixed costs.
4) A few generators offer at the ceiling price of 1000 $/MWh.

Curves with this typical shape are often dubbed "hockey sticks."

Figure 4.15 shows two demand curves built using data from the same website. The curve on the left is for the trading period ending at 3 a.m. (i.e. close to the minimum

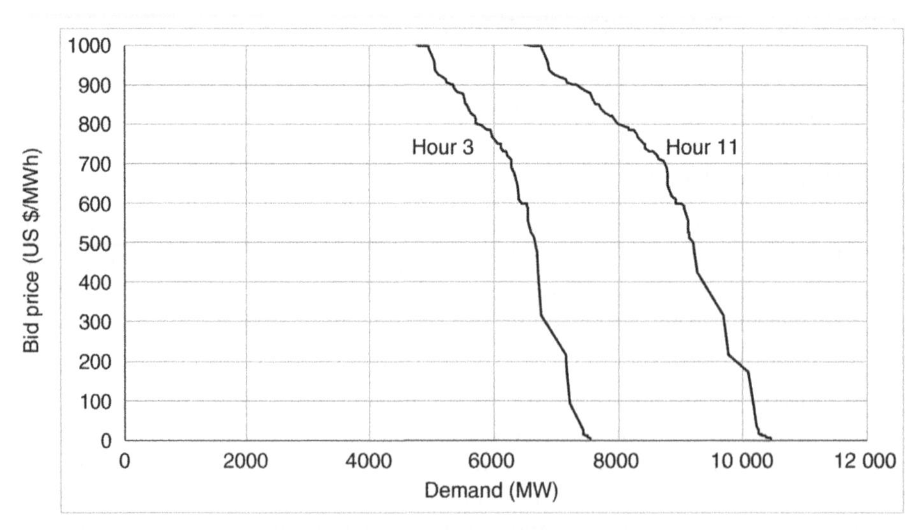

Figure 4.15 Bid curves for the ISO-New England day-ahead market for hours 3 and 11 of March 30, 2016.

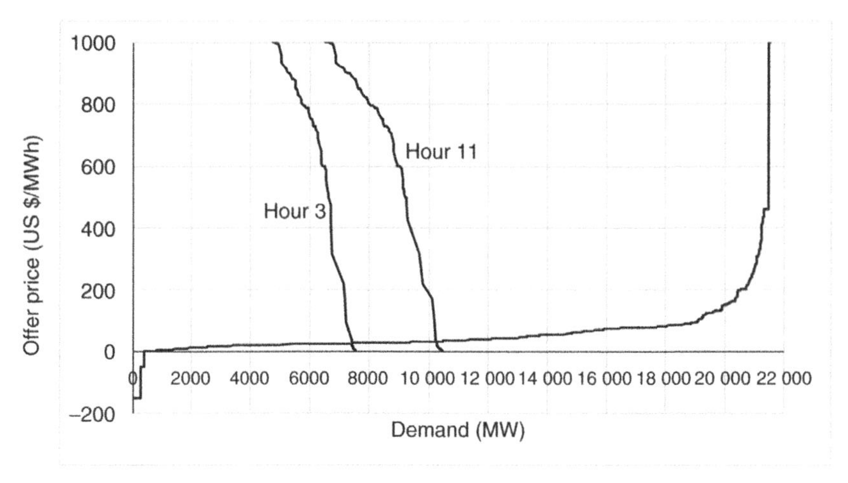

Figure 4.16 Market clearing for the ISO-New England day-ahead market for hours 3 and 11 of March 30, 2016.

demand for that day) while the curve on the right is for the trading period ending at 11 a.m. (i.e. close to the maximum demand for that day). No price was submitted for about 7550 MW of demand bids at hour 3 and 10 500 MW at hour 11, indicating that these consumers are not price-sensitive. On the other hand, since roughly 2800 and 4000 MW of price-sensitive bids were submitted for hours 3 and 11, respectively, these demand curves exhibit some price elasticity. However, a substantial part of this elasticity probably stems from virtual bids submitted in the day ahead market rather than from actual load flexibility.

If, as shown in Figure 4.16, we superimpose the offer and bid curves, we can see that the market clearing price for the various market periods of this particular day will vary within a relatively narrow range. This is to be expected because March 30, 2016 was a relatively mild spring day that did not require much electric heating or cooling. The generation capacity offered in the market was therefore much larger than the demand and none of the generating units that bid at a high price were needed. On the other hand, on a particularly hot or cold day, the demand curve shifts far to the right. Because of their relative shapes, the intersection of the supply and demand curves can shoot up to very high prices for a small increment in load at peak time, creating price spikes.

Figure 4.17 summarizes how the ISO-New England market cleared during the year 2015 using a price duration curve, i.e. a plot showing the percentage of hours during which the market clearing price exceeded a given value. By comparison, Figure 4.18 shows that prices on the electricity market of the Canadian province of Alberta generally vary over a narrower range but spike to much higher values a few percent of the time.

4.9.2 Exercising Market Power

Generators will often use economic or physical withholding to try to increase the market clearing price. As we discussed in Section 2.5.1, economic withholding entails offering some capacity at a high price in the expectation that this capacity will be needed to clear the market. Looking at Figures 4.14 and 4.16, this means making the steep part of the offer

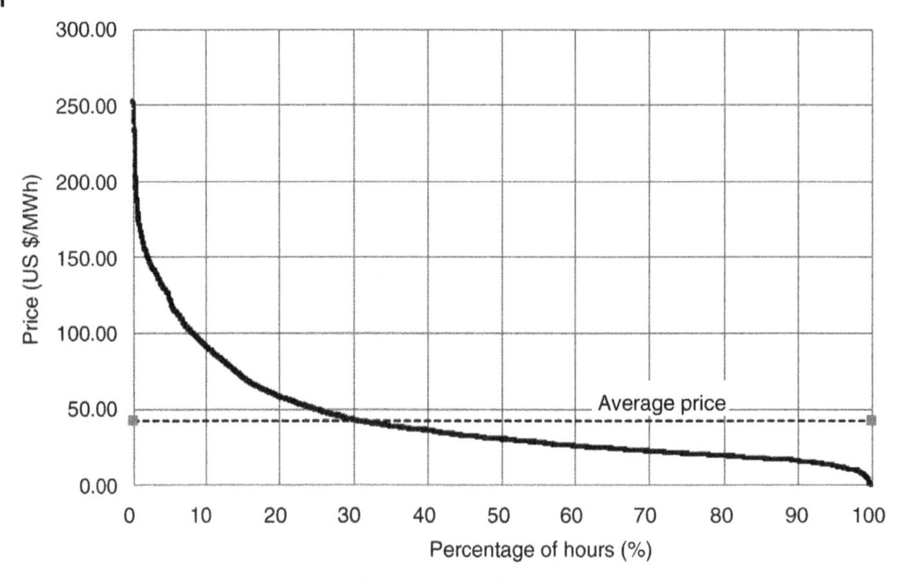

Figure 4.17 Day-ahead hourly price duration curve for the Boston area of ISO-New England for the year 2015. Prices are in US dollars.

curve even steeper to push up its intersection with the (also steep) demand curve. Physical withholding consists in not offering a substantial amount of generation into the market. Withholding capacity thus shifts the rest of the offer curve to the left, resulting again in a higher market clearing price. The shapes of the offer and demand curves provide an intuitive explanation of why the exercise of market power is more likely to be significant during periods of high demand. As we mentioned before, exercising market power is less effective if the price elasticity of the demand is higher. Figure 4.19 illustrates the fact that only a relatively small fraction of the total demand needs to be price-sensitive to significantly reduce the market price during periods of peak load.

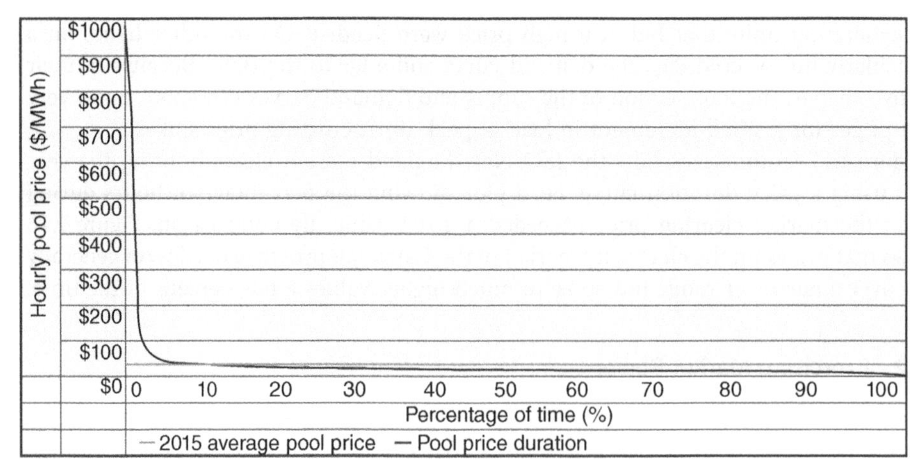

Figure 4.18 Price duration curve for the Alberta Electricity Market for 2015. Prices are in Canadian dollars. Source: www.aeso.ca.

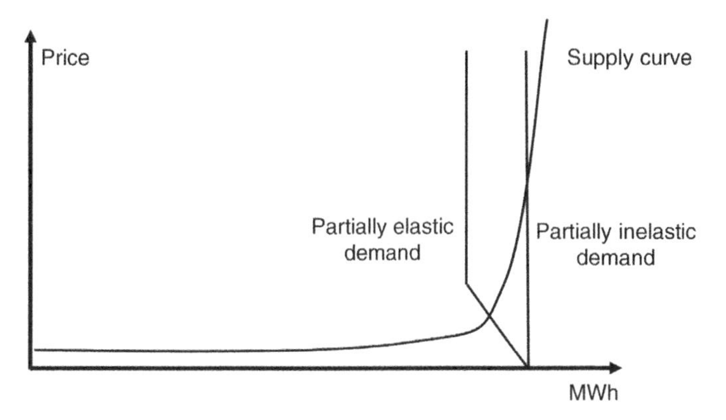

Figure 4.19 Effect of a partially elastic demand on market clearing.

In Chapter 5, we will discuss how limits on the capacity of the transmission network create additional opportunities to exercise market power.

4.9.3 Dealing with Market Power

It is not unusual for firms to form a *cartel* and collude to divvy the market and keep prices high. While exercising market power is not prohibited, collusion is illegal, and regulatory authorities impose substantial penalties when companies are caught in the act. However, collusion often takes a subtler, tacit form. Instead of discussing how to rig the market, firms that compete on a regular basis can send each other signals through published prices. While a firm could achieve a bigger profit in the short term by undercutting its competitors' prices, it may realize that in the long run it is in its own interest not to do so.

Because market power is a serious concern, designers of electricity markets have put in place various mechanisms to mitigate its effects. The simplest technique is to impose a price cap, i.e. automatically limit prices to a value set by the regulatory authority. This cap must be set relatively high because high prices are occasionally justified because they signal a need to invest in additional generation capacity and because they help generators recoup their fixed costs. Some markets also implement bid caps, i.e. limits on the offer price that generators are allowed to submit. Finally, some markets have developed bid mitigation techniques. When the exercise of market power is suspected, offending offers are replaced by standard offers based on the characteristics of the generating unit and the current fuel cost. Prices are then recalculated using these standard offers and compared to the original prices. If these new prices are significantly lower than the old prices, bids are capped at the standard offers.

In addition to these ex ante measures aimed at preventing the exercise of market power, regulatory authorities occasionally try to punish perpetrators. However, it is difficult to prove that an abuse of market power has occurred because generating companies can easily argue that they know their power plants much better than regulators. For example, where a regulator may argue that capacity has been withheld from the market to raise prices, the generator may retort that the power plant in question was in urgent need of maintenance or repair. Similarly, they can argue that the regulator underestimates the cost of running peaking plants. Nevertheless, the threat of large fines can deter the most blatant abuses of market power.

4.10 Problems

4.1 Cheapo Electrons is an electricity retailer. Table 4.14 shows the load that it forecast its consumers would use over a 6-h period. Cheapo Electrons purchased on the forward market and the power exchange exactly enough energy to cover this forecast. The table shows the average price that it paid for this energy for each hour. As one might expect, the actual consumption of its customers did not exactly match the load forecast and it had to purchase or sell the difference on the spot market at the prices indicated. Assuming that Cheapo Electrons sells energy to its customers at a flat rate of 24.00 \$/MWh, calculate the profit or loss that it made during this 6-h period. What would be the rate that it should have charged its customers to break even?

4.2 The input–output curve of a gas-fired generating unit is approximated by the following function:

$$H(P) = 120 + 9.3P + 0.0025P^2 \text{ (MJ/h)}$$

This unit has a minimum stable generation of 200 MW and a maximum output of 500 MW. The cost of gas is 1.20 \$/MJ. Over a 6-h period, the output of this unit is sold on a market for electrical energy at the prices shown in Table 4.15.

Assuming that this unit is optimally dispatched, is initially on-line, and cannot be shut down, calculate its operational profit or loss for this period.

4.3 Repeat the calculation of Problem 4.2 assuming that the cost curve is replaced by a three-segment piecewise linear approximation whose values correspond with those given by the quadratic function for 200, 300, 400, and 500 MW.

4.4 Assume that the unit of Problem 4.2 has a startup cost of \$500 and that it is initially shutdown. Given the same prices as in Problem 4.2, when should this unit be brought on-line and when should it be shutdown to maximize its operational

Table 4.14 Data for Problem 4.1.

Period	1	2	3	4	5	6
Load forecast (MWh)	120	230	310	240	135	110
Average cost (\$/MWh)	22.5	24.5	29.3	25.2	23.1	21.9
Actual load (MWh)	110	225	330	250	125	105
Spot price (\$/MWh)	21.6	25.1	32	25.9	22.5	21.5

Table 4.15 Data for Problem 4.2.

Period	1	2	3	4	5	6
Price (\$/MWh)	12.5	10	13	13.5	15	11

profit? Assume that dynamic constraints do not affect the optimal dispatch of this generating unit.

4.5 Repeat Problem 4.4 taking into account that the minimum up-time of this unit is 4 h.

4.6 Borduria Generation owns three generating units that have the following cost functions:

Unit A: $15 + 1.4\ P_A + 0.04\ P_A^2$ \$/h
Unit B: $25 + 1.6\ P_B + 0.05\ P_B^2$ \$/h
Unit C: $20 + 1.8\ P_C + 0.02\ P_C^2$ \$/h.

How should these units be dispatched if Borduria Generation must supply a load of 350 MW at minimum cost?

4.7 How would the dispatch of Problem 4.6 change if Borduria Generation had the opportunity to buy some of this energy on the spot market at a price of 8.20 \$/MWh?

4.8 If, in addition to supplying a 350 MW load, Borduria Generation had the opportunity to sell energy on the electricity market at a price of 10.20 \$/MWh, what is the optimal amount of power that it should sell? What profit would it derive from this sale?

4.9 Repeat Problem 4.8 if the outputs of the generating units are limited as follows:
$P_A^{max} = 100$ MW
$P_B^{max} = 80$ MW
$P_C^{max} = 250$ MW.

4.10 Consider a market for electrical energy that is supplied by two generating companies whose cost functions are as follows:

$$C_A = 36 \cdot P_A\ (\$/h)$$

$$C_B = 31 \cdot P_B\ (\$/h)$$

The inverse demand curve for this market is estimated to be:

$$\pi = 120 - D\ (\$/MWh)$$

Assuming a Cournot model of competition, use a table similar to the one used in Example 4.10 and calculate the equilibrium point of this market (price, quantity, production, and profit of each firm).

(*Hint:* Use a spreadsheet. A resolution of 5 MW is acceptable.)

4.11 Write and solve the optimality conditions for Problem 4.10.

4.12 Consider a pumped hydro plant with an energy storage capacity of 1000 MWh and an efficiency of 75%. Assume that it takes 4 h to completely empty or fill the upper reservoir of this plant if the plant operates at rated power and that the upper

Table 4.16 Data for Problem 4.12.

Period	1	2	3	4	5	6
Price ($/MWh)	40.92	39.39	39.18	40.65	45.42	56.34
Period	7	8	9	10	11	12
Price ($/MWh)	58.05	60.15	63.39	59.85	54.54	49.50

reservoir is initially empty. Suppose that the operator of this plant has decided to go through a full cycle during the 12-h period using a very simple strategy: pump water to the upper reservoir during the 4 h with the lowest energy prices and release it during the 4 h with the highest energy prices. Table 4.16 shows the prices during this 12-h period.

Calculate the profit or loss that this plant would make during this cycle of operation. Determine the value of the plant efficiency that would make the profit or loss equal to zero.

(*Hint:* Use a spreadsheet.)

4.13 Dragon Power Ltd is considering the construction of a new power plant in addition to the 50 MW power plant it already operates in the Syldavian electricity market. Possible capacities for the new plant are 50, 100, and 150 MW. The marginal cost of production of the existing plant and the new plant is 25 $/MWh. Dragon Power's analysis of the Syldavian market shows that the following:

- Its competitors currently operate 200 MW of generation capacity.
- The incremental cost of operation of the generation capacity of its competitors is 30 $/MWh.
- The demand can be represented by the demand curve $\pi = 450 - D$, where D is the total demand and π is the market price.

a Using a Cournot model, determine the capacity of the new plant that would maximize the total hourly operating profit of Dragon Power. What is the total operating profit of Dragon Power with this plant in service?

b What would be the effect of the construction of a new plant of the optimal capacity on the hourly profit of Dragon Power's competitors?

c Assuming that this plant is built and that the demand is 300 MW, estimate how much imperfect competition in the Syldavian market costs consumers at each hour.

d Another team of analysts estimates that the demand curve of the Syldavian market is $\pi = 440 - 1.2\,D$. How does this revised estimate affect the profitability of the optimal new plant? (Assume that the plant must make an hourly profit of 6000 $/h to cover its fixed costs.) What can you conclude about the robustness of the Cournot model?

Further Reading

Bunn (2000) surveys techniques for forecasting loads and prices, while Bushnell and Mansur (2001) provide some data about how consumers respond when faced with market prices for electricity. Techniques for maximizing the profits of a price-taking generator are discussed in Arroyo and Conejo (2001), while Conejo et al. (2010) discuss techniques for decision-making under uncertainty. Baldick et al. (2005) address the topic of market power mitigation. The reader interested in supply function equilibria will find discussions of this subject in Day et al. (2002). Kirschen (2003) discusses various aspects of demand-side participation in electricity markets. Bunn and Oliveira (2001) as well as Yu et al. describe how agent-based modeling can be used to study market designs. Wood and Wollenberg (2014) is the classic reference on generation scheduling.

Arroyo, J.M. and Conejo, A.J. (2000). Optimal response of a thermal unit to an electricity spot market. *IEEE Trans. Power Syst.* 15 (3): 1098–1104.

Baldick, R., Helman, U., Hobbs, B.F., and O'Neill, R.P. (2005). Design of efficient generation markets. *Proc. IEEE* 93 (11): 1998–2012.

Bunn, D.W. (2000). Forecasting loads and prices in competitive power markets. *Proc. IEEE* 88 (2): 163–169.

Bunn, D.W. and Oliveira, F.S. (2001). Agent-based simulation – an application to the new electricity trading arrangements of England and Wales. *IEEE Trans. Evol. Comput.* 5 (5): 493–503.

Bushnell, J.B. and Mansur, E.T. (2001). The impact of retail rate deregulation on electricity consumption in San Diego. Working Paper PWP-082, Program on Workable Energy Regulation. University of Californian Energy Institute, April 2001. https://ei.haas.berkeley.edu/research/papers/PWP/pwp082r.pdf.

Conejo, A.J., Carrión, M., and Morales, J.M. (2010). *Decision Making under Uncertainty in Electricity Markets*, International Series in Operations Research & Management Science. Springer.

Day, C.J., Hobbs, B.F., and Pang, J.-S. (2002). Oligopolistic competition in power networks: a conjectured supply function approach. *IEEE Trans. Power Syst.* 17 (3): 597–607.

Kirschen, D.S. (2003). A demand-side view of electricity markets. *IEEE Trans. Power Syst.* 18 (2): 520–527.

Wood, A.J. and Wollenberg, B.F. (2014). *Power Generation, Operation and Control*, 3e. Wiley.

Yu, N.P., Liu, C.C., and Price, J. (2010). Evaluation of market rules using a multi-agent system method. *IEEE Trans. Power Syst.* 25 (1): 470–479.

5

Transmission Networks and Electricity Markets

5.1 Introduction

In most, if not all, regions of the world, the assumption that electrical energy can be traded as if all generators and loads were connected to the same bus is not tenable. Transmission constraints and losses in the network connecting generators and loads can introduce gross distortions in the market for electrical energy if they are not taken into account properly. Furthermore, the creation of wholesale electricity markets spanning large regions encourages transactions between participants that are geographically distant and thus increases congestion in the transmission network.

In this chapter, we study the effects that transmission networks have on trading of electrical energy and the special techniques that can be used to hedge against the limitations and price fluctuations that are caused by these networks. In Chapter 6, we discuss in more detail what operators do to maintain the reliability of the system, including the stability of the transmission network. For the sake of simplicity, in this chapter, we assume that system stability is guaranteed if the active power flowing through each branch of the transmission network remains below a given limit. If the market participants agree upon a set of transactions that naturally satisfy these conditions (e.g. when the demand is very low), nothing needs to be done and the transmission network has no effect on the market outcome. On the other hand, if one or more stability limits would be violated, the market outcome is affected. In a bilateral or decentralized trading system, this means that some transactions have to be limited or curtailed. In a centralized market, injections must be adjusted and the market clearing price is no longer the same at every bus.

5.2 Decentralized Trading over a Transmission Network

In a decentralized or bilateral trading system, all transactions for electrical energy involve only two parties: a buyer and a seller. These two parties agree on a quantity, a price, and any other condition that they may want to attach to the trade. The system operator does not get involved in these transactions and does not set the prices at which transactions take place. Its role is limited to maintaining the balance and the operational reliability of the system. This involves the following:

- Buying or selling energy to balance the load and the generation. Under normal circumstances, the amounts involved in these balancing transactions should be small.

Fundamentals of Power System Economics, Second Edition. Daniel S. Kirschen and Goran Strbac.

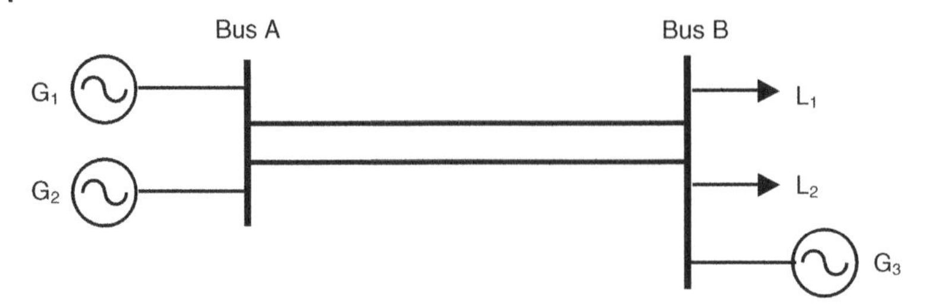

Figure 5.1 Bilateral trading in a two-bus power system.

- Limiting the transactions between generators and consumers if operational reliability cannot be maintained through other means.

Let us consider the two-bus power system shown in Figure 5.1 where trading in electrical energy operates on a bilateral basis. Let us suppose that generator G_1 has signed a contract for the delivery of 300 MW to load L_1 and generator G_2 has agreed to deliver 200 MW to load L_2. Since these transactions are bilateral, the agreed prices are a private matter between the buyer and the seller. On the other hand, the amount of power to be transmitted must be reported to the system operator because this power flows on the transmission network that is open to all parties. The system operator must check that the system can operate within its limits when all these transactions are implemented. In this case, this is not a problem as long as the capacity of the transmission lines connecting buses A and B is at least 500 MW. If the maximum amount of power that can be transmitted between buses A and B without affecting the reliability of the system is less than 500 MW, the network is said to be congested and the system operator has to intervene. Some of the bilateral transactions that were concluded between generators at bus A and loads at bus B must then be curtailed.

5.2.1 Physical Transmission Rights

With modern power system analysis software, determining that a set of transactions could result in a violation of the transmission limits can be computationally demanding but is conceptually simple. Deciding which transactions should be curtailed to remain within these limits is a much more complex question. Administrative procedures can be established to determine the order in which transactions should be cut back. Such transmission load relief procedures take into account the nature of the transactions (firm or nonfirm), the order in which they were registered with the system operator and possibly some historical factors. However, such administrative procedures do not factor in the relative economic benefits of the various transactions because a decentralized trading environment does not provide a framework for evaluating these benefits. Administrative curtailments are therefore economically inefficient and should be avoided.

Advocates of decentralized electricity trading believe that the parties considering transactions for electrical energy are best placed to decide whether they wish to use the transmission network. When they sign a contract, producers at bus A and consumers at bus B who do not wish to see their transaction denied because of congestion should therefore also purchase the right to use the transmission system for this transaction. Since these transmission rights are purchased at a public auction, the parties have the opportunity to decide whether this additional cost is justifiable.

For example, let us suppose that generator G_1 and load L_1 have agreed on a price of 30.00 \$/MWh while Generator G_2 and load L_2 agreed on 32.00 \$/MWh. At the same time, Generator G_3 offers energy at 35.00 \$/MWh. Load L_2 should therefore not agree to pay more than 3.00 \$/MWh for transmission rights because this would make the energy it purchases from G_2 more expensive than the energy it could purchase from G_3. The price of transmission rights would have to rise above 5.00 \$/MWh before L_1 reaches the same conclusion. The cost of transmission rights is also an argument that the consumers can use in their negotiations with the generators at bus B to convince them to lower their prices.

Transmission rights of this type are called *physical transmission rights* because they are intended to support the actual transmission of a certain amount of power over a given transmission link.

5.2.2 Problems with Physical Transmission Rights

Our simple example makes physical transmission rights appear simpler than they turn out to be. The first difficulty is practical and arises because the path that power takes through a network is determined by physical laws and not by the wishes of market participants. The second problem is that physical transmission rights have the potential to exacerbate the exercise of market power by some participants. Let us consider these two issues in turn.

5.2.2.1 Parallel Paths

Two fundamental laws govern current and power flows in electrical networks: Kirchhoff's current law (KCL) and Kirchhoff's voltage law (KVL). KCL specifies that the sum of all the currents entering a node must be equal to the sum of all the currents exiting this node. KCL implies that the active and reactive powers must both be in balance at each node. KVL specifies that the sum of the voltage drops across all the branches of any loop must be equal to zero or, equivalently, that the voltage drops along parallel paths must be equal. Since these voltage drops are proportional to the current flowing through the branch, KVL determines how the currents (and hence the active and reactive power flows) distribute themselves through the network. In the simple example shown in Figure 5.2, a current $\overline{I}$ can flow from node 1 to node 2 along two parallel paths of impedances z_A and z_B. The voltage difference between the two nodes is thus

$$\overline{V_{12}} = z_A \overline{I_A} = z_B \overline{I_B}$$

Since $\overline{I} = \overline{I_A} + \overline{I_B}$, we have:

$$\overline{I_A} = \frac{z_B}{z_A + z_B} \overline{I} \tag{5.1}$$

$$\overline{I_B} = \frac{z_A}{z_A + z_B} \overline{I} \tag{5.2}$$

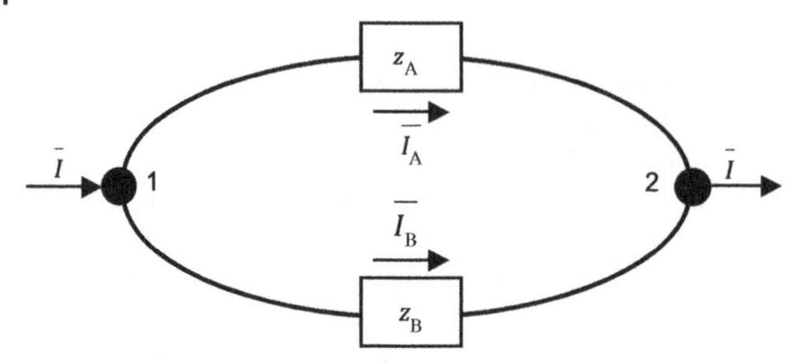

Figure 5.2 Illustration of Kirchhoff's voltage law.

Currents in parallel paths therefore divide themselves in inverse proportion of the impedance of each path. To simplify the following discussion, we assume that the resistance of any branch is much smaller than its reactance:

$$Z = R + jX \approx jX \tag{5.3}$$

We also neglect the flow of reactive power through the network and the losses. Under these assumptions, the system of Figure 5.2 can be depicted in terms of active power flows as shown in Figure 5.3. The active power flows in the parallel paths are related by the following expressions:

$$F^{A} = \frac{x_B}{x_A + x_B} P \tag{5.4}$$

$$F^{B} = \frac{x_A}{x_A + x_B} P \tag{5.5}$$

The factors relating the active power injections and the branch flows are called power transfer distribution factors (PTDF).

5.2.2.2 Example

A two-bus system does not illustrate the effect of KVL because there is only one path that the power can follow.[1] We must therefore consider a network with three buses and one loop. Figure 5.4 illustrates such a system and Table 5.1 gives its parameters. To keep

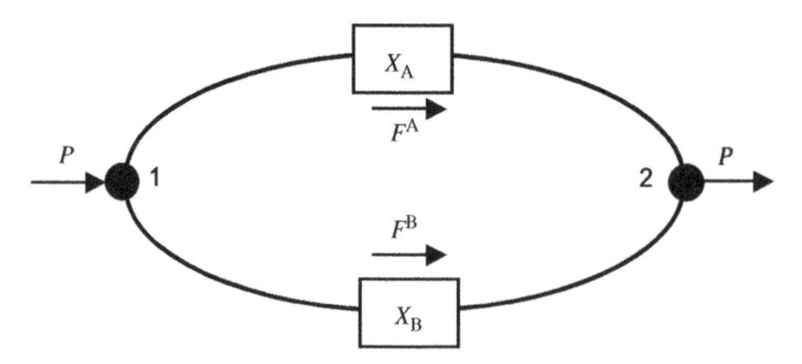

Figure 5.3 Active power flows on parallel paths.

1 For the sake of simplicity, we treat a line with two identical circuits as a single branch.

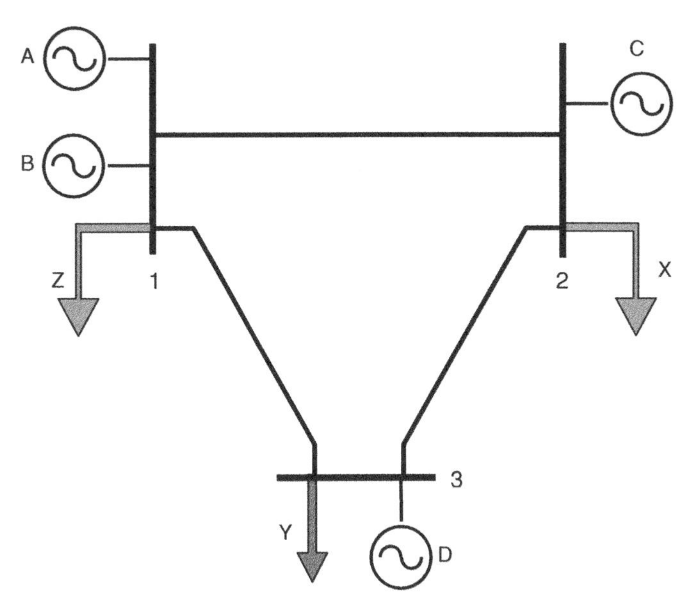

Figure 5.4 A simple three-bus system.

matters simple, we assume that network limitations take the form of constant capacity limits on the active power flowing in each line and that the resistance of the lines is negligible.

Let us suppose that generator B and load Y want to sign a contract for the delivery of 400 MW. If generator B injects these 400 MW at bus 1 and load Y extracts them at bus 3, this power flows along the two paths shown in Figure 5.5. The amounts of power flowing along paths I and II are given by:

$$F^{\mathrm{I}} = \frac{0.2}{0.2 + 0.3} \times 400 = 160\,\mathrm{MW}$$

$$F^{\mathrm{II}} = \frac{0.3}{0.2 + 0.3} \times 400 = 240\,\mathrm{MW}$$

To guarantee that this transaction can actually take place, the parties therefore need to secure 240 MW of transmission rights on line 1–3 as well as 160 MW of transmission rights on lines 1–2 and 2–3. This is clearly not possible if this transaction is the only one taking place in this network because the maximum capacity of lines 1–2 and 2–3 are 126

Table 5.1 Branch data for the three-bus system of Figure 5.4.

Branch	Reactance (p.u.)	Capacity (MW)
1–2	0.2	126
1–3	0.2	250
2–3	0.1	130

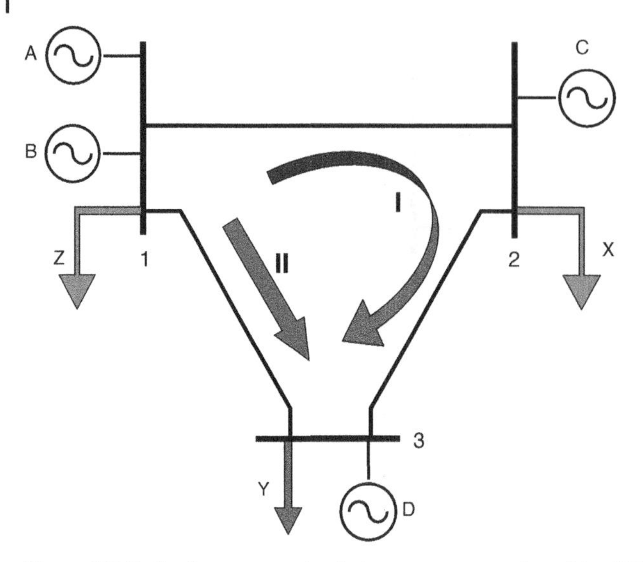

Figure 5.5 Paths for a transaction between generator B and load Y.

and 130 MW, respectively. In the absence of any other transaction, the maximum amount that A and Y can trade is limited by the capacity of line 1–2:

$$P^{\max} = \frac{0.5}{0.2} \times 126 = 315\,\text{MW}$$

However, suppose that load Z would like to purchase 200 MW from generator D. This power would flow along the paths shown in Figure 5.6 in the following proportions:

$$F^{\text{III}} = \frac{0.2}{0.2 + 0.3} \times 200 = 80\,\text{MW}$$

$$F^{\text{IV}} = \frac{0.3}{0.2 + 0.3} \times 200 = 120\,\text{MW}$$

Let us calculate what the flows in this network would be if both of these transactions were to take place at the same time. For this calculation, we can make use of the superposition theorem because our simplifying assumptions have linearized the relations between flows and injections. The flows in the various lines are thus given by:

$$F_{12} = F_{23} = F^{\text{I}} - F^{\text{III}} = 160 - 80 = 80\,\text{MW}$$

$$F_{13} = F^{\text{II}} - F^{\text{IV}} = 240 - 120 = 120\,\text{MW}$$

The transaction between generator D and load Z thus creates a counterflow that increases the power that generator D and load Y can trade.

If we do not want the transmission network to limit trading opportunities unnecessarily, the amount of physical transmission rights that is made available must take into account possible counterflows. In keeping with the bilateral or decentralized trading philosophy, the system operator should only check that the system would operate within its limits if all the proposed transactions were implemented. If this is not the case, the market participants have to adjust their position through further bilateral contracts

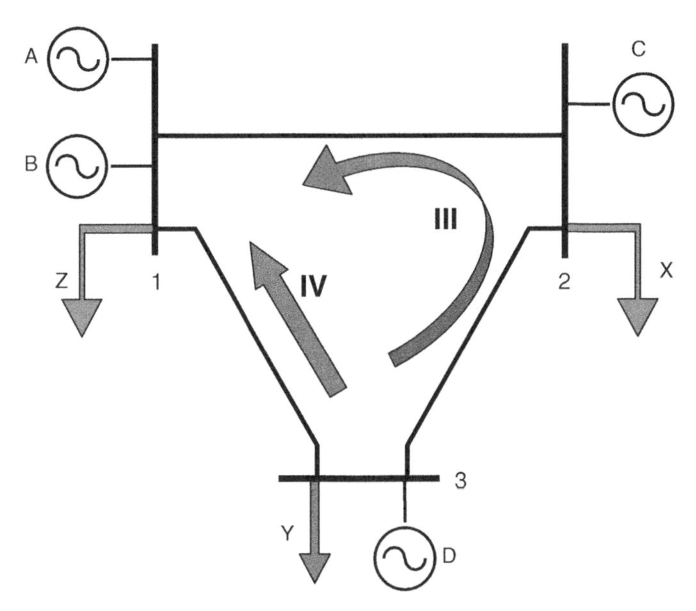

Figure 5.6 Paths for a transaction between generator D and load Z.

until an acceptable system operating state is achieved. Bilateral energy trading is therefore closely coupled with bilateral trading in physical transmission rights.

In theory, if the market is efficient, participants should be able to discover through repeated interactions a combination of bilateral trades in energy and transmission rights that achieves the economic optimum. In practice, in a power system with more than a few capacity constraints, the amount of information that needs to be exchanged is so large that it is unlikely that this optimum could be reached quickly enough through bilateral interactions.

5.2.2.3 Physical Transmission Rights and Market Power

We have defined physical transmission rights as giving their owner the right to transmit a certain amount of power for a certain time through a given branch of the transmission network. If physical transmission rights are treated like other types of property rights, their owners can use them or sell them. They can also decide to keep them but not sell them. In a perfectly competitive market, buying physical transmission rights but not using them would be an irrational decision. On the other hand, in a less than perfectly competitive market, physical transmission rights can enhance the ability of some participants to exert market power. Consider, for example, the two-bus power system of Figure 5.1. If generator G_3 is the only generator connected to bus B, it might want to purchase physical transmission rights for power flowing from bus A to bus B. If G_3 does not use or resell these rights, it effectively decreases the amount of power that can be sold at bus B by the other generators. This artificial reduction in transmission capacity enhances the market power that G_3 exerts at bus B and allows it to increase the profit margin on its production. It also has a detrimental effect on the economic efficiency of the overall system. See Joskow and Tirole (2000) for a comprehensive discussion of this issue.

To avoid this problem, it has been suggested that a "use them or lose them" provision be attached to physical transmission rights. Under this provision, transmission capacity that a participant has reserved but does not use is released to others who wish to use it. In theory, this approach should prevent market participants from hoarding transmission capacity for the purpose of enhancing market power. In practice, enforcing this condition is difficult because the unused transmission capacity may be released so late that other market participants are unable to readjust their trading positions.

5.3 Centralized Trading over a Transmission Network

In a centralized or pool-based trading system, producers and consumers submit their bids and offers to the system operator, who also acts as market operator. The system operator, which must be independent from all the other parties, selects the bids and offers that optimally clear the market while respecting the constraints imposed by the transmission network. As part of this process, the system operator also determines the market clearing prices. We shall show that, when losses and congestion in the transmission network are taken into account, the price of electrical energy depends on the bus where power is injected or extracted. The price that consumers and producers pay or are paid is the same for all participants connected to the same bus. This was not necessarily the case in a decentralized trading system where prices are determined by bilateral contracts. In a centralized trading system, the system operator thus has a much more active role than it does in the bilateral model. Economic efficiency is indeed achieved only if it optimizes the use of the transmission network.

5.3.1 Centralized Trading in a Two-Bus System

We begin our analysis of the effects of a transmission network on centralized trading of electrical energy using a simple example involving the fictitious countries of Borduria and Syldavia. After many years of hostility, these two countries have decided that the path to progress lies in economic cooperation. One of the projects that are under consideration is the reenergization of an existing electrical interconnection between the two national power systems. Before committing themselves to this project, the two governments have asked Bill, a highly regarded independent economist, to study the effect that this interconnection would have on their electricity markets and to evaluate the benefit that this interconnection would bring to both countries.

Bill begins his study by analyzing the power systems of both countries. He observes that both countries have developed centralized electricity markets that are quite competitive. The price of electrical energy in each market thus reflects closely its marginal cost of production. In both countries, the installed generation capacity exceeds the demand by a significant margin. Using regression analysis, Bill estimates the supply function for the electricity market in each country. Borduria's supply function is:

$$\pi_B = MC_B = 10 + 0.01P_B \ (\$/MWh) \tag{5.6}$$

While in Syldavia, it has the following form:

$$\pi_S = MC_S = 13 + 0.02P_S \ (\$/MWh) \tag{5.7}$$

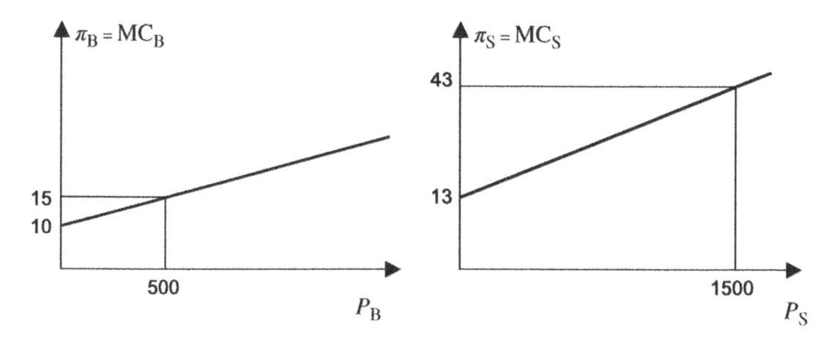

Figure 5.7 Supply functions for the electrical energy markets of Borduria and Syldavia.

Figure 5.7 shows that, like all such functions, these two supply functions increase monotonically with the demand for electrical energy. For the sake of simplicity, Bill assumes that the demands in Borduria and Syldavia are constant and equal to 500 and 1500 MW, respectively. He also assumes that these demands have a price elasticity of zero. When the two national electricity markets operate independently, the prices are thus as follows:

$$\pi_B = MC_B = 10 + 0.01 \times 500 = 15\,\$/\text{MWh} \tag{5.8}$$

$$\pi_S = MC_S = 13 + 0.02 \times 1500 = 43\,\$/\text{MWh} \tag{5.9}$$

Neither country is interconnected with other countries. Since the transmission infrastructure within each country is quite strong and very rarely affects the operation of the market for electrical energy, Bill decides that the simple model shown in Figure 5.8 is adequate for the study he needs to perform.

5.3.1.1 Unconstrained Transmission

Under normal operating conditions, the interconnection can carry 1600 MW. If all the generators in Syldavia were to be shut down, the entire load of that country could therefore still be supplied from Borduria through the interconnection. The capacity of this link is thus larger than the power that could possibly need to be transmitted.

Equations (5.8) and (5.9) show that electricity prices in Borduria are significantly lower than in Syldavia. One could therefore envision that Bordurian generators might supply not only their domestic demand but also all the Syldavian demand. We would then have:

$$P_B = 2000\,\text{MW} \tag{5.10}$$

$$P_S = 0\,\text{MW} \tag{5.11}$$

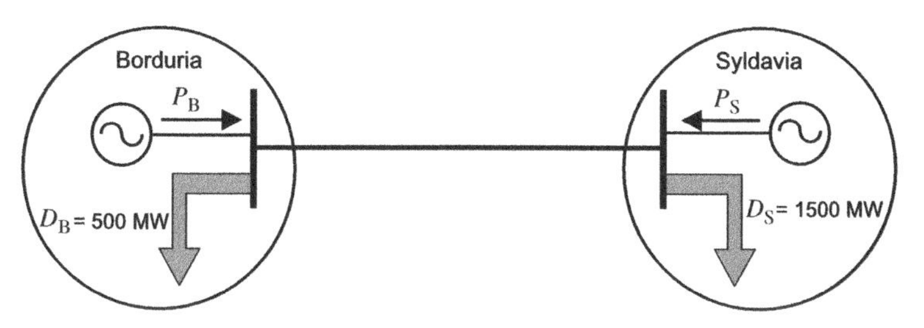

Figure 5.8 Model of the Borduria–Syldavia interconnection.

Replacing these values in (5.6) and (5.7), we find that the marginal cost of producing electrical energy in the two systems would be the following:

$$MC_B = 30\,\$/\text{MWh} \tag{5.12}$$

$$MC_S = 13\,\$/\text{MWh} \tag{5.13}$$

This situation is clearly not tenable because Bordurian generators would demand 30 \$/MWh, while Syldavian generators would be willing to sell energy at 13 \$/MWh. The Bordurian generators are thus not able to capture the entire market because a process of price equalization would take place. In other words, the interconnection forces the markets for electrical energy in both countries to operate as a single market, where the same price applies to the energy consumed in both countries:

$$\pi = \pi_B = \pi_S \tag{5.14}$$

Generators from both countries compete to supply the total demand, which is equal to the sum of the two national demands:

$$P_B + P_S = D_B + D_S = 500 + 1500 = 2000\,\text{MW} \tag{5.15}$$

Since the generators in both countries are willing to produce up to the point where their marginal cost of production is equal to the market clearing price, Equations (5.6) and (5.7) are still applicable. To determine the market equilibrium, Bill solves the system of equations (5.6), (5.7), (5.14), and (5.15) and obtains the following solution:

$$\pi = \pi_B = \pi_S = 24.30\,\$/\text{MWh} \tag{5.16}$$

$$P_B = 1433\,\text{MW} \tag{5.17}$$

$$P_S = 567\,\text{MW} \tag{5.18}$$

The flow of power in the interconnection is equal to the surplus of generation over load in the Bordurian system and the deficit in the Syldavian system:

$$F_{BS} = P_B - D_B = D_S - P_S = 933\,\text{MW} \tag{5.19}$$

A flow of power from Borduria to Syldavia makes intuitive sense because the price of electricity in Borduria is lower than in Syldavia when the interconnection is not in service.

Figure 5.9 offers a graphical representation of the operation of this single market. The productions of the Bordurian and Syldavian generators are plotted respectively from left to right and right to left. Since the two vertical axes are separated by the total load in the system, any point on the horizontal axis represents a feasible dispatch allocation of this load between generators in the two countries. This diagram also shows the supply curves of the two national markets. The prices in Borduria and Syldavia are measured along the left and right axes, respectively.

When the two systems operate as a single market, the prices in both systems must be identical. Given the way this diagram is constructed, the intersection of the two supply curves gives this operating point. The diagram then shows the production in each country and the flow on the interconnection.

5.3.1.2 Constrained Transmission

Over the course of a year, various components of the transmission system must be taken out of service for maintenance. These components include not only lines and

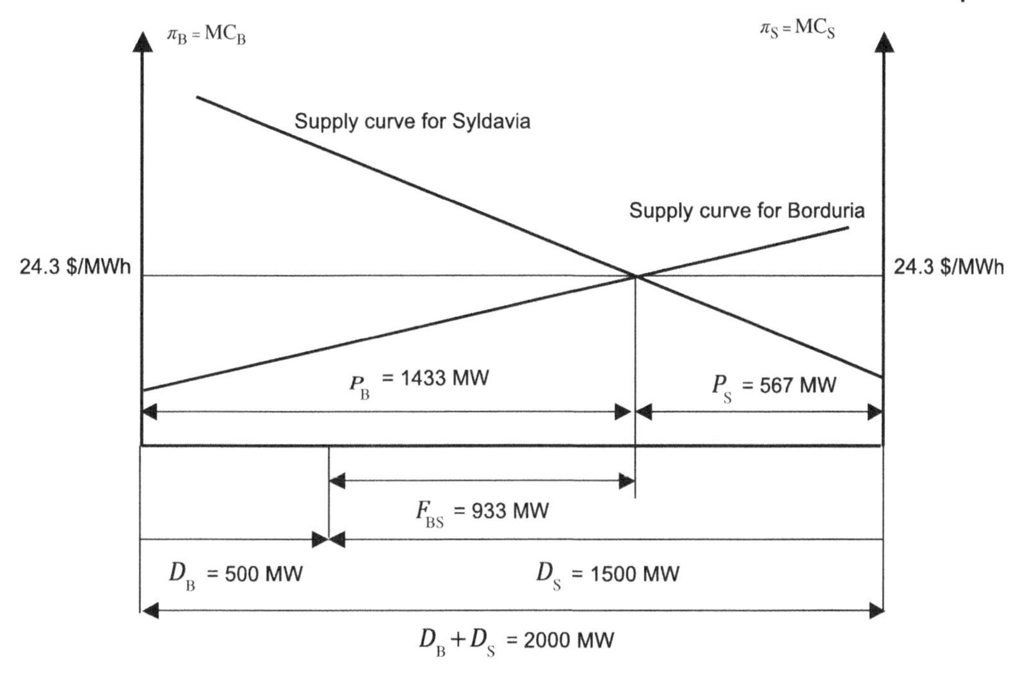

Figure 5.9 Graphical representation of the combination of the Syldavian and Bordurian electricity markets into a single market.

transformers but also some generating plants that provide essential reactive support services. The Borduria/Syldavia interconnection is therefore not always able to carry its nominal 1600 MW capacity. After consulting transmission engineers, Bill estimates that, during a significant part of each year, the interconnection is only able to carry a maximum of 400 MW. He therefore needs to study how the system behaves under these conditions.

When the capacity of the interconnection is limited to 400 MW, the production in Borduria must be reduced to 900 MW (500 MW of local load and 400 MW sold to consumers in Syldavia). The production in Syldavia is then 1100 MW. Using Equations (5.6) and (5.7), we find that:

$$\pi_B = MC_B = 10 + 0.01 \times 900 = 19\,\$/\text{MWh} \tag{5.20}$$

$$\pi_S = MC_S = 13 + 0.02 \times 1100 = 35\,\$/\text{MWh} \tag{5.21}$$

Figure 5.10 illustrates this situation. The constraint on the capacity of the transmission corridor creates a difference of 16 $/MWh between the prices of electrical energy in Borduria and Syldavia. If electricity were a normal commodity, traders would spot a business opportunity in this price disparity: if they could find a way of shipping more power from Borduria to Syldavia, they could make money by buying energy in one market and selling it in the other. However, this opportunity for *spatial arbitrage* cannot be realized because the interconnection is the only way to transmit power between the two countries and it is already fully utilized. This price difference persists as long as the capacity of the interconnection remains below the capacity needed to ensure free interchanges. Limits on the flows in the transmission network thus divide what should be a single market into separate markets. Because of this congestion, an additional

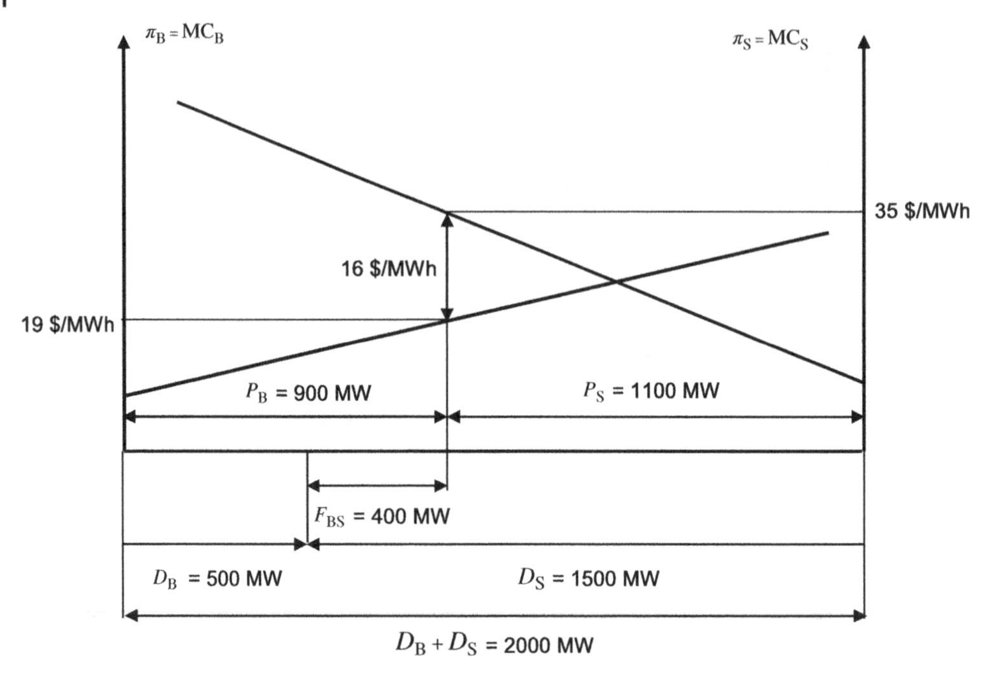

Figure 5.10 Graphical representation of the effect of congestion on the Syldavian and Bordurian electricity markets.

megawatt of load in each country has to be provided solely by the local generators. The marginal cost of producing electrical energy is therefore different in each country. If these separate markets are still sufficiently competitive, the prices are still equal to the marginal costs. We thus have what is called *locational marginal pricing* because the marginal cost depends on the location where the energy is produced or consumed. If a different price is defined at each bus or node in the system, locational marginal pricing is called *nodal pricing*. Our example shows that locational marginal prices are higher in areas that normally import power and lower in areas that export power.

Bill summarizes his findings so far in Table 5.2 using the following notations: R represents the revenue accruing to a group of generators from the sale of electrical energy; E represents the payment made by a group of consumers for the purchase of electrical energy; the subscripts B and S denote, respectively, Borduria and Syldavia. F_{BS} represents the power flowing on the interconnection. This quantity is positive if power flows from Borduria to Syldavia.

Table 5.2 shows that the biggest beneficiaries of the reenergization of the interconnection are likely to be the Bordurian generators and the Syldavian consumers. Bordurian consumers would see an increase in the price of electrical energy. Syldavian generators would lose a substantial share of their market. Overall, the interconnection would have a positive effect because it would reduce the total amount of money spent by consumers on electrical energy. This saving arises because the energy produced by less efficient generators is replaced by energy produced by more efficient ones. Congestion on the interconnection reduces the benefit that it provides. Note that this congestion partially shields Syldavian generators from the competition of their Bordurian counterparts.

Table 5.2 Operation of the Borduria–Syldavia interconnection as separate markets, as a single market and as a single market with congestion.

	Separate markets	Single market	Single market with congestion
P_B (MW)	500	1 433	900
π_B (\$/MWh)	15	24.33	19
R_B (\$/h)	7 500	34 865	17 100
E_B (\$/h)	7 500	12 165	9 500
P_S (MW)	1 500	567	1 100
π_S (\$/MWh)	43	24.33	35
R_S (\$/h)	64 500	13 795	38 500
E_S (\$/h)	64 500	36 495	52 500
F_{BS} (MW)	0	933	400
$R_{TOTAL} = R_B + R_S$	72 000	48 660	55 600
$E_{TOTAL} = E_B + E_S$	72 000	48 660	62 000

We have assumed so far that the markets are perfectly competitive. If competition were less than perfect, congestion in the interconnection would allow Syldavian generators to raise their prices above their marginal cost of production but would intensify competition in the Bordurian market. On the other hand, by linking more participants, the interconnection would make the resulting single market more competitive.

5.3.1.3 Congestion Surplus

Bill decides that it would be interesting to quantify the effect that congestion on the interconnection would have on producers and consumers in both countries. He calculates the prices in Borduria and Syldavia as a function of the amount of power flowing on the interconnection:

$$\pi_B = MC_B = 10 + 0.01\,(D_B + F_{BS}) \tag{5.22}$$

$$\pi_S = MC_S = 13 + 0.02\,(D_S - F_{BS}) \tag{5.23}$$

Bill assumes that consumers pay the going price in their local market independently of where the energy they consume is produced. The total payment made by consumers is thus given by:

$$E_{TOTAL} = \pi_B \cdot D_B + \pi_S \cdot D_S \tag{5.24}$$

Combining Equations (5.22)–(5.24), Figure 5.11 shows how this payment varies as a function of F_{BS}. As Bill expected, this payment decreases monotonically as the flow between the two countries increases. The curve does not extend beyond $F_{BS} = 933$ MW because we saw earlier that a greater interchange does not make economic sense.

Similarly, Bill assumes that generators are paid the going price in their local market for the electrical energy they produce, independently of where this energy is consumed. The total revenue collected by the generators from the sale of electrical energy in both markets

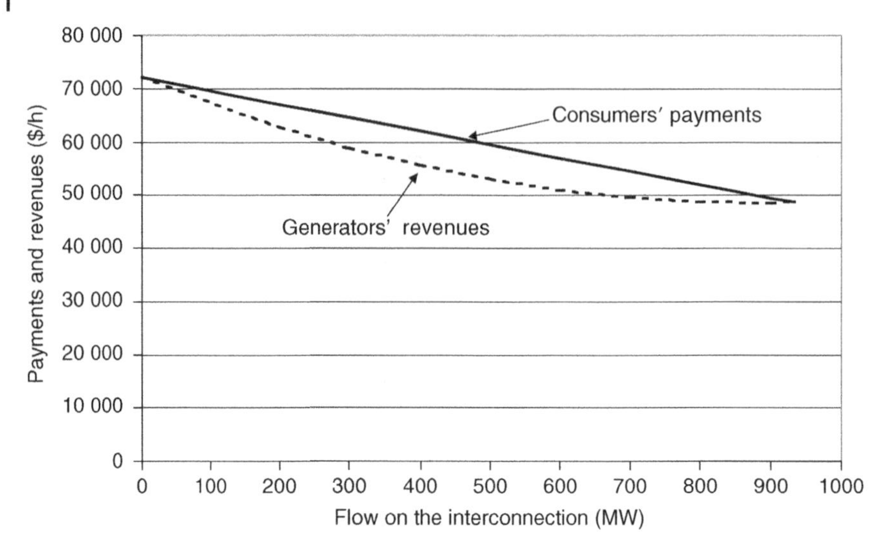

Figure 5.11 Consumers' payments (solid line) and generators' revenue (dashed line) as a function of the flow on the interconnection between Borduria and Syldavia.

is thus given by:

$$R_{\text{TOTAL}} = \pi_B \cdot P_B + \pi_S \cdot P_S = \pi_B \cdot (D_B + F_{BS}) + \pi_S \cdot (D_S - F_{BS}) \tag{5.25}$$

This quantity has also been plotted in Figure 5.11 as a function of the power flowing on the interconnection. We observe that this revenue is less than the payment made by the consumers except when the interconnection is not congested ($F_{BS} = 933$ MW) or when it is not in service ($F_{BS} = 0$MW). Combining Equations (5.24) and (5.25) while recalling that the flow on the interconnection is equal to the surplus of production over consumption in each country, we can write:

$$\begin{aligned} E_{\text{TOTAL}} - R_{\text{TOTAL}} &= \pi_S \cdot D_S + \pi_B \cdot D_B - \pi_S \cdot P_S - \pi_B \cdot P_B \\ &= \pi_S \cdot (D_S - P_S) + \pi_B \cdot (D_B - P_B) \\ &= \pi_S \cdot F_{BS} + \pi_B \cdot (-F_{BS}) \\ &= (\pi_S - \pi_B) \cdot F_{BS} \end{aligned} \tag{5.26}$$

This difference between payments and revenues is called the *merchandizing surplus* and is equal to the product of the difference between the prices in the two markets and the flow on the interconnection. In this case, since this surplus is due to the congestion in the network, it is also called the *congestion surplus*.

In particular, for the case where the flow on the interconnection is limited to 400 MW, we have:

$$E_{\text{TOTAL}} - R_{\text{TOTAL}} = (\pi_S - \pi_B) \cdot F_{BS} = (35 - 19) \cdot 400 = \$6400 \tag{5.27}$$

Note that this amount is identical to the quantity we obtain if we take the difference between the total payment and the total revenue given in the last column of Table 5.2.

In a pool system where all market participants buy or sell at the centrally determined price applicable to the network node where they are connected, this congestion surplus is collected by the market operator. It should not, however, be kept by this market operator because this would it give a perverse incentive to increase congestion or at least not work very hard to reduce congestion. On the other hand, simply returning the congestion surplus to the market participants would blunt the effect of nodal marginal pricing, which is designed to encourage efficient economic behavior. We will return to this issue when we discuss the management of congestion risks and financial transmission rights (FTRs) later in this chapter.

5.3.2 Centralized Trading in a Three-Bus System

In our discussion of decentralized or bilateral trading, we already mentioned that KCL and KVL dictate how power flows in a transmission network with more than two buses. We therefore need to explore the effect that these physical laws have on centralized trading. We carry out this investigation using the same three-bus system that we used when we discussed bilateral trading. Figure 5.12 shows the configuration of this system and Table 5.3 gives its parameters. We again assume that network limitations take the form of constant capacity limits on the active power flowing in each line and that the resistance of the lines is negligible.

When we analyzed this system in the context of bilateral trading, we did not need to consider price or cost information because this data remains private to the parties involved in each bilateral transaction. On the other hand, in a centralized trading system, producers and consumers submit their offers and bids to the system operator, who uses this information to optimize the operation of the system. Since we are taking the perspective of the system operator, we assume that we have access to the data given in Table 5.4. We also assume that, since the market is perfectly competitive, the generators' bids are equal to their marginal cost. For the sake of simplicity, the marginal cost of each generator is assumed constant over its generation capacity and the demand side is represented by the constant loads shown in Figure 5.12.

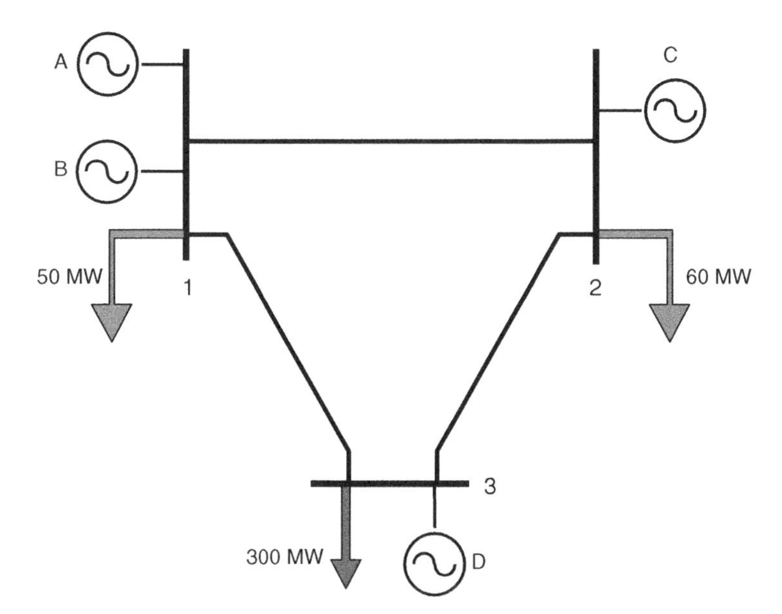

Figure 5.12 Simple three-bus system used to illustrate centralized trading.

Table 5.3 Branch data for the three-bus system of Figure 5.12.

Branch	Reactance (p.u.)	Capacity (MW)
1–2	0.2	126
1–3	0.2	250
2–3	0.1	130

5.3.2.1 Economic Dispatch

If we ignore the constraints that the network might impose, the total load of 410 MW should be dispatched solely on the basis of the offers or marginal cost of the generators in a way that minimizes the total cost of supplying the demand. Since we have assumed that these generators have a constant marginal cost over their entire range of operation and that the demand is not price sensitive, this dispatch is easy to compute: the generators are ranked in order of increasing marginal cost and loaded up to their capacity until the load is satisfied. We get:

$$\begin{aligned} P_A &= 125\,\text{MW} \\ P_B &= 285\,\text{MW} \\ P_C &= 0\,\text{MW} \\ P_D &= 0\,\text{MW} \end{aligned} \tag{5.28}$$

The total cost of the economic dispatch is:

$$C_{ED} = MC_A \cdot P_A + MC_B \cdot P_B = 2647.50\,\$/\text{h} \tag{5.29}$$

We must check whether this dispatch would cause one or more flows to exceed the capacity of a line. In a large network, we would calculate the branch flows using a power flow program. For a simple system like this one, we can do this computation by hand and gain a more intuitive understanding of the way power flows through the network. Given the assumed flow directions shown in Figure 5.13, we can write the power balance equation at each bus or node:

$$\text{Node 1}: F_{12} + F_{13} = 360 \tag{5.30}$$

$$\text{Node 2}: F_{12} - F_{23} = 60 \tag{5.31}$$

$$\text{Node 3}: F_{13} + F_{23} = 300 \tag{5.32}$$

Table 5.4 Generator data for the three-bus system of Figure 5.12.

Generator	Capacity (MW)	Marginal cost ($/MWh)
A	140	7.5
B	285	6
C	90	14
D	85	10

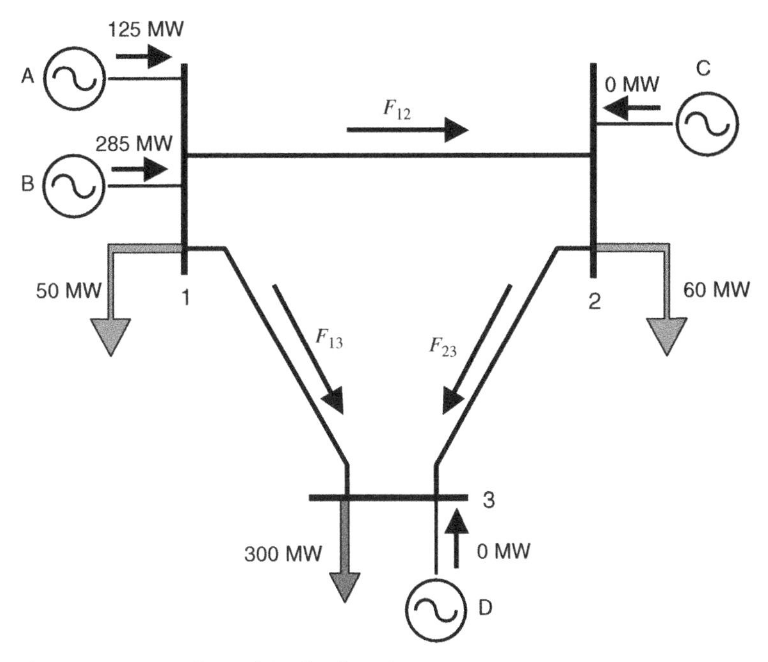

Figure 5.13 Basic dispatch in the three-bus system.

In this case, we get three equations in three unknowns. However, these equations are linearly dependent because the power balance also holds for the system as a whole. For example, subtracting Equation (5.31) from Equation (5.30) gives Equation (5.32). Since one of these equations can be eliminated with no loss of information, we are left with two equations and three unknowns. This is hardly surprising because we have not taken into account the impedances of the branches.

Rather than simply add an equation based on KVL, let us again make use of the superposition theorem. Figure 5.14 shows how our original problem can be decomposed into two simpler problems. If we succeed in determining the flows in these two simpler problems, we can easily find the flows in the original problem because we know from the superposition theorem that:

$$F_{12} = F_1^{\mathrm{A}} + F_2^{\mathrm{A}} \tag{5.33}$$

$$F_{13} = F_1^{\mathrm{B}} + F_2^{\mathrm{B}} \tag{5.34}$$

$$F_{23} = F_1^{\mathrm{A}} - F_2^{\mathrm{B}} \tag{5.35}$$

Let us consider the first problem. Three hundred megawatt is injected at bus 1 and extracted at bus 3. Since this power can flow along two paths (A and B), we have:

$$F_1^{\mathrm{A}} + F_1^{\mathrm{B}} = 300 \tag{5.36}$$

The reactances of paths A and B are, respectively, as follows:

$$x_1^{\mathrm{A}} = x_{12} + x_{23} = 0.3\,\text{p.u.} \tag{5.37}$$

$$x_1^{\mathrm{B}} = x_{13} = 0.2\,\text{p.u.} \tag{5.38}$$

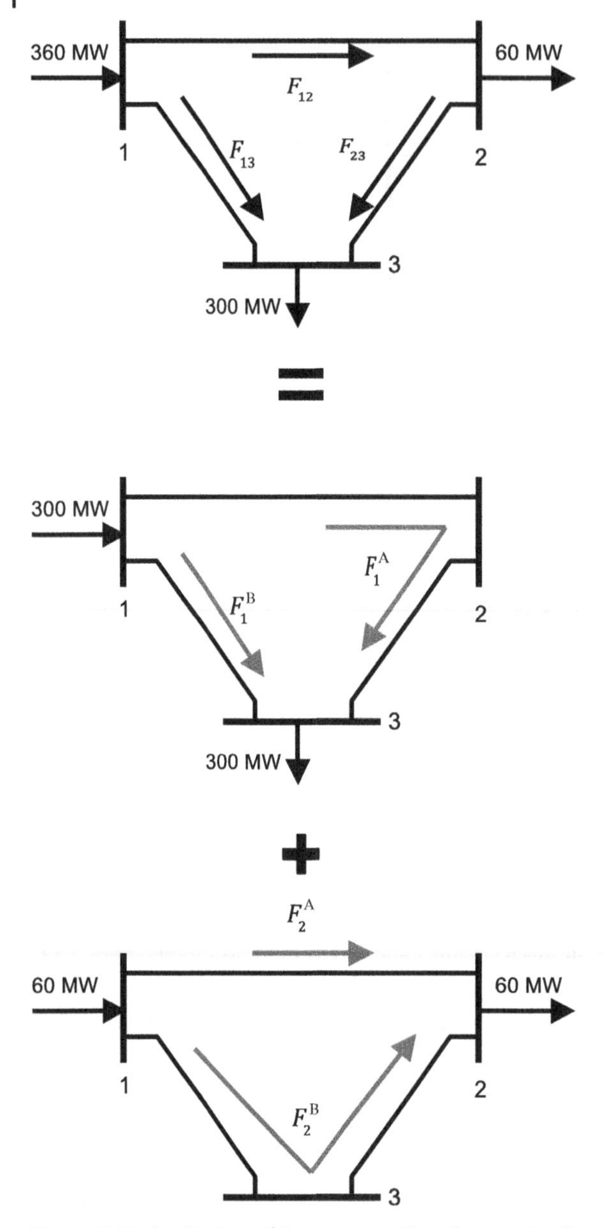

Figure 5.14 Application of the superposition theorem to the calculation of the line flows in the three-bus system.

Since these 300 MW divide themselves between the two paths in accordance with Equations (5.4) and (5.5), we have:

$$F_1^A = \frac{0.2}{0.3 + 0.2} \cdot 300 = 120\,\text{MW} \tag{5.39}$$

$$F_1^B = \frac{0.3}{0.3 + 0.2} \cdot 300 = 180\,\text{MW} \tag{5.40}$$

Similarly, for the second circuit, 60 MW is injected at bus 1 and extracted at bus 2. In this case, the impedances of the two paths are:

$$x_2^{\mathrm{A}} = x_{12} = 0.2\,\text{p.u.} \tag{5.41}$$

$$x_2^{\mathrm{B}} = x_{13} + x_{23} = 0.3\,\text{p.u.} \tag{5.42}$$

Hence:

$$F_2^{\mathrm{A}} = \frac{0.3}{0.3 + 0.2} \cdot 60 = 36\,\text{MW} \tag{5.43}$$

$$F_2^{\mathrm{B}} = \frac{0.2}{0.3 + 0.2} \cdot 60 = 24\,\text{MW} \tag{5.44}$$

Equations (5.33)–(5.35) give the flows in the original system:

$$F_{12} = F_1^{\mathrm{A}} + F_2^{\mathrm{A}} = 120 + 36 = 156\,\text{MW} \tag{5.45}$$

$$F_{13} = F_1^{\mathrm{B}} + F_2^{\mathrm{B}} = 180 + 24 = 204\,\text{MW} \tag{5.46}$$

$$F_{23} = F_1^{\mathrm{A}} - F_2^{\mathrm{B}} = 120 - 24 = 96\,\text{MW} \tag{5.47}$$

Figure 5.15 illustrates this solution. From these results, we conclude that the economic dispatch would overload branch 1–2 by 30 MW because it would have to carry 156 MW when its capacity is only 126 MW. This is clearly not acceptable.

5.3.2.2 Correcting the Economic Dispatch

While the economic dispatch minimizes the total production cost, this solution is not viable because it does not respect the transmission capacity limits. We must therefore determine the least cost modifications that remove the line overload. We begin by noting that the economic dispatch concentrates all the generation at bus 1. To reduce the flow on branch 1–2, we can increase the generation either at bus 2 or at bus 3. Let us first consider what happens when we increase the generation at bus 2 by 1 MW. Since we neglect losses, we must also reduce the generation at bus 1 by 1 MW. Figure 5.16 illustrates this incremental redispatch.

Since the incremental flow ΔF^{A} is in the opposite direction as the flow F_{12}, increasing the generation at bus 2 and reducing it at bus 1 reduces the overload on this branch. To

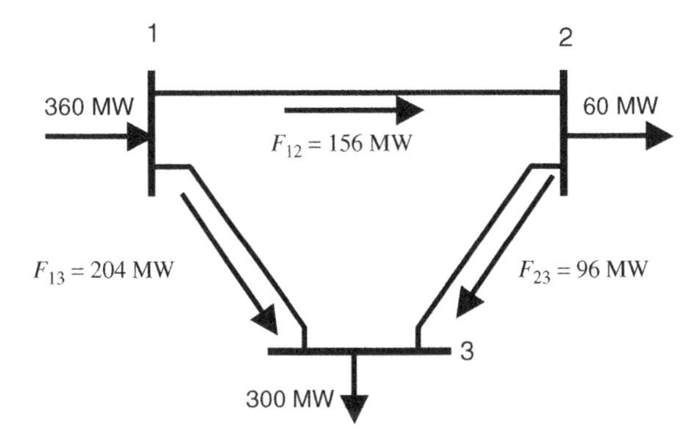

Figure 5.15 Flows for the basic dispatch in the three-bus system.

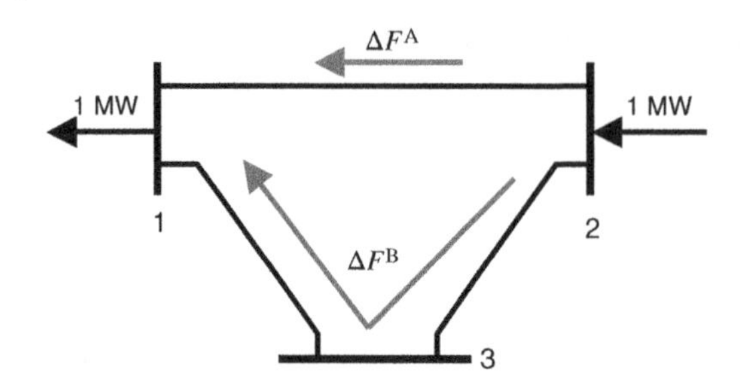

Figure 5.16 Effect of an incremental change in the generation at bus 2.

quantify this effect, we can again make use of the superposition theorem. Since the reactances of paths A and B are, respectively,

$$x^A = x_{12} = 0.2\,\text{p.u.} \tag{5.48}$$

$$x^B = x_{23} + x_{13} = 0.3\,\text{p.u.} \tag{5.49}$$

and since the sum of the two flows must be equal to 1 MW, we get:

$$\Delta F^A = 0.6\,\text{MW} \tag{5.50}$$

$$\Delta F^B = 0.4\,\text{MW} \tag{5.51}$$

Every megawatt injected at bus 2 and extracted at bus 1 thus reduces the flow on branch 1–2 by 0.6 MW. Given that this line is overloaded by 30 MW, a total of 50 MW of generation must be shifted from bus 1 to bus 2 to satisfy the line capacity constraint. Figure 5.17 illustrates this redispatch and its superposition with the economic dispatch to yield what we call the constrained dispatch. We observe that the flow on branch 1–3 has also been reduced by this redispatch but that the flow on branch 2–3 has increased. However, this increase is tolerable because this flow remains smaller than the capacity specified in Table 5.3. To implement this constrained dispatch, the generators connected at bus 1 must produce a total of 360 MW to meet the local load of 50 MW and inject a net 310 MW into the network. At the same time, the generator at bus 2 must produce 50 MW. An additional 10 MW is taken from the network to supply the local load of 60 MW. Under these conditions, the least cost generation dispatch is:

$$\begin{aligned} P_A &= 75\,\text{MW} \\ P_B &= 285\,\text{MW} \\ P_C &= 50\,\text{MW} \\ P_D &= 0\,\text{MW} \end{aligned} \tag{5.52}$$

where we have reduced the output of generator A rather than the output of generator B because generator A has a higher marginal cost. The total cost of this constrained dispatch is:

$$C_2 = MC_A \cdot P_A + MC_B \cdot P_B + MC_C \cdot P_C = 2972.50\,\$/\text{h} \tag{5.53}$$

This cost is necessarily higher than the cost of the economic dispatch that of Equation (5.29). The difference represents the cost of respecting the transmission limits.

We mentioned that we could also relieve the overload on branch 1–2 by increasing the output of generator D connected at bus 3. Let us calculate the extent and the cost of this alternative redispatch using the same procedure. Figure 5.18 shows the two paths along which an extra MW injected at bus 3 and extracted at bus 1 would divide itself.

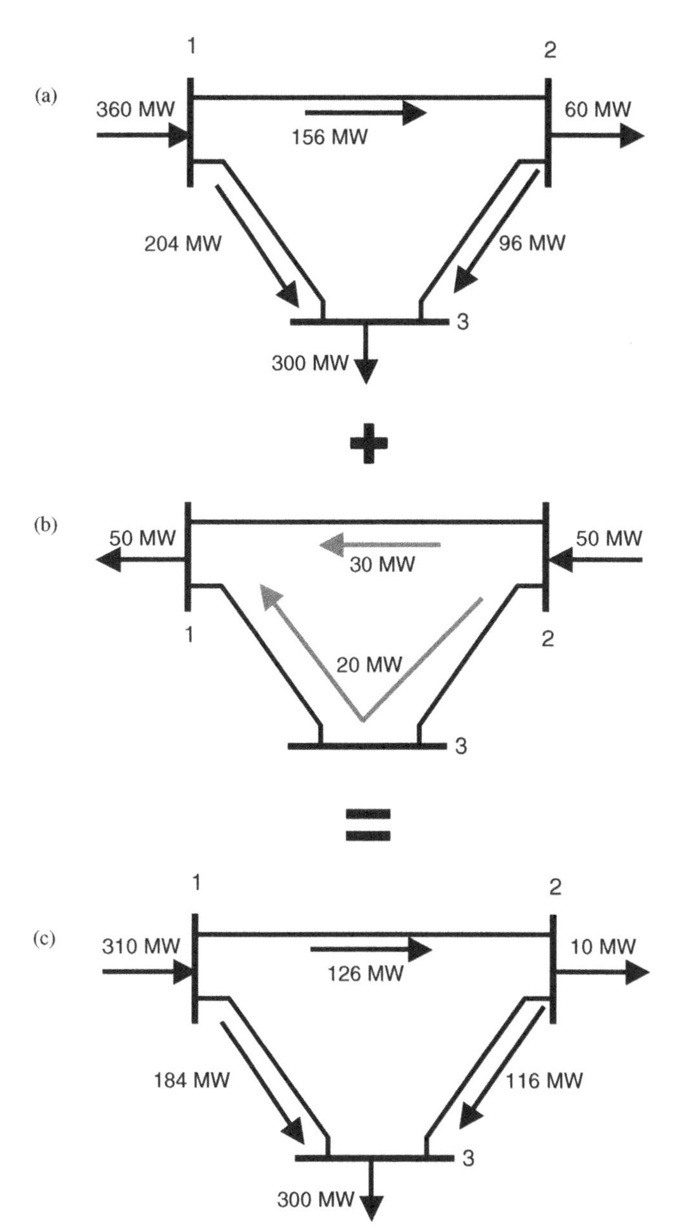

Figure 5.17 Superposition of the redispatch of generation from bus 1 to bus 2 (b) on the economic dispatch (a) to produce a constrained dispatch that meets the constraints on line flows (c).

Given that the reactances of paths A and B are, respectively,

$$x^{\mathrm{A}} = x_{23} + x_{12} = 0.3\,\text{p.u.} \tag{5.54}$$

$$x^{\mathrm{B}} = x_{13} = 0.2\,\text{p.u.} \tag{5.55}$$

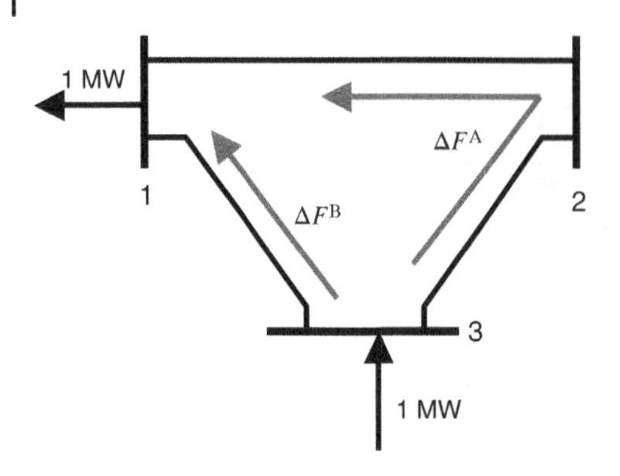

Figure 5.18 Effect of an incremental change in the generation at bus 3.

and that the sum of the two flows must be equal to 1 MW, we get:

$$\Delta F^{A} = 0.4\,\text{MW} \tag{5.56}$$

$$\Delta F^{B} = 0.6\,\text{MW} \tag{5.57}$$

Every MW injected at bus 3 and extracted at bus 1 thus reduces the flow on branch 1–2 by 0.4 MW. This means that we need to shift 75 MW of generation from bus 1 to bus 3 to reduce the flow on branch 1–2 by 30 MW and remove the overload. Figure 5.19 shows how superposing this redispatch on the economic dispatch reduces the flows through all branches of the network. As expected, the flow on branch 1–2 is equal to the maximum capacity of that branch. Since the total power to be produced at bus 1 is now reduced by 75 MW, the generation dispatch for this case is:

$$\begin{aligned} P_A &= 50\,\text{MW} \\ P_B &= 285\,\text{MW} \\ P_C &= 0\,\text{MW} \\ P_D &= 75\,\text{MW} \end{aligned} \tag{5.58}$$

The total cost of this constrained dispatch is:

$$C_3 = \text{MC}_A \cdot P_A + \text{MC}_B \cdot P_B + \text{MC}_D \cdot P_D = 2835\,\$/\text{h} \tag{5.59}$$

Let us now compare these two ways of removing the overload on branch 1–2. If we make use of the generation at bus 3, we need to redispatch 75 MW. On the other hand, if we call upon the generation at bus 2, we need to shift only 50 MW. This is because the flow on branch 1–2 is less sensitive to an increase in generation at bus 3 than to an increase at bus 2. However, since the marginal cost of generator D is less than the marginal cost of generator C, increasing the generation at bus 3 is the cheaper solution. The cost of the transmission constraints is thus equal to the difference between the cost of this constrained dispatch and the cost of the economic dispatch:

$$C_S = C_3 - C_{ED} = 2835.00 - 2647.50 = 187.50\,\$/\text{h} \tag{5.60}$$

5.3.2.3 Nodal Prices

We have already alluded to the concept of nodal marginal price when we discussed the Borduria–Syldavia interconnection. We are now in a position to clarify this concept. The nodal marginal price is equal to the cost of supplying an additional megawatt of load at the node under consideration by the cheapest possible means while respecting the constraints imposed by the network capacity limits.

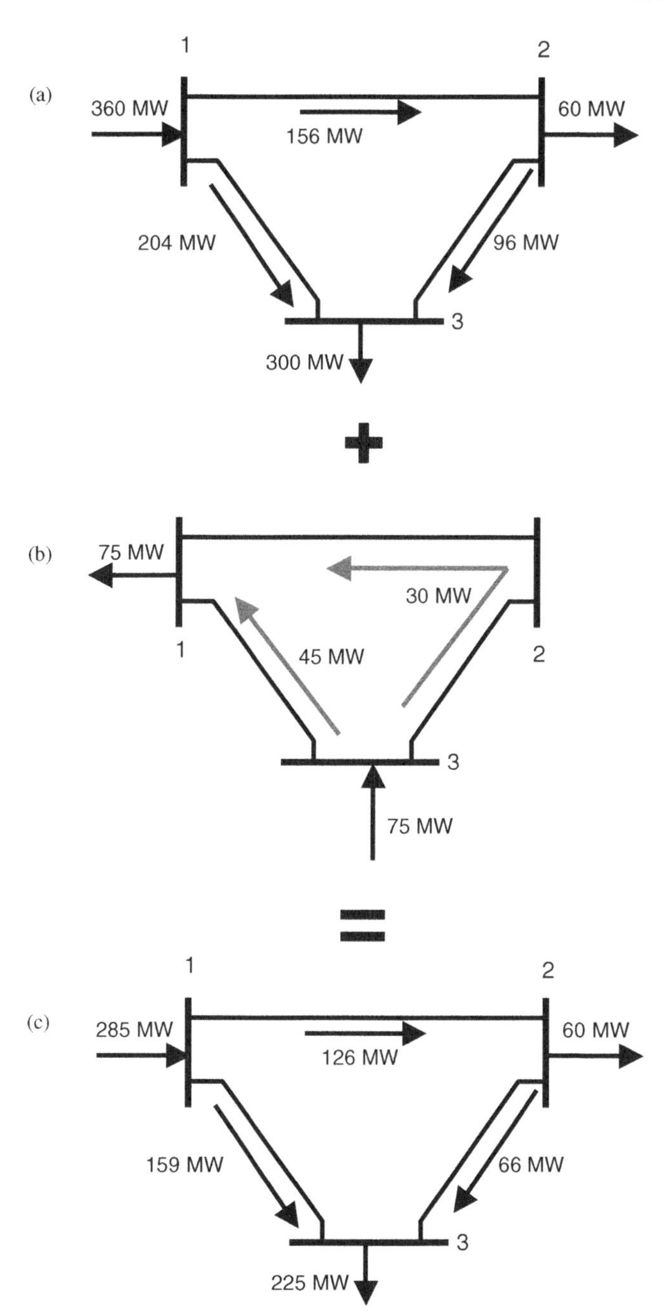

Figure 5.19 Superposition of the redispatch of generation from bus 1 to bus 3 (b) on the economic dispatch (a) to produce a constrained dispatch that meets the constraints on line flows (c).

In our three-bus example, this means that we do not start from the economic dispatch but from the constrained dispatch given by Equation (5.58). The output of generator D has thus been increased to remove the overload on branch 1–2. At node 1, it is clear that an additional megawatt of load should be produced by generator A. The marginal cost of generator A is indeed lower than the marginal cost of generators C and D. While it is higher than the marginal cost of generator B, this generator is unable to produce an additional megawatt because it is already loaded up to its maximum capacity. The network has no impact on the marginal price at this node because the additional megawatt is both produced and consumed locally. The nodal marginal price at bus 1 is therefore:

$$\pi_1 = MC_A = 7.50\,\$/\text{MWh} \tag{5.61}$$

What is the cheapest way of supplying an additional megawatt at bus 3? Generator A has the lowest marginal cost and is not fully loaded. Unfortunately, increasing the generation at bus 1 would inevitably overload branch 1–2. The next cheapest option is to increase the output of generator D. Since this generator is located at bus 3, this additional megawatt would not flow through the network. Therefore, we have:

$$\pi_3 = MC_D = 10\,\$/\text{MWh} \tag{5.62}$$

Supplying an additional megawatt at bus 2 is a more complex matter. We could obviously generate it locally using generator C, but this looks rather expensive because at 14 $/MWh the marginal cost of this generator is much higher than the marginal cost of the other generators. If we choose to adjust the output of the generators at the other buses, we must consider what might happen in the network. Figure 5.20 shows how an additional megawatt of load at bus 2 would flow through the network if it were produced at bus 1 or bus 3. We can see that in both cases we increase the flow on branch 1–2. Since

(a)

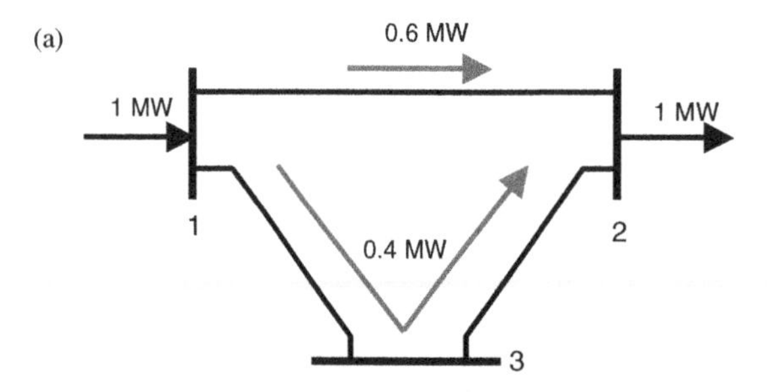

(b)

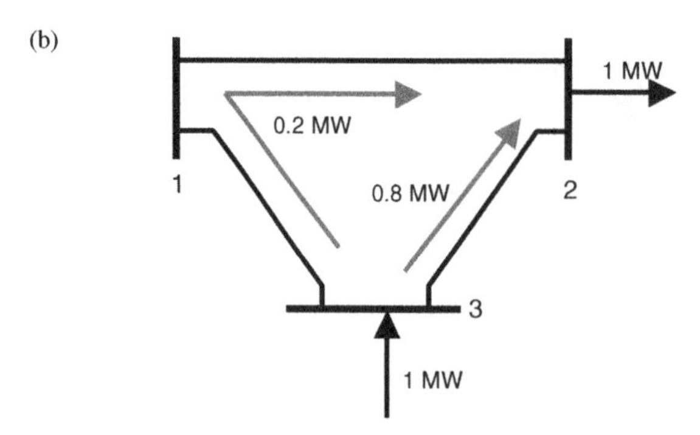

Figure 5.20 Incremental flows in the network due to an additional megawatt of load at bus 2 when this megawatt is produced at bus 1 (a) or at bus 3 (b).

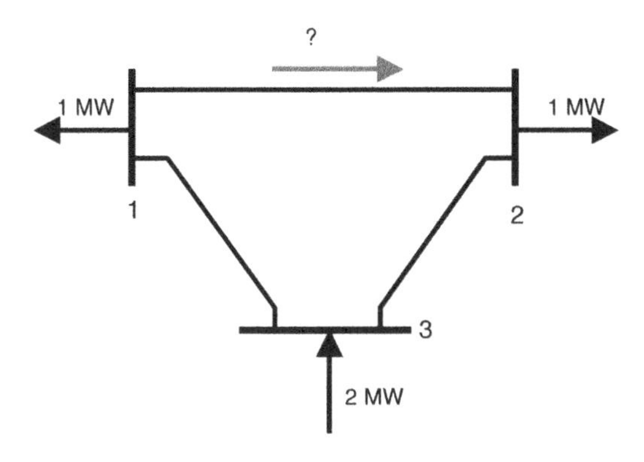

Figure 5.21 Supplying an additional megawatt of load at bus 2 by increasing production at bus 3 and reducing it at bus 1.

the flow on this branch is already at its maximum value, neither solution is acceptable. Any combination of generation increases at buses 1 and 3 would also be unacceptable.

We could, however, increase generation at bus 3 and reduce it at bus 1. For example, as shown in Figure 5.21, we could increase the generation by 2 MW at bus 3 and reduce it by 1 MW at bus 1. The net increase is then equal to the additional load at bus 2. We can then, once again, use superposition to determine the resulting incremental flows. The first diagram in Figure 5.22 shows that if 1 MW is injected at bus 3 and extracted at bus 1, the flow on branch 1–2 would decrease by 0.4 MW. The other diagram shows that another

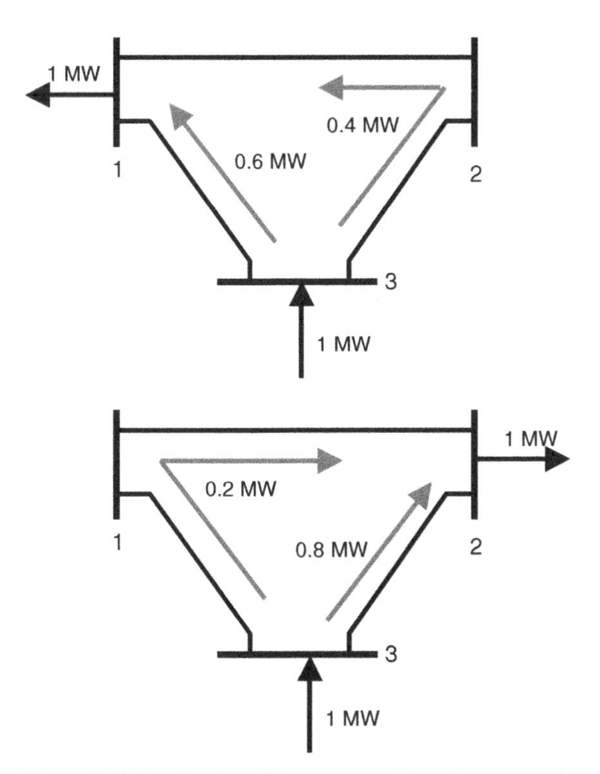

Figure 5.22 Application of the superposition theorem to the analysis of the conditions illustrated in Figure 5.17.

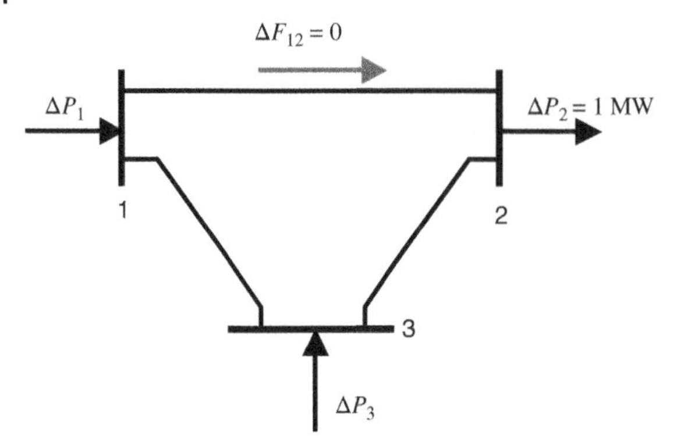

Figure 5.23 Formulation of the problem of supplying an additional megawatt of load at bus 2 without changing the flow on branch 1–2.

1 MW injected at bus 3 but extracted at bus 2 increases this flow by 0.2 MW. Overall, the flow on branch 1–2 decreases by 0.2 MW. Supplying an additional megawatt at bus 2 by increasing the production at bus 3 by 2 MW and decreasing it at bus 1 by 1 MW is thus acceptable because it keeps the flow on branch 1–2 below the maximum capacity of this branch. But is it optimal? This combination of injections has not simply kept the flow on branch 1–2 at its maximum: it has decreased it to 0.2 MW below this maximum. This means that we have reduced too much the output of the generation connected at bus 1, which is cheaper than generation at bus 3.

Figure 5.23 illustrates the formulation that we use to determine how we can supply an additional megawatt at bus 2 by redispatching generation at buses 1 and 3 without overloading branch 1–2. We must have:

$$\Delta P_1 + \Delta P_3 = \Delta P_2 = 1\,\text{MW} \tag{5.63}$$

Using the sensitivities shown in Figure 5.20, we can also write:

$$0.6\Delta P_1 + 0.2\Delta P_3 = \Delta F_{12} = 0\,\text{MW} \tag{5.64}$$

Solving these equations, we get:

$$\Delta P_1 = -0.5\,\text{MW} \tag{5.65}$$

$$\Delta P_3 = 1.5\,\text{MW} \tag{5.66}$$

Supplying at minimum cost an additional megawatt at bus 2 therefore requires that we increase the output of generator D by 1.5 MW and reduce the output of generator A by 0.5 MW. The cost of this megawatt, and hence the nodal marginal price at bus 2, is thus:

$$\pi_2 = 1.5 \cdot \text{MC}_\text{D} - 0.5 \cdot \text{MC}_\text{A} = 11.25\,\$/\text{MWh} \tag{5.67}$$

In summary, we observe the following:

- Generator A sets a price of 7.50 $/MWh at bus 1. Generator B has a lower marginal cost (6.00 $/MWh) but has no influence on the prices because it operates at its upper limit.
- Generator D sets a nodal marginal price of 10.00 $/MWh at bus 3.
- At bus 2 the price is set at 11.25 $/MWh by a combination of the prices of the marginal generators A and D.

These observations can be generalized to more complex networks. In a system without transmission constraints, if we model all generators as having constant marginal costs, all generators, except one, produce either their minimum or their maximum output. The exception is the marginal generator, whose output is such that the total generation is equal to the total load. Such a generator is said to be part-loaded. The marginal cost of this generator sets the price for the entire system because it provides the hypothetical additional megawatt that determines the marginal price. When a transmission limit constrains the economic dispatch, another generator becomes marginal in the sense that it is neither at its maximum or its minimum output. In general, if there are m binding transmission constraints in the system, there are $m + 1$ marginal generators. Each of these part-loaded generators sets the marginal price at the bus where it is connected. Nodal marginal prices at the other buses are determined by a combination of the prices of the marginal generators. This combination depends on the application of KVL to the constrained network. We will see shortly that this can lead to flows and prices that do not behave in an intuitively obvious manner.

5.3.2.4 Congestion Surplus

Before looking at these counterintuitive situations, let us summarize the economic operation of this three-bus system. Table 5.5 shows the load and generation as well as the nodal price at each bus. It also shows the payments made by consumers and the revenues collected by generators if energy is bought and sold at nodal marginal prices. All of these quantities are calculated for one hour of operation at constant loads.

If we compare the sum of the consumer payments at all the buses and the sum of the generator revenues at all the buses, we notice that these two quantities do not match. More money is collected from the consumers than is paid to the generators. This difference is the merchandising surplus that we already encountered in our two-bus Borduria–Syldavia example. This surplus is again caused by congestion in the network. If the capacity of branch 1–2 were larger than 156 MW, we could implement the unconstrained economic dispatch. The marginal prices would then be identical at all nodes and the total amount collected by generators would be equal to the total amount paid by consumers.

5.3.2.5 Economically Counterintuitive Flows

Locational differences in the price of producing goods are quite common in economics. The most intuitive example is probably the production of fruits and vegetables, which are

Table 5.5 Summary of the economic operation of the three-bus system.

	Bus 1	Bus 2	Bus 3	System
Consumption (MW)	50	60	300	410
Production (MW)	335	0	75	410
Nodal marginal price ($/MWh)	7.50	11.25	10.00	—
Consumer payments ($/h)	375.00	675.00	3000.00	4050.00
Producer revenues ($/h)	2512.50	0.00	750.00	3262.50
Congestion surplus ($/h)				787.50

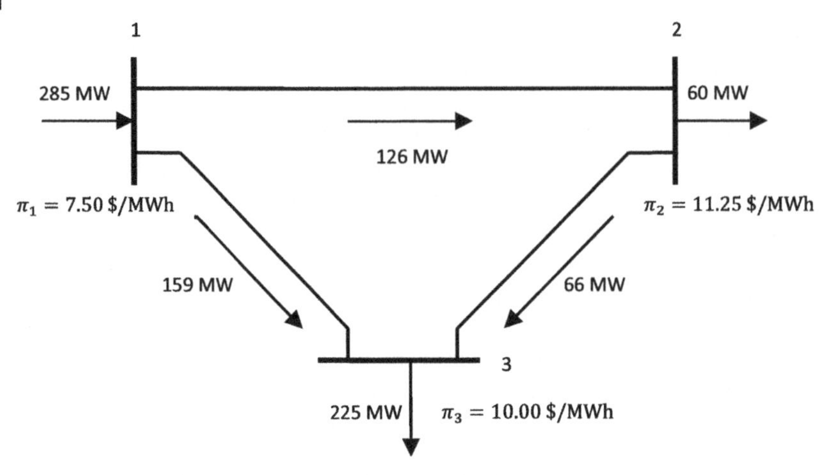

Figure 5.24 Nodal marginal prices and flows in the three-bus system. Power in branch 2–3 flows from a higher marginal price to a lower marginal price.

cheaper to grow outdoors in a warm climate than in greenhouses in a cold climate. If markets are competitive, the price of these goods is lower in the warmer regions and higher in the colder regions. If trade between these regions is free, fruits and vegetables are shipped from the regions with low prices to the regions with high prices. No rational trader would transport grapes from Alaska to California and hope to make a profit. In electricity networks however, such economically counterintuitive transportation does occur, even when operation is optimal. Figure 5.24 shows the flows and the nodal prices for the constrained dispatch of our three-bus system. The flows in branches 1–2 and 1–3 carry energy from a node with a lower marginal price to nodes with higher marginal prices. On the other hand, power flows from a higher price node to a lower price node on branch 2–3. This phenomenon does not occur because somebody behaves irrationally but because the laws of physics (KVL in this case) take precedence over the "laws" of the market.

Table 5.6 shows the surplus that each branch generates by carrying power in our three-bus network. In the case of line 2–3, this amount is negative because the power flows from a node with a higher price to a node with a lower price. The sum over all the lines, however, is equal to the merchandising or congestion surplus that we calculated on a node-by-node basis in Table 5.5.

Table 5.6 Contribution of each branch to the merchandising surplus of the three-bus system.

Branch	Flow (MW)	"From" price ($/MWh)	"To" price ($/MWh)	Surplus ($/h)
1–2	126	7.50	11.25	472.50
1–3	159	7.50	10.00	397.50
2–3	66	11.25	10.00	–82.50
Total				787.50

5.3.2.6 Economically Counterintuitive Prices

In our three-bus example, we have assumed so far that the flow on branch 1–2 could not exceed 126 MW. As we will see in Chapter 6, in a real system, the maximum flow allowable on a line is not necessarily fixed. If this flow is constrained by the thermal rating of the line, the limit depends on the weather conditions because wind and cold ambient temperature reduce the temperature rise inside the conductors. If the limit on the line flow arises from stability considerations, it is determined by the configuration of the rest of the system. Studying how the maximum flow on branch 1–2 affects the nodal marginal prices is thus not just a mathematical curiosity.

Table 5.7 summarizes the effect of the maximum flow on this branch has on the operation and the economics of the three-bus system. Each row of this table corresponds to a different value of this flow. For each of these values, we have calculated the constrained dispatch and the nodal prices using the same procedure as before. We have also calculated the amounts collected by the generators and paid by the consumers, as well as the cost of producing the energy, the generator profits and the congestion surplus. The last row of the table shows that a limit of 160 MW or more does not constrain the dispatch. Generator A is then the only marginal generator and the nodal prices are uniform across the network. Under these conditions, the network does not produce a surplus. On the other hand, for a limit of less than 70 MW, there is no generation dispatch that can supply the load without violating the flow constraint on branch 1–2.

For limits between 70 and 90 MW, generator A does not produce energy, generators B and C are part-loaded, and generator D is fully loaded. The nodal prices at buses 1 and 2 are thus 6.00 and 14.00 \$/MWh, respectively, while the nodal price at bus 3 is 11.33 \$/MWh. This last value falls between the prices at the other two nodes. However, it must be above 10.00 \$/MWh because generator D is fully loaded.

One would expect that further increasing the line capacity should result in lower prices because the system would be less constrained. Table 5.7 shows that this is not necessarily the case. If we raise the limit up to 120 MW, the prices at nodes 1 and 3 increase while the price at node 2 remains constant. The system is not being operated inefficiently, however, because the cheaper generators (A and B) produce more power while the more expensive generator (C) produces less. Overall, the cost to generators of producing electrical energy decreases while the consumer payments, the generators' profits and the congestion surplus increase. In this case, measures to increase the transmission capacity benefit producers at the expenses of consumers. Why did this happen? Increasing the flow on branch 1–2 allowed us to increase the output of the generators connected to bus 1. At some point, generator B reached its maximum capacity and generator A became the marginal generator, raising the nodal price at bus 1 to 7.50 \$/MWh. Using superposition, we can then check that the price at bus 3 is given by:

$$\pi_3 = \frac{1}{3}\pi_1 + \frac{2}{3}\pi_2 = 11.83\,\$/\text{MWh} \tag{5.68}$$

If we increase the limit beyond 120 MW, generator C produces nothing. The price at node 2 becomes a combination of the prices at nodes 1 and 3 because generator D is a marginal generator. Nodal prices, generator revenues, generator profits, consumer payments, and congestion surplus all decrease until we reach the noncongested state for a limit of 156 MW.

Table 5.7 Effect of the maximum flow on branch 1–2 on the operation of the three-bus system.

F_{12}^{max}	P_A	P_B	P_C	P_D	π_1	π_2	π_3	Generator costs	Generator revenues	Generator profits	Consumer payments	Congestion surplus
70	0.00	238.33	86.67	85.00	6.00	14.00	11.33	3493.33	3606.67	113.33	4540.00	933.33
80	0.00	255.00	70.00	85.00	6.00	14.00	11.33	3360.00	3473.33	113.33	4540.00	1066.67
90	0.00	271.67	53.33	85.00	6.00	14.00	11.33	3226.67	3340.00	113.33	4540.00	1200.00
100	3.33	285.00	36.67	85.00	7.50	14.00	11.83	3098.33	3681.67	583.33	4765.00	1083.33
110	20.00	285.00	20.00	85.00	7.50	14.00	11.83	2990.00	3573.33	583.33	4765.00	1191.67
120	36.67	285.00	3.33	85.00	7.50	14.00	11.83	2881.67	3465.00	583.33	4765.00	1300.00
130	60.00	285.00	0.00	65.00	7.50	11.25	10.00	2810.00	3237.50	427.50	4050.00	812.50
140	85.00	285.00	0.00	40.00	7.50	11.25	10.00	2747.50	3175.00	427.50	4050.00	875.00
150	110.00	285.00	0.00	15.00	7.50	11.25	10.00	2685.00	3112.50	427.50	4050.00	937.50
160	125.00	285.00	0.00	0.00	7.50	7.50	7.50	2647.50	3075.00	427.50	3075.00	0.00

5.3.2.7 More Economically Counterintuitive Prices

Let us now consider what happens if the capacity of branch 2–3 is reduced to 65 MW, for example, because of maintenance work on one of the two circuits of that branch. Under these conditions, the minimum cost constrained dispatch is:

$$\begin{aligned} P_A &= 47.5\,\text{MW} \\ P_B &= 285\,\text{MW} \\ P_C &= 0\,\text{MW} \\ P_D &= 77.5\,\text{MW} \end{aligned} \tag{5.69}$$

This dispatch produces the following flows:

$$\begin{aligned} F_{12} &= 125\,\text{MW} \\ F_{13} &= 157.5\,\text{MW} \\ F_{23} &= 65\,\text{MW} \end{aligned} \tag{5.70}$$

The flow on line 2–3 is thus the only one that is constrained. The marginal generators are A and D because generator B is producing its maximum output and generator C is not producing at all. Generator A thus sets π_1 at 7.50 \$/MWh while generator D sets π_3 at 10.00 \$/MWh. To calculate the marginal price at node 2, we need to calculate the cost of an additional megawatt of load at that node. Since the marginal generators would supply this megawatt, we have:

$$\Delta P_1 + \Delta P_3 = 1 \tag{5.71}$$

These increments in generation must be such that they keep the flow on branch 2–3 at its limit. Considering the relative reactances of the paths, we have:

$$-0.4\Delta P_1 - 0.8\Delta P_3 = 0 \tag{5.72}$$

The negative signs arise because increasing the generation at either bus 1 or bus 3 while increasing the load at bus 2 decreases the flow on branch 2–3. Solving Equations (5.71) and (5.72), we get:

$$\begin{aligned} \Delta P_1 &= 2\,\text{MW} \\ \Delta P_3 &= -1\,\text{MW} \end{aligned} \tag{5.73}$$

An additional megawatt at bus 2 would therefore cost us the price of two megawatts at bus 1 but we would save the price of one megawatt at bus 3. We therefore have:

$$\pi_2 = 2 \times 7.50 - 1 \times 10 = 5.00\,\$/\text{MWh} \tag{5.74}$$

The marginal price at node 2 is thus lower than the price at either of the other buses, i.e. lower than the marginal cost of any marginal generator!

5.3.2.8 Nodal Pricing and Market Power

We have assumed so far that the markets are perfectly competitive and that the nodal price is equal to the marginal cost when an incremental amount of energy would be produced using local generators. While this assumption greatly simplifies the analysis, it is highly questionable in practice, especially when the transmission network is congested. We will now show that KVL can make strategic bidding easy and profitable. Let us go back to our three-bus example with, as in the previous section, a constraint on branch 2–3 rather than branch 1–2. Suppose that generator C at bus 2 desperately wants to produce some power. Such a situation could happen if the startup cost of C is high and its owner decides that it is cheaper to produce at a loss for a while rather than having to restart the unit later. A similar situation arises if C is a cogeneration plant that must run to produce the steam required for an industrial process. The owner of generator C realizes that, if the plant is to run, she must bid below the current nodal marginal price of 5.00 \$/MWh. She therefore decides to bid at 3.00 \$/MWh. If the other generators bid at their marginal cost, the economic dispatch is then:

$$\begin{aligned} P_A &= 35\,\text{MW} \\ P_B &= 285\,\text{MW} \\ P_C &= 90\,\text{MW} \\ P_D &= 0\,\text{MW} \end{aligned} \tag{5.75}$$

However, this dispatch must be modified as follows to satisfy the constraint on branch 2–3:

$$\begin{aligned} P_A &= 32.5\,\text{MW} \\ P_B &= 285\,\text{MW} \\ P_C &= 7.5\,\text{MW} \\ P_D &= 85\,\text{MW} \end{aligned} \tag{5.76}$$

Since generators A and C are marginal, they set the nodal prices at buses 1 and 2 at 7.50 and 3.00 \$/MWh, respectively. Generator D is operating at its upper limit and therefore does not set the price at bus 3. Using the same technique as above, we find that to supply an additional megawatt at bus 3, we would have to increase the output of generator A by 2 MW and decrease the output of generator C by 1 MW. The marginal price at node 3 is thus given by:

$$\pi_3 = 2\pi_1 - \pi_2 = 12.00\,\$/\text{MWh} \tag{5.77}$$

The submission of a low bid at bus 2 increases the price at bus 3 from 10.00 to 12 \$/MWh and the output of the generator at that bus from 77.5 to 85 MW. Generator C's low bid thus has the counterintuitive consequence of being very profitable for generator D!

This fact is unlikely to go unnoticed by the owner of generator D who may decide to see what happens if he raises his own bid to 20.00 \$/MWh. Under these conditions, the constrained dispatch becomes:

$$\begin{aligned} P_A &= 47.5\,\text{MW} \\ P_B &= 285\,\text{MW} \\ P_C &= 0.0\,\text{MW} \\ P_D &= 77.5\,\text{MW} \end{aligned} \tag{5.78}$$

The marginal generators set the nodal prices at buses 1 and 3:

$$\begin{aligned} \pi_1 &= 7.50\,\$/\text{MWh} \\ \pi_3 &= 20.0\,\$/\text{MWh} \end{aligned} \tag{5.79}$$

On the other hand, supplying an additional megawatt at bus 2 would require increasing the injection at bus 1 by 2 MW and decreasing the injection at bus 3 by 1 MW. We therefore have:

$$\pi_2 = 2\pi_1 - \pi_3 = 2 \times 7.50 - 1 \times 20.00 = -5.00\,\$/\text{MWh} \tag{5.80}$$

Since the price at bus 2 is negative, consumers connected to that bus would be paid to consume and generators would have to pay for the privilege of producing energy! Besides making life miserable for generator C, generator D increases its profit by raising its bid, even though its output decreases:

$$\Delta\Omega_D = 77.50 \times 20 - 85 \times 10 = \$700$$

Generator D is able to exert market power because it is in a very favorable position with respect to the constraint on branch 2–3. In fact, given the loads in the system, the output of generator D cannot be reduced below 77.5 MW without violating the constraint. No matter what generator D bids, its output does not drop below that level. Generator D thus enjoys a locational monopoly.

In general, network constraints increase opportunities for strategic bidding because not all generators are connected at locations where they can relieve a given constraint. In many cases, the number of generators that can effectively alleviate a constraint is likely to be small. Congestion in the transmission network can therefore transform a reasonably competitive global market into a collection of smaller local energy markets. Since these smaller markets inevitably have fewer active participants than the global market, some of them are likely to be able to exert market power. Such scenarios are not easy to detect or analyze. See Day et al. (2002) for a discussion of techniques that can be used to model strategic bidding when network constraints are taken into consideration.

5.3.2.9 A Few Comments on Nodal Marginal Prices

Our collection of small examples has demonstrated that nodal prices at buses without a marginal generator can be higher, lower, or in between the prices at buses with marginal generators. A nodal price can even be negative! We have also shown that unlike normal commodities, electrical energy can flow from a high price to a low price. All of these effects are a consequence of the interactions between economics and KVL. They demonstrate the wisdom of the statement[2]: "Never trust a technique proven on the basis of a two-bus system."

These results may run against economic common sense, but they are mathematically correct. Efficiently trading electrical energy in a centralized market that spans a constrained transmission network requires the use of nodal marginal prices computed using an optimization procedure that maximizes the global welfare.[3] Unfortunately, as we saw, these prices are dictated not only by economics but also by KVL. Even in our simple three-bus examples, understanding these prices takes time and effort. In a real system, the analysis is even more difficult. This puts electricity traders in the position of "having to take the computer's word for it," which is not entirely satisfactory compared to trading in normal commodities.

Using a very small example allowed us to explain the factors driving the nodal prices in detail. Skeptical readers may be pardoned for suspecting that the phenomena that we have described are an artifact of this small network and would not occur in a real system.

---

2 Attributed to L.A. Dale.

3 The optimization procedure that we have used in our examples minimizes the production cost. Since we assume in these examples that the price elasticity of demand is zero, minimizing cost is equivalent to maximizing welfare. See Hogan (1992) for the generalization.

This is unfortunately not the case. Counterintuitive prices have been observed in several systems.

5.3.3 Losses in Transmission Networks

Transmitting electrical power through a network inevitably results in losses of energy. Since one or more generators must produce this lost energy and since these generators expect to be paid for all the energy they produce, a mechanism must be devised to take losses and their cost into account in electricity markets.

5.3.3.1 Types of Losses

Before going further, we should make a distinction between the three different types of losses that are encountered in power systems. The first type is the *variable losses*. These losses are caused by the current flowing through the lines, cables, and transformers of the network. Variable losses are also called load losses, series losses, copper losses, or transport-related losses. As Equation (5.81) shows, these losses are proportional to the resistance R of the branch and to the square of the current in this branch. They can also be expressed as a function of the apparent power S or the real and reactive powers P and Q flowing through the branch. Since the voltage in a power system does not normally deviate much from its nominal value and since the active power flow is usually much larger than the reactive power flow, these variable losses can, as a first approximation, be treated as a quadratic function of the active power flow:

$$L^{\text{variable}} = I^2 R \approx \left(\frac{S}{V}\right)^2 R = \frac{P^2 + Q^2}{V^2} \cdot R \approx \frac{R}{V^2} \cdot P^2 = K \cdot P^2 \tag{5.81}$$

Note that Equation (5.81) is ambiguous because the power at the receiving end of the line is not the same as the power at the beginning of the line because of the variable losses!

The second type of losses is the *fixed losses*. Most of these losses are caused by hysteresis and eddy current losses in the iron core of the transformers. The rest is due to the corona effect in transmission lines. Fixed losses are proportional to the square of the voltage and independent of the power flows. However, since the voltage varies relatively little from its nominal value, as a first approximation, these losses can be treated as constant. Fixed losses are also called no-load losses, shunt losses, or iron losses.

The third type of losses is called *nontechnical losses*. This euphemism covers energy that is stolen from the power system.

Because of their quadratic dependence on the power flows, variable losses are much more significant during periods of peak load. Averaged over a whole year, in western European countries, 1–3% of the energy produced is lost in the transmission system and 4–9% in the distribution system. In the remainder of this discussion, we consider only the variable losses because they are typically much larger than the fixed losses.

5.3.3.2 Marginal Cost of Losses

Figure 5.25 shows a two-bus system where a generator connected at bus 1 supplies a load connected at bus 2 through a line of resistance R. For the sake of simplicity, we assume that the load is purely active and we neglect the effect of the reactive power flow on the losses in this line. We also assume that the voltage is equal to its nominal value at both buses.

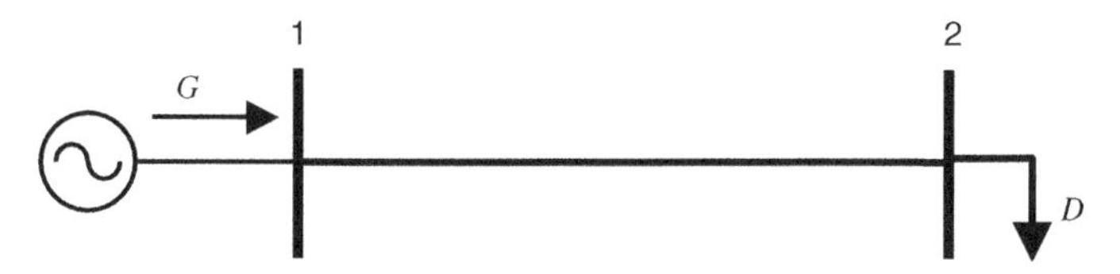

Figure 5.25 Two-bus system illustrating the calculation of the marginal cost of losses.

These assumptions allow us to express the losses as follows:

$$L = K \cdot D^2 \tag{5.82}$$

where D is the load at bus 2 and $K = {}^{R}/_{V^2}$. The generation at bus 1 is thus given by:

$$G(D) = D + L = D + K \cdot D^2 \tag{5.83}$$

If the load increases from D to $D + \Delta D$, the generation must increase by:

$$\Delta G = G(D + \Delta D) - G(D) = \Delta D + 2\Delta D \cdot D \cdot K = (1 + 2D \cdot K)\Delta D \tag{5.84}$$

where we have neglected the second-order term in ΔD. If the marginal cost of generation at bus 1 is c, the increase in the cost of generation due to an increase in load ΔD at bus 2 is:

$$\Delta C = c(1 + 2D \cdot K)\Delta D$$

and the marginal cost at bus 2 is:

$$\frac{\Delta C}{\Delta D} = c(1 + 2D \cdot K)$$

If we assume that competition is perfect in this system, the prices of energy at buses 1 and 2 are given by:

$$\pi_1 = c \tag{5.85}$$

$$\pi_2 = \pi_1(1 + 2D \cdot K) \tag{5.86}$$

The difference in price between the two buses thus increases linearly with the line flow because the losses are a quadratic function of the load.

Because of the losses, the total amount paid by consumers at bus 2 exceeds the amount received by generators at bus 1. A merchandizing surplus MS thus arises in the network. This surplus is equal to the value of the energy sold at bus 2 minus the cost of purchasing the energy produced at bus 1:

$$\mathrm{MS} = \pi_2 D - \pi_1\left(D + K \cdot D^2\right) \tag{5.87}$$

Using the expressions for the prices given in Equations (5.85) and (5.86), we get:

$$\begin{aligned}\mathrm{MS} &= c(1 + 2 \cdot K \cdot D)D - c\left(D + K \cdot D^2\right)\\ &= c \cdot K \cdot D^2\end{aligned} \tag{5.88}$$

While less energy is consumed at bus 2 than is produced at bus 1, the difference in price between these two buses is sufficient to ensure that this surplus is always positive. In this case, the merchandising surplus is equal to the cost of supplying the losses because there is only one generator with a defined marginal cost. In a more complex network, one cannot obtain a closed-form expression similar to Equation (5.88). It is therefore impossible to establish a rigorous method for quantifying the cost of losses. The point of Equation (5.88) is thus to show that the merchandising surplus is a rough indication of the cost of losses.

5.3.3.3 Effect of Losses on Generation Dispatch

Let us go back to the Borduria–Syldavia interconnection that we introduced at the beginning of this chapter to study the effect of losses on the dispatch of the generating units. To keep matters simple, we first assume that the interconnection is not congested and that its coefficient $K = {}^{R}/_{V^2} = 0.00005\,\text{MW}^{-1}$.

Using Equations (5.6) and (5.7), the variable costs of producing energy in Borduria and Syldavia are given by:

$$C_B(P_B) = \int_0^{P_B} MC_B(P)dP = 10P_B + \frac{1}{2}\cdot 0.01P_B^2 \tag{5.89}$$

$$C_S(P_S) = \int_0^{P_S} MC_S(P)dP = 13P_S + \frac{1}{2}\cdot 0.02P_S^2 \tag{5.90}$$

If the joint Borduria–Syldavia electricity market operates efficiently and competitively, at equilibrium, it minimizes the total variable cost of producing electrical energy:

$$\min(C_B + C_S) = \min\left(10P_B + \frac{1}{2}\cdot 0.01P_B^2 + 13P_S + \frac{1}{2}\cdot 0.02P_S^2\right) \tag{5.91}$$

This minimization is subject to the power balance constraint. In other words, the power generated in Borduria and Syldavia must be equal to the sum of the load and the losses:

$$P_B + P_S = D_B + D_S + K\cdot F_{BS}^2 \tag{5.92}$$

where K represents the resistance of the interconnection between the two countries and F_{BS} the active power flow at the Syldavia end of the interconnection. We again make the assumption that the voltage at all buses is kept at nominal value. To solve this optimization problem, we adopt an empirical approach where we vary the flow F_{BS} and then calculate the productions in Syldavia and Borduria using:

$$P_S = D_S - F_{BS} \tag{5.93}$$

$$P_B = D_B + F_{BS} + K\cdot F_{BS}^2 \tag{5.94}$$

The total production cost can then be computed using Equations (5.89) and (5.90). Figure 5.26 shows how this total cost varies as a function of the flow on the interconnection when we do and do not consider the cost of the losses in the interconnector. This figure clearly shows that the losses reduce the optimal power transfer from 933 to 853 MW. Table 5.8 gives the details of these two optimal solutions. The losses make the Bordurian generators somewhat less competitive because a fraction of the energy that they produce is lost during its transfer to Syldavian customers. Production therefore decreases in Borduria and increases in Syldavia. It is worth noting that the size of this redispatch is significantly larger than the amount of losses. Because of this redispatch, the

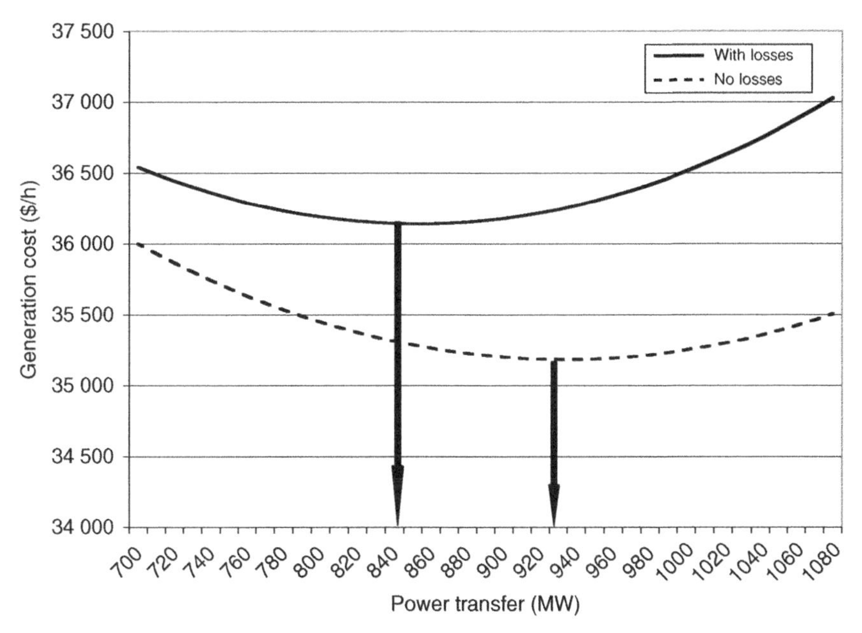

Figure 5.26 Total generation cost in the Borduria–Syldavia interconnection as a function of the flow on the interconnector when the losses in this interconnector are and are not taken into consideration. The coefficient $K = {}^{R}/_{V^2}$ of the interconnector is 0.000 05 MW^{-1}. The resistances of the internal lines of the Bordurian and Syldavian networks have been neglected. The demands in Borduria and Syldavia are 500 and 1500 MW, respectively.

marginal costs of production (and hence the local energy prices) are no longer equal in Borduria and Syldavia. A price differential of about 2.00 $/MWh arises. Syldavian consumers are indifferent between buying from local generators at 25.94 $/MWh or from Bordurian producers at 23.89 $/MWh and paying a 2.00 $/MWh transmission charge. Similarly, Bordurian consumers are indifferent between buying from local producers or from more expensive Syldavian generators because they get rewarded for entering into a transaction that reduces the losses.

Table 5.8 Effect of losses on the operation of the Borduria–Syldavia interconnection.

	Without losses	With losses
P_B (MW)	1 433	1 389
P_S (MW)	567	647
Losses (MW)	0	36
Power transfer (MW)	933	853
MC_B ($/MWh)	24.33	23.89
MC_S ($/MWh)	24.33	25.94
Total generation cost ($/h)	35 183	36 134

Table 5.9 Operation of the Borduria–Syldavia system when losses in the interconnection are taken into consideration.

	Borduria	Syldavia	System
Consumption (MW)	500	1 500	2 000
Production (MW)	1 389	647	2 036
Nodal marginal price ($/MWh)	23.89	25.94	—
Consumer payments ($/h)	11 945.00	38 910.00	50 855.00
Producer revenues ($/h)	33 183.21	16 783.18	49 966.39
Merchandising surplus ($/h)			888.61

5.3.3.4 Merchandising Surplus

Table 5.9 summarizes the operation of the Borduria–Syldavia interconnection when losses in the tie-line are taken into consideration. Consumers and producers buy and sell energy at their local price, which is assumed equal to the local marginal cost of production.

The presence of losses thus creates a merchandising surplus of 888.61 $/h. We get the same result if we treat this surplus as the "profit" made by the operator of the interconnection if it were to buy energy in Borduria and sell it in Syldavia. The quantity bought in Borduria would be 889 MW (that is, 1389 – 500 MW) and the price would be 23.89 $/MWh. The quantity sold in Syldavia would be 853 MW (that is, 1500 – 647 MW) and the price 25.94 $/MWh. The profit or surplus would then be:

$$853 \times 25.94 - 889 \times 23.89 = 888.61\,\$/\text{h}$$

Note that this is not the same thing as multiplying the quantity transported by the price differential because of the losses in the interconnection.

5.3.3.5 Combining Losses and Congestion

Losses occur whether or not the system is congested. Let us consider the case where the flow on the interconnection is constrained at 600 MW. The generators in Syldavia therefore produce 900 MW to meet the local load of 1500 MW. The nodal price (which we assume is equal to the marginal cost) in Syldavia is then:

$$\pi_S = MC_S = 13 + 0.02 P_S = 31.00\,\$/\text{MWh} \tag{5.95}$$

Using Equation (5.94), we find the production of the generators in Borduria:

$$P_B = D_B + F_{BS} + K \cdot F_{BS}^2 = 500 + 600 + 18 = 1118\,\text{MW} \tag{5.96}$$

The marginal cost and the nodal price in Borduria are then:

$$\pi_B = MC_B = 10 + 0.01 P_B \;\; = 21.18\,\$/\text{MWh} \tag{5.97}$$

The price differential is 9.82 $/MWh and is primarily due to the constraint. Table 5.10 summarizes the operation of the interconnection under these conditions.

Since the constraint on the interconnection reduces the flow, it also decreases the losses.

Table 5.10 Operation of the Borduria–Syldavia system when both losses and congestion in the interconnection are taken into consideration.

	Borduria	Syldavia	System
Consumption (MW)	500	1 500	2 000
Production (MW)	1 118	900	2 018
Nodal marginal price ($/MWh)	21.18	31.00	—
Consumer payments ($/h)	10 590	46 500	57 090
Producer revenues ($/h)	23 679	27 900	51 579
Merchandising surplus ($/h)			5 511

5.3.3.6 Handling of Losses Under Bilateral Trading

Because losses are not a linear function of the flows in the transmission system, the losses caused by a transaction are not simply a function of the amount of power traded and the location of the parties involved in the transaction. These losses depend also on all the other transactions taking place in the network. Allocating the losses or their cost between all the market participants is thus a problem that does not have a rigorous solution. Nevertheless, this cost must be paid and shared fairly. A fair mechanism is one where participants that contribute more to losses (for example, remote generators and consumers) pay a larger share than the others. See Conejo et al. (2002) for a discussion of various methods that have been proposed to allocate the cost of losses on an approximately fair basis.

5.3.4 Mathematical Formulation of Nodal Pricing

In an actual power system, the size and complexity of the network are such that the prices of electrical energy obviously cannot be computed in the ad hoc manner that we have used in the examples of the previous section. A centralized market operator needs a mathematical formulation that can be used to calculate these prices in a systematic manner. This market operator receives bids and offers from producers and consumers. It must then set prices to clear the market and select the accepted bids and offers. These decisions must maximize the global welfare generated by the system while respecting the limits imposed by the transmission network. In the following sections, we consider four, progressively more complex, variants of this constrained optimization problem. Once again, to keep matters simple, we assume that competition is perfect throughout the network. The bids submitted by the generators are thus equal to their marginal costs.

5.3.4.1 Network with a Single Busbar

Let us first take a step backward and see how we can formalize trading in electrical energy when the demand and the production are connected to the same busbar. Such a trivial "network" does not cause losses and does not limit the transfer of power between generation and load.

The economic welfare is equal to the difference between the benefit that consumers derive from the consumption of electrical energy and the cost of producing this energy.

We assume that the consumers' benefit is given by a function $B(D)$ of the total demand D and that the hourly cost of electrical energy is given by the function $C(P)$ of the total power P produced by the generators. This cost function $C(P)$ represents either the actual cost of production or the offers that the generators have submitted. As we have mentioned before, in a perfectly competitive market these two functions differ only by a constant. Obviously, to maintain the stability of the system, the generation must be equal to the load. We can thus formulate the operation of this system as the following constrained optimization problem:

Maximize $[B(D) - C(P)]$ subject to: $P - D = 0$.

The Lagrangian function of this problem is:

$$\ell(D, P, \pi) = B(D) - C(P) + \pi(P - D) \tag{5.98}$$

where we have chosen, for reasons that will soon be obvious, to represent the Lagrange multiplier by π. The optimality conditions are given by setting the partial derivatives of the Lagrangian to zero:

$$\frac{\partial \ell}{\partial D} \equiv \frac{dB}{dD} - \pi = 0 \tag{5.99}$$

$$\frac{\partial \ell}{\partial P} \equiv -\frac{dC}{dP} + \pi = 0 \tag{5.100}$$

$$\frac{\partial \ell}{\partial \pi} \equiv P - D = 0 \tag{5.101}$$

From Equations (5.99) and (5.100), we get:

$$\frac{dB}{dD} = \frac{dC}{dP} = \pi \tag{5.102}$$

Equation (5.102) formalizes a point that we discussed in Chapter 2, namely that consumers demand energy up to the point where the marginal benefit they derive from this consumption equals the price they pay. Similarly, generators produce up to the point where their marginal cost is equal to the price they receive. At equilibrium, in a perfectly competitive market, the price is equal to the value of the Lagrange multiplier of the optimization problem.

5.3.4.2 Network of Infinite Capacity with Losses

Let us now consider the case where demand and generation are connected to various nodes of a network. Since we assume that this network has an infinite capacity, transmission constraints are nonexistent and have therefore no effect on the prices of electricity. On the other hand, we take into consideration the effect that the distribution of the generations and loads has on the losses in the network.

Instead of treating generation and load separately, it is convenient to consider the *net power injection* at each node. If both generators and consumers are connected to a particular node, this net injection is positive when the local production exceeds the demand and negative when the opposite holds. If we denote by I_k the net injection at node k, we have:

$$I_k = P_k - D_k \tag{5.103}$$

If the various nodes were not connected by a network, the net injection at each node would have to be equal to zero and the economic optimization would need to be carried out independently at each node, as we discussed above. A network creates economic welfare by allowing trades between nodes with positive net injections and nodes with negative net injections.

At each node, we define a function $W_k(I_k)$. If I_k is negative, this function is equal to the benefit to consumers at node k of this net injection. If I_k is positive, it is equal to minus the cost of producing this net injection. Summing over all the nodes, we get the overall welfare created by the network:

$$W = \sum_{k=1}^{n} W_k(I_k) \tag{5.104}$$

As the objective of this optimization problem, we could choose to maximize this total welfare:

$$\max_{I_k}(W) = \max_{I_k}\left[\sum_{k=1}^{n} W_k(I_k)\right] \tag{5.105}$$

Since maximizing a function is equivalent to minimizing its opposite, we can also define the objective function as follows:

$$\min_{I_k}(-W) = \min_{I_k}\left\{\sum_{k=1}^{n}[-W_k(I_k)]\right\} = \min_{I_k}\left\{\sum_{k=1}^{n} C_k(I_k)\right\} \tag{5.106}$$

The second formulation is preferable because it is consistent with the traditional definition of the *optimal power flow problem*.[4] In this problem, the demands are assumed to be completely insensitive to prices and fixed loads are specified at each node. The benefit accruing to consumers is thus constant and does not need to be taken into consideration in the optimization. Under these conditions, Equation (5.106) represents the minimization of the total cost of producing energy:

$$\min_{I_k}(-W) = \min_{I_k}\left\{\sum_{k=1}^{n} C_k(I_k)\right\} \tag{5.107}$$

Since we assume that the network has an infinite capacity, the only constraint on this optimization is the need to maintain a power balance. The sum of the net injections at all nodes must therefore be equal to the power losses in the branches of the network:

$$\sum_{k=1}^{n} I_k = L(I_1, I_2, \ldots, I_{n-1}) \tag{5.108}$$

The power losses depend on the flows in the branches and thus on the net injections as shown by the function L in Equation (5.108). This function cannot depend on the injections at all the nodes. If it did, there would be no way to satisfy the power balance because any adjustment in the injections would cause a change in the losses. To get

4 The constrained economic dispatch that we encountered in the three-bus example above is a simplified version of the optimal power flow problem.

around this difficulty, one bus in the system is designated as the *slack bus* and the injection at this bus is omitted from the variables of the function L. Given all the other net injections, the injection at the slack bus can then be adjusted to satisfy Equation (5.108). Since the concept of slack bus is purely mathematical and has no physical implications, the choice of a slack bus is entirely arbitrary. In Equation (5.108) and the rest of this chapter, we have chosen bus n as the slack bus.

We combine Equations (5.107) and (5.108) to build the Lagrangian function of this optimization problem:

$$\ell = \sum_{k=1}^{n} C_k(I_k) + \pi \left[L(I_1, I_2, \ldots, I_{n-1}) - \sum_{k=1}^{n} I_k \right] \tag{5.109}$$

The conditions for optimality are then:

$$\frac{\partial \ell}{\partial I_k} \equiv \frac{\mathrm{d}C_k}{\mathrm{d}I_k} + \pi \left(\frac{\partial L}{\partial I_k} - 1 \right) = 0, \quad k = 1, \ldots, n-1 \tag{5.110}$$

$$\frac{\partial \ell}{\partial I_n} \equiv \frac{\mathrm{d}C_n}{\mathrm{d}I_n} - \pi = 0 \tag{5.111}$$

$$\frac{\partial \ell}{\partial \pi} \equiv L(I_1, I_2, \ldots, I_{n-1}) - \sum_{k=1}^{n} I_k = 0 \tag{5.112}$$

Combining Equations (5.110) and (5.111), we get:

$$\frac{\mathrm{d}C_k}{\mathrm{d}I_k} = \frac{\mathrm{d}C_n}{\mathrm{d}I_n} \left(1 - \frac{\partial L}{\partial I_k} \right) = \pi \left(1 - \frac{\partial L}{\partial I_k} \right), \quad k = 1, \ldots, n-1 \tag{5.113}$$

The Lagrange multiplier π thus represents the marginal cost or marginal benefit of an injection of power at the slack bus. In a competitive context, this is the nodal price at the slack bus. The nodal prices at the other buses are related to the price at the slack bus by Equation (5.113). If an increase in the net injection at node k adds to the losses, we have:

$$\frac{\partial L}{\partial I_k} > 0 \tag{5.114}$$

Hence, we get:

$$\frac{\mathrm{d}C_k}{\mathrm{d}I_k} < \frac{\mathrm{d}C_n}{\mathrm{d}I_n} \tag{5.115}$$

The nodal price paid to generators at node k is thus smaller than the nodal price at the slack bus to penalize them for the additional losses that they would cause by injecting an increment of power in the network at that node. On the other hand, consumers at node k pay a lower price because a load increase at that bus would reduce the losses. The opposite holds true if an increase in the net injection at node k reduces the losses. Finally, if the losses are neglected, the nodal prices at all buses are equal.

5.3.4.3 Network of Finite Capacity with Losses

As we will discuss in Chapter 6, system operators must consider not only the limits imposed by the thermal capacity of lines and cables but also how to guarantee the transient and voltage stability of the power system in the face of faults and outages. For the purpose of calculating prices, these considerations are translated into limits on the flow of power on certain lines or groups of lines. We model all these constraints as follows:

$$F_l(I_1, I_2, \ldots, I_{n-1}) \leq F_l^{\max}, \qquad l = 1, \ldots, m \tag{5.116}$$

where F_l is the flow on branch l and $F_l^{\max}$ is the maximum value that this flow is allowed to take, and m is the total number of branches in the network. Note that the net injection at the slack bus is not included in the expressions for the branch flows to avoid creating an overdetermined problem.

We take these inequality constraints into account by adding them to the Lagrangian function of the previous optimization problem (Equation (5.109)):

$$\begin{aligned} \ell = \sum_{k=1}^{n} C_k(I_k) + \pi \left[L(I_1, I_2, \ldots, I_{n-1}) - \sum_{k=1}^{n} I_k \right] \\ + \sum_{l=1}^{m} \mu_l \left[F_l^{\max} - F_l(I_1, I_2, \ldots, I_{n-1}) \right] \end{aligned} \tag{5.117}$$

The optimality conditions become:

$$\frac{\partial \ell}{\partial I_k} \equiv \frac{\mathrm{d}C_k}{\mathrm{d}I_k} + \pi \left(\frac{\partial L}{\partial I_k} - 1 \right) - \sum_{l=1}^{m} \mu_l \frac{\partial F_l}{\partial I_k} = 0, \qquad k = 1, \ldots, n-1 \tag{5.118}$$

$$\frac{\partial \ell}{\partial I_n} \equiv \frac{\mathrm{d}C_n}{\mathrm{d}I_n} - \pi = 0 \tag{5.119}$$

$$\frac{\partial \ell}{\partial \pi} \equiv L(I_1, I_2, \ldots, I_{n-1}) - \sum_{k=1}^{n} I_k = 0 \tag{5.120}$$

$$\frac{\partial \ell}{\partial \mu_l} \equiv F_l^{\max} - F_l(I_1, I_2, \ldots, I_{n-1}) = 0, \qquad l = 1, \ldots, m \tag{5.121}$$

$$\mu_l \cdot \left[F_l^{\max} - F_l(I_1, I_2, \ldots, I_{n-1}) \right] = 0; \mu_l \geq 0 \quad l = 1, \ldots, m \tag{5.122}$$

We can gain a better understanding of the implications of these equations by considering the special case where the flow on only one line (say line i) is constrained. Since all the Lagrange multipliers μ_l are then equal to zero except μ_i, we get:

$$\frac{\mathrm{d}C_k}{\mathrm{d}I_k} = \pi \left(1 - \frac{\partial L}{\partial I_k} \right) + \mu_i \frac{\partial F_i}{\partial I_k}, \qquad k = 1, \ldots, n-1 \tag{5.123}$$

$$\frac{\mathrm{d}C_n}{\mathrm{d}I_n} = \pi \tag{5.124}$$

$$\sum_{k=1}^{n} I_k = L(I_1, I_2, \ldots, I_{n-1}) \tag{5.125}$$

$$F_i(I_1, I_2, \ldots, I_{n-1}) = F_l^{\max}; \quad \mu_i > 0 \tag{5.126}$$

Equation (5.123) shows that the nodal price at every node (except the slack bus) is affected by a binding flow constraint on any line. The effect on the price at node k depends on the shadow cost of the constraint (the Lagrange multiplier μ_i) and the sensitivity $\partial F_i / \partial I_k$ of the flow on branch i to the net injection at node k.

5.3.4.4 Network of Finite Capacity, DC Power Flow Approximation

Formulation

Solving Equations (5.123)–(5.126) is computationally difficult not only because they implicitly involve the solution of the power flow equations but also because they are nonlinear. Instead of using a full and accurate AC model, this optimization is typically carried out using a linearized model called a DC power flow. The equations for the DC power flow are derived from the equation of the AC power flow by making the following simplifying assumptions:

- The resistance of each branch is negligible compared to the reactance.
- The magnitude of the voltage at every bus is equal to its nominal value.
- The differences in voltage angles across each branch are sufficiently small to allow the following approximations:

$$\cos\left(\theta_i - \theta_j\right) \approx 1$$
$$\sin\left(\theta_i - \theta_j\right) \approx \theta_i - \theta_j$$

Under these assumptions, the flow of reactive power in the network is negligible and the net active power injections are related to the bus voltage angles through the following set of equations:

$$I_i = \sum_{j=1}^{n} y_{ij}\left(\theta_i - \theta_j\right), \quad i = 1, \ldots, n \tag{5.127}$$

where y_{ij} represents the inverse of the reactance of the branch between nodes i and j and θ_i represents the voltage angle at node i. The flow of active power between nodes i and j is then given by:

$$F_{ij} = y_{ij}\left(\theta_i - \theta_j\right), \quad i, j = 1, \ldots, n \tag{5.128}$$

Since the DC power flow neglects the resistance of the branches and thus ignores the losses, we no longer have to take into consideration an equality constraint similar to Equation (5.108). We have, however, introduced a new set of variables θ_i, which is counterbalanced by the new set of equations (5.127). The constraints on the branch flows are given by:

$$y_{ij}\left(\theta_i - \theta_j\right) \le F_{ij}^{\max}, \quad i, j = 1, \ldots, n \tag{5.129}$$

Note that this formulation distinguishes two constraints for each branch: one on the flow from node i to node j and one on the flow from node j to node i. Obviously, only one of these two constraints can be binding at any time.

The Lagrangian function of this optimization problem is:

$$\begin{aligned} \ell = & \sum_{i=1}^{n} C_i(I_i) + \sum_{i=1}^{n} \pi_i \left[\left\{ \sum_{j=1}^{n} y_{ij}\left(\theta_i - \theta_j\right) \right\} - I_i \right] \\ & + \sum_{i=1}^{n} \sum_{j=1}^{n} \mu_{ij} \left[F_{ij}^{\max} - y_{ij}\left(\theta_i - \theta_j\right) \right] \end{aligned} \tag{5.130}$$

Taking the partial derivatives of this function with respect to the variables, we get the following optimality conditions:

$$\frac{\partial \ell}{\partial I_i} \equiv \frac{\mathrm{d}C_i}{\mathrm{d}I_i} - \pi_i = 0, \quad i = 1, \ldots, n \tag{5.131}$$

$$\frac{\partial \ell}{\partial \theta_i} \equiv \sum_{j=1}^{n} y_{ij}\left(\pi_i - \pi_j + \mu_{ij} - \mu_{ji}\right) = 0, \quad i = 1, \ldots, n-1 \tag{5.132}$$

$$\frac{\partial \ell}{\partial \pi_i} \equiv \left\{\sum_{j=1}^{n} y_{ij}(\theta_i - \theta_j)\right\} - I_i = 0, \quad i = 1, \ldots, n \tag{5.133}$$

$$\frac{\partial \ell}{\partial \mu_{ij}} \equiv F_{ij}^{\max} - y_{ij}(\theta_i - \theta_j) \geq 0, \quad i,j = 1, \ldots, n \tag{5.134}$$

$$\mu_{ij} \cdot \left[F_{ij}^{\max} - y_{ij}(\theta_i - \theta_j)\right] = 0; \mu_{ij} \geq 0, \quad i,j = 1, \ldots, n \tag{5.135}$$

Note that there are only $n - 1$ equations like (5.132) because the voltage angle at one of the nodes (typically the slack bus) is taken as a reference and is thus not a variable. Equations (5.134) and (5.135) exist only for the pairs ij that correspond to network branches.

Equation (5.131) shows that with this formulation the Lagrange multipliers π_i are equal to the nodal prices. Let us define $C^{\min}$ as the value of the cost at the optimum. This cost is a function of the flow limit on branch ij. Using Equation (5.130), we get:

$$\frac{\partial C^{\min}}{\partial F_{ij}^{\max}} = \mu_{ij} \tag{5.136}$$

The Lagrange multiplier μ_{ij} thus represents the marginal cost of this constraint. It is expressed in \$/MWh because it is equal to the saving that would accrue each hour if the flow in branch ij could be increased by 1 MW.

Implementation

In practice, operators must deal with a multitude of operational reliability issues that we will discuss in Chapter 6. Dealing with these issues involves much more than placing fixed limits on the active power flows on some lines. While the DC power flow approximation is convenient and computationally efficient, it would be foolish to think that it provides a sound basis for actually running a power system. However, it can be used to determine nodal marginal prices that reflect with sufficient accuracy the marginal effect that producers and consumers at each node have on the cost of operating the system.

If the offer functions are piecewise linear, we can solve the problem defined by Equations (5.131)–(5.135) using a linear programming package to get the optimal active power injections and the resulting voltage angle at each bus. In addition to the value of these primal variables, we also get the value of the dual variables:

- The Lagrange multipliers associated with the nodal load balance constraints (i.e. the nodal prices).
- The Lagrange multipliers associated with the binding line flow constraints.

It is interesting to compare how this rigorous mathematical formulation compares with the ad hoc method that we used to calculate nodal prices in Section 5.3.2. Linear programming dispatches generators at their minimum or maximum output, or at an elbow point of their offer curve, except for those that have to be redispatched to satisfy a constraint. As we argued in our three-bus example, if there are m constraints, there are $m+1$ such marginal generators.[5] Equation (5.131) determines the price of electrical energy at the buses where marginal generators are connected but not for generators that are operating at a breakpoint because the derivative of this function is not defined at those points. This equation is also of no use at buses where an offer or bid function is not available.

If there are m active constraints, we thus have:

- $m+1$ known prices π_i
- $n-m-1$ unknown prices π_i
- m unknown Lagrange multipliers μ_{ij}.

To find the value of these $n-1$ unknown variables, we have the $n-1$ equations (5.132). If we denote by K and U the sets of buses where the prices are respectively known and unknown, we can rearrange these equations as follows to have all the unknown variables are on the left-hand side:

$$Y_{ii}\pi_i - \sum_{j\in U} y_{ij}\pi_j + \sum_{j=1}^{n} y_{ij}\left(\mu_{ij} - \mu_{ji}\right) = \sum_{j\in K} y_{ij}\pi_j, \quad i \in U; i \neq \text{slack bus} \tag{5.137}$$

$$-\sum_{j\in U} y_{ij}\pi_j + \sum_{j=1}^{n} y_{ij}\left(\mu_{ij} - \mu_{ji}\right) = -Y_{ii}\pi_i + \sum_{j\in K} y_{ij}\pi_j, \quad i \in K; i \neq \text{slack bus} \tag{5.138}$$

where Y_{ii} represents the ith diagonal element of the admittance matrix of the network. Remember that μ_{ij} is nonzero only if the flow on the branch between nodes i and j is equal to its limits. The Lagrange multipliers μ_{ij} and μ_{ji} cannot be nonzero simultaneously because they correspond to flows on the same branch but in opposite directions. Even though we have written summations covering all the buses, the only nonzero terms are those for which bus j is at the opposite end of a branch connected to node i.

Example

Let us recalculate the nodal marginal prices for the three-bus example of Section 5.3.2 using this formulation given the constrained optimal dispatch of Equation (5.58). Generators A and D, located, respectively, at buses 1 and 3, are not operating at one of their limits. The price of electrical energy at these buses is thus known and equal to the marginal cost of these generators:

$$\pi_1 = \frac{dC_A}{dP_A} = 7.5\,\$/\text{MWh} \tag{5.139}$$

$$\pi_3 = \frac{dC_D}{dP_D} = 10.0\,\$/\text{MWh} \tag{5.140}$$

5 If the demand side takes an active part in the operation of the system by submitting bids to increase or decrease load, we could also have marginal loads.

On the other hand, the price at bus 2 is unknown. We thus have:

$$K = \{1, 3\}$$
$$U = \{2\}$$

The shadow cost μ_{12} of the constraint on the flow from bus 1 to bus 2 is also unknown. The other Lagrange multipliers μ_{ij} are equal to zero because the corresponding constraints are not binding. If we choose arbitrarily bus 3 as our slack bus, we can use the template provided by (5.137) and (5.138) to write the following equations:

$$i = 1: \quad -y_{12}\pi_2 + y_{12}\mu_{12} = -Y_{11}\pi_1 + y_{13}\pi_3 \tag{5.141}$$

$$i = 2: \quad Y_{22}\pi_2 - y_{12}\mu_{12} = y_{21}\pi_1 + y_{23}\pi_3 \tag{5.142}$$

Since the admittance matrix of this network is:

$$Y = \begin{pmatrix} -10 & 5 & 5 \\ 5 & -15 & 10 \\ 5 & 10 & -15 \end{pmatrix} \tag{5.143}$$

Equations (5.141) and (5.142) become:

$$\begin{cases} 5\pi_2 - 5\mu_{12} = 25 \\ -15\pi_2 + 5\mu_{12} = -137.5 \end{cases} \tag{5.144}$$

Solving these equations gives:

$$\begin{cases} \pi_2 = 11.25\,\$/\text{MWh} \\ \mu_{12} = 6.25\,\$/\text{MWh} \end{cases} \tag{5.145}$$

The nodal price at bus 2 is identical to the value given in Equation (5.67). Note that the shadow cost of the constraint on branch 1–2 is not equal to the difference between the marginal prices at nodes 1 and 2 because there is more than one path between these two nodes.

5.3.4.5 AC Modeling

The DC approximation discussed in the previous section has the advantage of supporting a linear and thus computationally reliable model of the system. However, it suffers from several limitations:

- It only models the flows of active power. An accurate check of branch current limits should also take into account the flows of reactive power.
- It does not model the voltage magnitudes and therefore ignores constraints on these variables.
- It considers only the active power injections and thus ignores the effect that reactive power injections and transformer tap changes have or could have on the state of the system.
- Because of these inaccuracies, operators impose conservative limits on line flows that hinder market efficiency.

A rigorous and accurate representation of the effects of the transmission network on electricity trading would therefore require the use of a full AC power flow model. Unfortunately, integrating such a highly nonlinear model into the market clearing process would require the use of nonlinear optimization techniques that are much slower than linear programming and for which convergence is not guaranteed. Since ensuring a timely market clearing is much more important than the accuracy of the solution, no AC optimization model has yet been implemented in an actual centralized market.

5.3.5 Managing Transmission Risks in a Centralized Trading System

We have already mentioned in previous chapters that it is unusual for producers and consumers of commodities to sell or buy entirely through the spot market. In Chapter 4, we saw how participants in centralized electricity markets use contracts for difference to manage their exposure to the risks associated with fluctuations in the spot price. In that chapter, however, we assumed that the transmission network did not affect trading in electrical energy. We have now seen how transmission capacity constraints limit the amount of power that can be transmitted across the network and create locational price differences. We must therefore consider the effect that congestion has on the feasibility of these contracts and what contractual tools are needed to manage the risks associated with this congestion. While losses also create differences in nodal marginal prices, these differences are smaller and more predictable than the differences caused by congestion. We therefore focus our discussion on the consequences of congestion. Our results can be generalized to cover the effect of losses.

5.3.5.1 The Need for Network-Related Contracts

By definition, in a centralized trading system, all the energy produced and consumed is traded physically through the system. Producers and consumers inject or extract power into the network according to the instructions of the system operator. In return, they receive or pay the centrally determined price in effect at the location where they are connected. However, all market participants are allowed to enter into bilateral financial contracts to protect themselves against the vagaries of the nodal prices. Let us examine what might happen when Borduria Power enters into a simple contract for difference with Syldavia Steel. This contract provides for the continuous delivery of 400 MW at a price of 30 \$/MWh. As before, we assume that there is no congestion within each of these two countries. There is thus a single nodal marginal price for Borduria (at which Borduria Power sells all its production) and a single nodal marginal price for Syldavia (at which Syldavia Steel still buys all its consumption).

As long as there is no congestion on the interconnection, these two nodal marginal prices are equal. Generators in Borduria thus see the same price as consumers in Syldavia. In particular, if the spot price is 24.30 \$/MWh, the contract between Borduria Power and Syldavia Steel is settled as follows:

- Borduria Power sells 400 MW at 24.30 \$/MWh and receives $400 \times 24.30 = \$9720$ in payment.
- Syldavia Steel buys 400 MW at 24.30 \$/MWh and pays $400 \times 24.30 = \$9720$.
- Syldavia Steel pays $400 \times (30 - 24.30) = \$2280$ to Borduria Power to settle the contract for difference.
- Borduria Power and Syldavia Steel have thus effectively traded 400 MW at 30 \$/MWh.
- If the nodal prices had been higher than 30 \$/MWh, Borduria Power would have made a payment to Syldavia Steel to settle the contract for difference.

Let us now consider what happens when the capacity of the interconnection is limited to 400 MW. We saw earlier that in this case, the nodal price in Borduria is 19.00 \$/MWh while it rises to 35.00 \$/MWh in Syldavia. Under these conditions:

- Borduria Power sells 400 MW at 19.00 \$/MWh and receives $400 \times 19.00 = \$7600$ in payment. According to the contract, it was supposed to collect $400 \times 30 = \$12\,000$. It is thus \$4400 short (\$12 000 – \$7600) and expects Syldavia Steel to pay this amount to settle the contract.
- Syldavia Steel buys 400 MW at 35.00 \$/MWh and pays $400 \times 35.00 = \$14\,000$. According to the contract, it was supposed to pay only $400 \times 30 = \$12\,000$. Syldavia Steel therefore expects Borduria Power to pay \$2000 to settle the contract.

These expectations are clearly incompatible. This contract for difference does not work when the transmission system is congested because it does not specify a location for the delivery of the power. If the two parties had agreed that the price to be compared to the contract price was the price in Syldavia, there would have been no ambiguity and the contract would be settled as follows:

- Borduria Power sells 400 MW at 19.00 \$/MWh for which it collects $400 \times 19.00 = \$7600$ from the system operator.
- Syldavia Steel buys 400 MW at 35.00 \$/MWh for which it pays $400 \times 35.00 = \$14\,000$ to the system operator.
- To settle the contract for difference, Borduria Power pays $400 \times (35.00 - 30.00) = \$2000$ to Syldavia Steel.

This contract thus works very well for Syldavia Steel, while Borduria Power suffers a substantial loss because it carried the risk associated with a price difference between Syldavia and Borduria. Parties wishing to protect themselves against price risks must therefore contract not only for the energy that they produce or consume but also for the ability of the transmission system to deliver this energy.

5.3.5.2 Financial Transmission Rights

Bill, the economist who has been asked to study the electrical reconnection between Borduria and Syldavia realizes, that, without the insurance provided by contracts, the full benefits of the interconnection are unlikely to be realized.

While pondering the example discussed in the previous section, he calculates the total shortfall in the contract for difference, i.e. the total amount of money that both parties should have received to settle the contract:

$$\$4400 + \$2000 = \$6400$$

He notices that this amount is exactly equal to the congestion surplus generated in the market, i.e. the difference between the total amount paid by the consumers and the total amount collected by the generators (see Table 5.2):

$$\$62\,000 - \$55\,600 = 6400\,\$/\text{h}$$

Bill realizes that if Borduria Power and Syldavia Steel were given access to this congestion surplus, they would be able to settle equitably their contract for difference. To convince himself that this is not just a coincidence, Bill develops an analytical representation of the settlement of a contract for difference in the presence of congestion.

He adopts the sign convention that a positive amount represents a revenue or a surplus and a negative amount an expense or a deficit. Given a contract for difference with a strike price π_C and an amount F, the total amount that a consumer such as Syldavia Steel expects to pay is:

$$E_C = -F \cdot \pi_C \tag{5.146}$$

Conversely, the total amount that a producer such as Borduria Power expects to receive is:

$$R_C = F \cdot \pi_C \tag{5.147}$$

The amounts that the consumer and producer pay and collect on the spot market are respectively:

$$E_M = -F \cdot \pi_S \tag{5.148}$$

and

$$R_M = F \cdot \pi_B \tag{5.149}$$

where Bill has taken into account the fact that the sale and the purchase are concluded at different nodal prices.

The amounts that the consumer and producer expect to pay or receive to settle the contract for difference are thus:

$$E_T = E_M - E_C = -F \cdot \pi_S - (-F \cdot \pi_C) = F(\pi_C - \pi_S) \tag{5.150}$$

and

$$R_T = R_M - R_C = F \cdot \pi_B - F \cdot \pi_C = -F(\pi_C - \pi_B) \tag{5.151}$$

If the producer and consumer trade are connected to the same node or if there is no congestion in the system, we have $\pi_S = \pi_B$ and the contract can be settled because:

$$E_T = -R_T \tag{5.152}$$

On the other hand, if $\pi_S \neq \pi_B$ both parties expect a payment and we have a total shortfall given by:

$$E_T + R_T = F(\pi_B - \pi_S) \tag{5.153}$$

Bill then compares Equation (5.153) with the expression for the congestion surplus given in Equation (5.26). He observes that both of these two expressions involve the product of a power transfer with a difference in price between two markets. The congestion surplus involves the maximum power that can be transferred between the two locations while the shortfall given in Equation (5.153) pertains to a specific transaction. The congestion surplus should therefore be able to cover the shortfalls for contracts up to the maximum power transfer between the two markets.

Bill concludes that problems with contracts for differences can be solved if the parties acquired what is called Financial Transmission Rights or FTRs. FTRs are defined between any two nodes in the network and entitle their holders to a revenue equal to the product of the amount of transmission rights bought and the price differential between the two nodes. Formally, the holder of FTRs for F MWh between locations B and S is entitled to the following amount taken from the congestion surplus:

$$R_{FTR} = F(\pi_S - \pi_B) \quad (5.154)$$

This amount is exactly what is needed to ensure that a contract for difference concluded between a producer at location B and a consumer at location S can be settled. Note that if there is no congestion in the transmission system, there is no price difference between locations B and S and the holder of FTRs receives no revenue. In this case, however, contracts for differences balance without problem.

Finally, Bill observes that holders of FTRs are indifferent about the origin or destination of the energy they consume or produce. For example, a consumer in Syldavia who owns F MWh worth of FTRs between Borduria and Syldavia can either:

- Buy F MWh of energy on the Bordurian market for a price π_B and use its transmission rights to have it delivered “for free” in Syldavia. In this case, it effectively pays $F \cdot \pi_B$.
- Buy the F MWh of energy on the Syldavian market for a price π_S and use its share of the congestion surplus to offset the higher price it paid for the energy. In this case, it pays $F \cdot \pi_S$ but receives $F \cdot (\pi_S - \pi_B)$.

In conclusion, FTRs completely isolate their holders from the risk associated with congestion in the transmission network. They provide a perfect hedge.

Bill must address one more question: how will producers and consumers obtain FTRs? Bill suggests that these rights should be auctioned. For each market period, the system operator would determine the amount of power that can be transmitted over the interconnection. FTRs for this amount of power would then be auctioned to the highest bidders. This auction would be open to all generators, consumers and even to speculators hoping to make a profit from locational differences in the price of electrical energy. The holder of these rights would be able to use them or resell them to another party. How much should bidders pay for an FTR? This depends on their expectations of the price differentials that might arise between the locations where these rights are defined. In the case of our example, if Bill’s estimates of the energy prices in Borduria and Syldavia and of the transmission capacity of the interconnection during periods of congestion are correct, the auction should result in a maximum price of

$$35.00\,\$/\text{MWh} - 19.00\,\$/\text{MWh} = 16.00\,\$/\text{MWh}$$

5.3.5.3 Point-to-Point Financial Transmission Rights

An important aspect of the definition of FTRs does not come out clearly from our two-bus Borduria/Syldavia example. FTRs are defined from any point in the network to any other point. These points do not have to be connected directly by a branch. The advantage of this approach from the perspective of a producer and a consumer who wish to enter into a transaction is that they do not have to concern themselves with the intricacies of the network. All they need to know is the bus where the power is injected and the bus where it is extracted. As far as they are concerned, the path that this power takes through the network is of no importance.

To illustrate this point, let us check how point-to-point FTRs work in our three-bus example. We consider first the basic conditions that we analyzed in Sections 5.3.2.2–5.3.2.4. Figure 5.27 summarizes the constrained economic operation of this system. Suppose that one of the consumers at bus 3 has signed a contract for difference with a generator connected to bus 1. This contract is for the supply of 100 MW at 8.00 $/MWh.

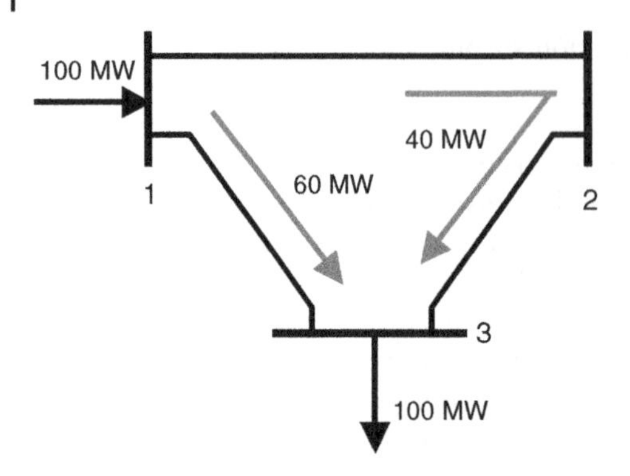

Figure 5.27 Transmission rights that must be acquired for a 100 MW transaction between buses 1 and 3.

The reference price for this contract is the nodal price at bus 1. As part of its risk management strategy, this consumer has also purchased 100 MW of FTRs from bus 1 to bus 3. As we saw, the nodal prices at buses 1 and 3 turn out to be 7.50 and 10.00 \$/MWh, respectively. This contract is settled as follows:

- The consumer pays $100 \times 10.00 = \$1000$ to the market operator for extracting 100 MW at bus 3.
- The generator receives $100 \times 7.50 = \$750$ from the market operator for injecting 100 MW at bus 1.
- The consumer pays $100 \times (8.00 - 7.50) = \$50$ to the generator to settle the contract for difference.
- The consumer collects $100 \times (10.00 - 7.50) = \$250$ from the market operator for the FTRs it owns between buses 1 and 3.

The consumer thus pays a total of \$800 for 100 MW, which is equivalent to a price of 8.00 \$/MWh.

As we mentioned earlier, the money that the market operator needs to pay the owners of FTRs comes from the congestion surplus that it collects because of network congestion. The market operator should therefore not sell FTRs for more capacity than the network can physically provide. Table 5.11 shows three combinations of FTRs that meet

Table 5.11 Some feasible combinations of point-to-point financial transmission rights for the three-bus example.

	Transmission rights			Settlement			
Combination	From bus	To bus	Amount (MW)	From bus price (\$/MWh)	To bus price (\$/MWh)	Revenue (\$)	Total (\$)
A	1	3	225	7.50	10.00	562.50	787.50
	1	2	60	7.50	11.25	225.00	
B	1	3	285	7.50	10.00	712.50	787.50
	3	2	60	10.00	11.25	75.00	
C	1	3	275	7.50	10.00	687.50	787.50
	1	2	10	7.50	11.25	37.50	
	3	2	50	10.00	11.25	62.50	

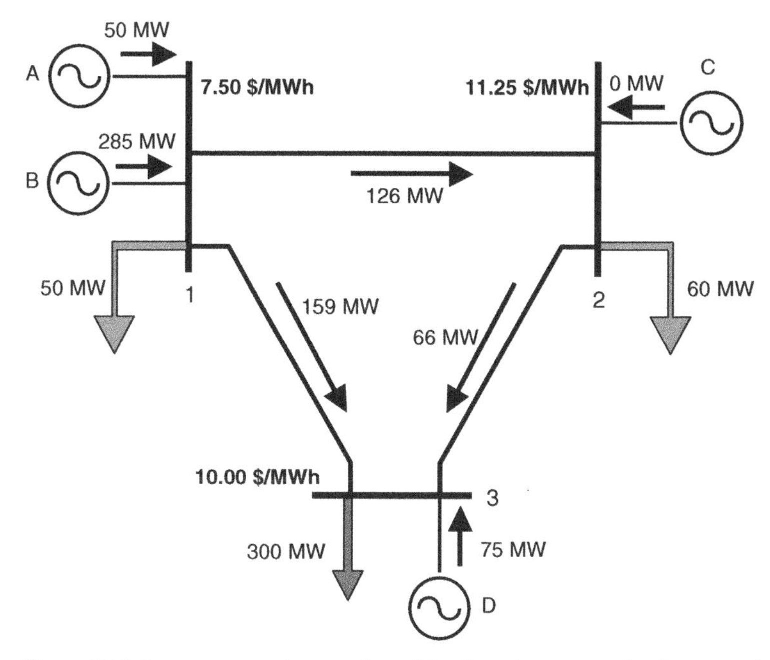

Figure 5.28 Secure economic operation of the three-bus system for the original network conditions.

this simultaneous feasibility condition for the three-bus example. Each of these combinations uses exactly the maximum transmission capacity illustrated in Figure 5.28.

Note that in each case the sum of the revenues that the holders of the rights collect based on the nodal prices is equal to the merchandising surplus collected by the market operator (see Table 5.5).

Let us see what happens if, as we explored in Section 5.3.2.7, the capacity of line 2–3 is limited to 65 MW. Figure 5.29 summarizes the operation of the system under these conditions. Table 5.12 summarizes the settlement of the three combinations of FTRs that are shown in Table 5.11.

Note that some of these FTRs have a negative value under these conditions. The holders of these rights therefore owe an additional amount to the market operator. This is a bit surprising because they actually paid to obtain these rights. However, it is not as bad as it may sound because contracts for difference can still be settled. Suppose for example that the load at bus 2 has signed a contract for difference with a generator at bus 1 for the delivery of 60 MW at 8.00 $/MWh. The reference price for this contract is again the nodal price at bus 1. This consumer had also purchased 60 MW of transmission rights between nodes 1 and 2. This contract would be settled as follows:

- The consumer pays $60 \times 5.00 = \$300$ to the market operator for extracting 60 MW at bus 2.
- The generator receives $60 \times 7.50 = \$450$ from the market operator for injecting 60 MW at bus 1.
- The consumer pays $60 \times (8.00 - 7.50) = \$30$ to the generator to settle the contract for difference.
- The consumer pays $60 \times (7.50 - 5.00) = \$150$ to the market operator for the FTRs it owns between buses 1 and 2.

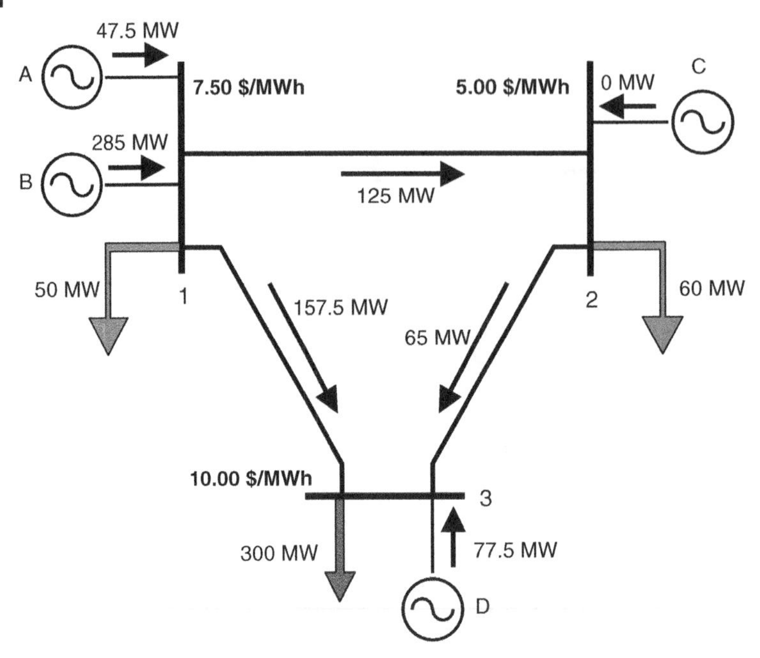

Figure 5.29 Secure economic operation of the three-bus system when the capacity of line 2–3 is limited to 65 MW.

This consumer thus pays a total of $480, which is equivalent to the 8.00 $/MWh strike price in its contract for difference.

A simple calculation, similar to the one we performed in Table 5.5, shows that under these operating conditions, the market operator collects a congestion surplus of $406.25. Unlike the previous case, this is somewhat short of the $412.50 shown in the last column of Table 5.12, which is the amount that the market operator must disburse to settle the FTRs. This discrepancy arises because the system operator was not able to deliver the point-to-point transmission capacity that it assumed when the FTRs were auctioned.

Table 5.12 Settlement of combinations of point-to-point financial transmission rights for the three-bus example when the capacity of line 2–3 is limited to 65 MW.

	Transmission rights			Settlement			
Combination	From bus	To bus	Amount (MW)	From bus price ($/MWh)	To bus price ($/MWh)	Revenue ($)	Total ($)
A	1	3	225	7.50	10.00	562.50	412.50
	1	2	60	7.50	5.00	–150.00	
B	1	3	285	7.50	10.00	712.50	412.50
	3	2	60	10.00	5.00	–300.00	
C	1	3	275	7.50	10.00	687.50	412.50
	1	2	10	7.50	5.00	–25.00	
	3	2	50	10.00	5.00	–250.00	

Note that the market operator must collect money from the FTRs that have a negative value to be able to roughly balance its account book.

This form of point-to-point FTR is therefore an obligation because the owner of the FTR may have to pay the system operator if the price difference is not in the expected direction. If such an arrangement is unpalatable, market participants can also purchase FTR options where the contract is exercised only if it is profitable for its holder.

5.3.5.4 Flowgate Rights

Instead of being defined from point to point, FTRs can be attached to a branch or flowgate in the network. They are then called flowgate rights (FGRs). FGRs operate like FTRs except that the value of these rights is not tied to the difference in nodal prices but to the value of the Lagrange multiplier or shadow cost associated with the maximum capacity of the flowgate. When a flowgate is not operating at its maximum capacity, the corresponding inequality constraint is not binding and the corresponding Lagrange multiplier μ has a value of zero. The only FGRs that produce revenues are thus those that are associated with congested branches.

For a few years, there was intense debate over the relative merits of point-to-point and flowgate transmission rights. To the best of our knowledge, all markets that support FTRs have chosen the point-to-point variety.

5.4 Problems

5.1 Consider the power system shown in Figure P5.1. Assuming that the only limitations imposed by the network are imposed by the thermal capacity of the transmission lines and that reactive power flows are negligible, check that the sets of transactions shown in Table P5.1 are simultaneously feasible.

Table P5.1 Sets of simultaneous transactions for Problem 5.1.

	Seller	Buyer	Amount
Set 1	B	X	200
	A	Z	400
	C	Y	300
Set 2	B	Z	600
	A	X	300
	A	Y	200
	A	Z	200
Set 3	C	X	1000
	X	Y	400
	B	C	300
	A	C	200
	A	Z	100

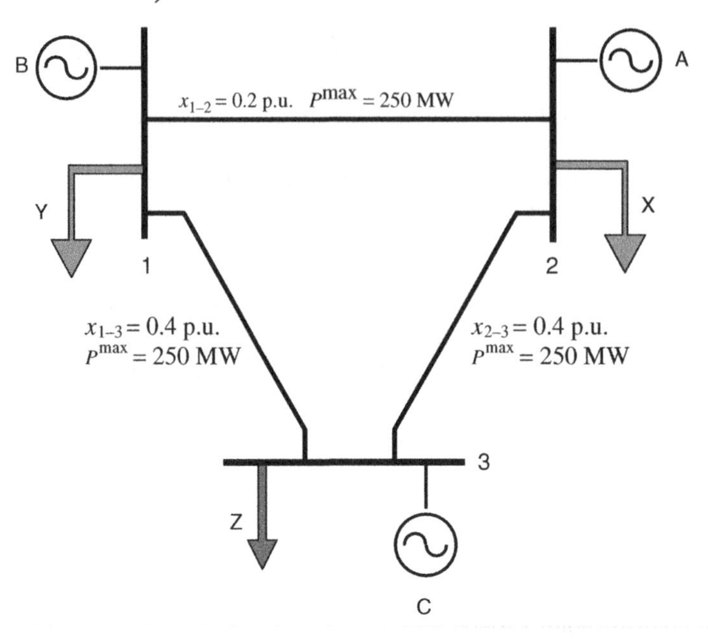

Figure P5.1 Three-bus power system for Problem 5.1.

5.2 Consider the two-bus power system shown in Figure P5.2. The marginal cost of production of the generators connected to buses A and B are given, respectively, by the following expressions:

$$\mathrm{MC_A} = 20 + 0.03P_\mathrm{A}\ (\$/\mathrm{MWh})$$
$$\mathrm{MC_B} = 15 + 0.02P_\mathrm{B}\ (\$/\mathrm{MWh})$$

Assume that the demand is constant and insensitive to price and that energy is sold at its marginal cost of production and that there are no limits on the output of the generators. Calculate the price of electricity at each bus, the production of each generator and the flow on the line for the following cases:

a The line between buses A and B is disconnected.
b The line between buses A and B is in service and has an unlimited capacity.
c The line between buses A and B is in service and has an unlimited capacity, but the maximum output of generator B is 1500 MW.
d The line between buses A and B is in service and has an unlimited capacity, but the maximum output of generator A is 900 MW. The output of generator B is unlimited.
e The line between buses A and B is in service but its capacity is limited to 600 MW. The output of the generators is unlimited.

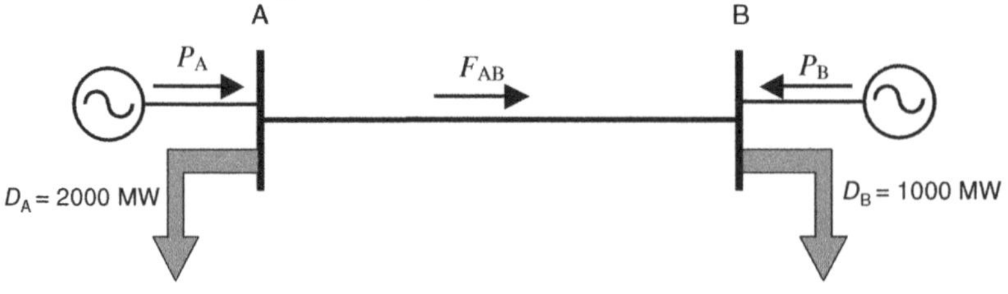

Figure P5.2 Two-bus power system for Problems 5.2–5.4, 5.10, and 5.11.

5.3 Calculate the generator revenues and the consumer payments for all the cases considered in Problem 5.2. Who benefits from the line connecting these two buses?

5.4 Calculate the congestion surplus for case (e) of Problem 5.2. Check your answer using the results of Problem 5.3. For what values of the flow on the line between buses A and B is the congestion surplus equal to zero?

5.5 Consider the three-bus power system shown in Figure P5.5. Table P5.5 shows the data about the generators connected to this system. Calculate the unconstrained economic dispatch and the nodal prices for the loading conditions shown in Figure P5.5.

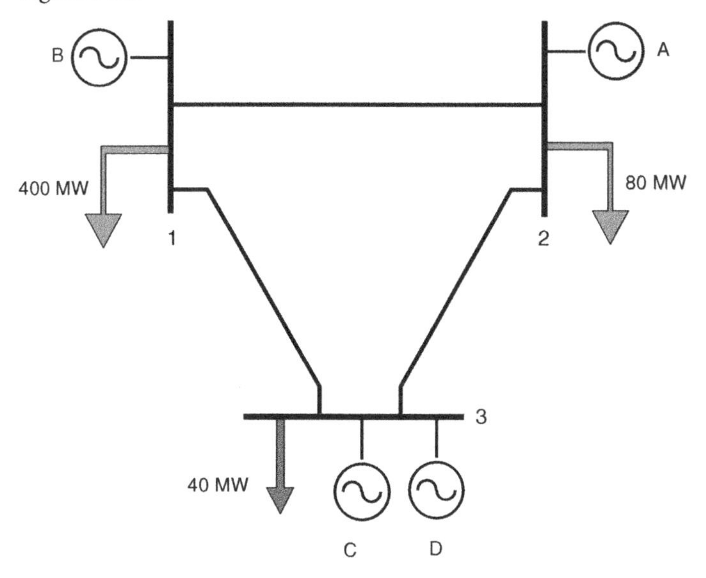

Figure P5.5 Three-bus power system for Problems 5.5–5.9 and 5.12–5.17.

Table P5.5 Characteristics of the generators for Problem 5.5.

Generator	Capacity (MW)	Marginal cost ($/MWh)
A	150	12
B	200	15
C	150	10
D	400	8

5.6 Table P5.6 gives the branch data for the three-bus power system of Problem 5.5. Using the superposition principle, calculate the flows that would result if the generating units were dispatched as calculated in Problem 5.5. Identify all the violations of transmission constraints.

Table P5.6 Characteristics of the branches for Problem 5.6.

Branch	Reactance (p.u.)	Capacity (MW)
1–2	0.2	250
1–3	0.3	250
2–3	0.3	250

5.7 Determine two ways of removing the constraint violations that you identified in Problem 5.6 by redispatching generating units. Which redispatch is preferable?

5.8 Calculate the nodal prices for the three-bus power system of Problems 5.5 and 5.6 when the generating units have been optimally redispatched to relieve the constraint violations identified in Problem 5.6 and corrected in Problem 5.7. Calculate the merchandising surplus and show that it is equal to the sum of the surpluses of each line.

5.9 Consider the three-bus power system described in Problems 5.5 and 5.6. Suppose that the capacity of branch 1–2 is reduced to 140 MW, while the capacity of the other lines remains unchanged. Calculate the optimal dispatch and the nodal prices for these conditions.

(*Hint:* The optimal solution involves a redispatch of generating units at all three buses.)

5.10 Consider the two-bus power system of Problem 5.2. Given that $K = {}^{R}/_{V^2} = 0.0001\ \text{MW}^{-1}$ for the line connecting buses A and B and that there is no limit on the capacity of this line, calculate the value of the flow that minimizes the total variable cost of production. Assuming that a competitive electricity market operates at both buses, calculate the nodal marginal prices and the merchandising surplus.

(*Hint:* Use a spreadsheet.)

5.11 Repeat Problem 5.10 for several values of K ranging from 0 to 0.0005. Plot the optimal flow and the losses in the line, as well as the marginal cost of electrical energy at both buses. Discuss your results.

5.12 Using the linearized mathematical formulation (DC power flow approximation), calculate the nodal prices and the marginal cost of the inequality constraint for the optimal redispatch that you obtained in Problem 5.7. Check that your results are identical to those that you obtained in Problem 5.8. Use bus 3 as the slack bus.

5.13 Show that the choice of slack bus does not influence the nodal prices for the DC power flow approximation by repeating Problem 5.12 using bus 1 and then bus 2 as the slack bus.

5.14 Using the linearized mathematical formulation (DC power flow approximation), calculate the marginal costs of the inequality constraints for the conditions of Problem 5.9.

5.15 Consider the three-bus system shown in Figure P5.5. Suppose that generator D and a consumer located at bus 1 have entered into a contract for difference for the delivery of 100 MW at a strike price of 11.00 $/MWh with reference to the nodal price at bus 1, Show that purchasing 100 MW of point-to-point financial rights between buses 3 and 1 provides a perfect hedge to generator D for the conditions of Problem 5.8.

5.16 What flowgate rights should generator D purchase to achieve the same perfect hedge as in Problem 5.15?

5.17 Repeat Problems 5.15 and 5.16 for the conditions of Problem 5.9.

5.18 Determine whether trading is centralized or decentralized in your region or country or in another area for which you have access to sufficient information. Determine also the type(s) of transmission rights that are used to hedge against the risks associated with network congestion.

5.19 Determine how the cost of losses is allocated in your region or country or in another area for which you have access to sufficient information.

5.20 Consider the small power system shown in Figure P5.20.

Assume that:

- Generating units A and B have the following constant marginal production costs:

$$MC_A = 20\,\$/\text{MWh}$$
$$MC_B = 40\,\$/\text{MWh}$$

- The no-load and start-up costs are neglected.
- All three transmission lines have the same reactance and a negligible resistance.
- The DC power flow assumption is valid.
- The capacity of each generator is 500 MW.
- The capacity of the lines depends on the weather conditions. Under cold weather conditions, each line is capable of carrying 400 MW. Under hot weather conditions, this capacity is reduced to 240 MW.
- This system is operated under the $N-0$ security criterion, i.e. we do not have to consider line or generator outages.

a Calculate the optimal power flow under cold weather conditions.
b Calculate the optimal power flow under hot weather conditions.
c Calculate the hourly cost of security under cold weather conditions.
d Calculate the hourly cost of security under hot weather conditions.

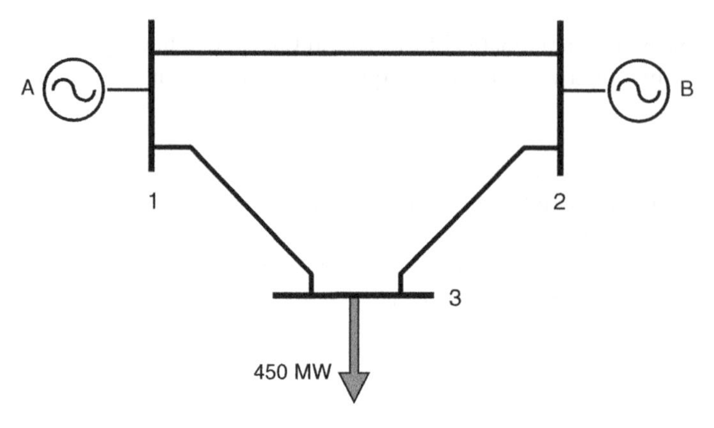

Figure P5.20 Three-bus power system for Problems 5.20–5.22.

5.21 Repeat Problem 5.20 but assume that the reactance of the line between buses 1 and 2 is two times the reactance of the other two lines.

5.22 Consider the small power system shown in Figure P5.20 and assume that:

- The load at bus 3 is now 300 MW.
- Generating units A and B have the following constant marginal production costs:

$$MC_A = 10\,\$/\text{MWh}$$
$$MC_B = 20\,\$/\text{MWh}$$

- All three transmission lines have the same impedance.

a Calculate the unconstrained optimal dispatch for these conditions.
b Calculate the hourly cost of this unconstrained dispatch.
c Calculate the power that would flow in each line if this dispatch was implemented.
d What is the marginal cost of energy at each node under these conditions?
e How should this unconstrained dispatch be modified if the flow in line 1–3 is limited to 150 MW for operational reliability reasons?
f Calculate the hourly cost of this constrained dispatch and the hourly cost of security.
g What is the marginal cost of energy at each node when the constraint on the flow on line 1–3 is taken into consideration?
h Identify an economic paradox[6] in this system.
i Assume that the Independent System Operator sells only point-to-point FTRs from bus 1 to bus 3. What is the maximum amount of transmission rights that it could sell without losing money?

5.23 Consider the three-bus power system shown in Figure P5.23a. Two identical circuits of equal impedance and equal MW capacity connect each pair of buses. The reactance and MW capacity of each circuit are given in Table P5.23.

---

6 *Paradox* (Noun): A statement, proposition, or situation that seems to be absurd or contradictory, but in fact is or may be true.

Table P5.23 Characteristics of the circuits in Problem 5.23.

From bus	To bus	Circuit reactance (p.u.)	Circuit capacity (MW)
1	2	0.2	120
1	3	0.4	180
2	3	0.2	250

Figure P5.23b shows the marginal cost curves of the two generators. Startup and no-load costs are assumed negligible. A DC (linear) transmission model is deemed acceptable.

a Calculate the economic dispatch ignoring network constraints.

b Check that this economic dispatch does not violate any transmission constraint when all the circuits are in service.

c Assuming that all generators bid at their marginal cost, what is the locational marginal price at each node under these conditions?

The system operator must operate this system in a way that guarantees $N-1$ security, i.e. there should be no line overload in the event of the outage of any transmission circuit. (We do not consider generation contingencies.)

d For each branch, determine the single circuit contingencies that would result in a flow constraint violation if this economic dispatch were implemented. (Take into account the fact that the transmission capacity and reactance of a branch change if one of its circuit is disconnected.)

Identify the contingency that would cause the worst violation of a flow constraint and the circuit or branch that would be overloaded.

e Determine the least cost generation dispatch that would avoid a line overload for the critical contingency identified in part (d). Assume that no postcontingency redispatch is allowed.

f What is the locational marginal price at each node under these conditions?

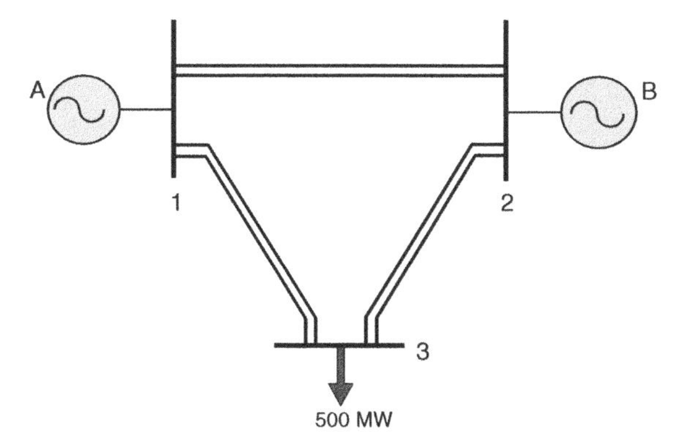

Figure P5.23a Three-bus power system for Problem 5.23.

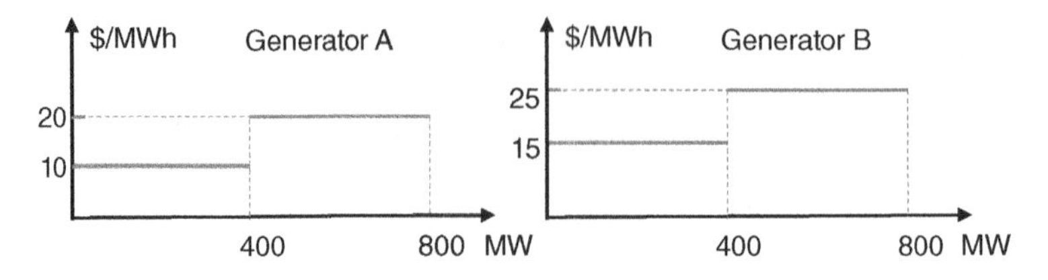

Figure P5.23b Marginal cost curves of the generators of Problem 5.23.

References

Conejo, A.J., Arroyo, J.M., Alguacil, N., and Guijarro, A.L. (2002). Transmission loss allocation: a comparison of different practical algorithms. *IEEE Trans. Power Syst.* 17 (3): 571–576.

Day, C.J., Hobbs, B.F., and Pang, J.-S. (2002). Oligopolistic competition in power networks: a conjectured supply function approach. *IEEE Trans. Power Syst.* 17 (3): 597–607.

Hogan, W.W. (1992). Contract networks for electric power transmission. *J. Regul. Econ.* 4 (3): 211–242.

Joskow, P. and Tirole, J. (2000). Transmission rights and market power on electric power networks. *RAND J. Econ.* 31 (3): 450–487.

Further Reading

Momoh (2000) provides a very readable discussion of the optimization techniques used in power systems. Hsu (1997) and Wu et al. (1996) are useful references on the principles of nodal pricing. Conejo et al. (2002) discuss the various methods used to handle losses. The seminal work on financial transmission rights was published by Hogan (1992). Joskow and Tirole (2000) discuss in detail market power issues with physical transmission rights. Day et al. (2002) present a numerical method to analyze market power issues in large networks.

Hsu, M. (1997). An introduction to the pricing of electric power transmission. *Util. Policy* 6 (3): 257–270.

Momoh, J.A. (2000). *Electric Power System Applications of Optimization*. Marcel Dekker.

Wu, F., Varaiya, P., Spiller, P., and Oren, S. (1996). Folk theorems on transmission access: proofs and counterexamples. *J. Regul. Econ.* 10 (1): 5–23.

6

Power System Operation

6.1 Introduction

6.1.1 The Need for Operational Reliability

When a buyer and a seller trade a commodity besides electricity, they implicitly or explicitly agree on how the goods will be delivered and who is responsible for their delivery. This can be as simple as the seller handing out a bag of apples across a market stall or as complex as arranging for transportation half-way around the world. If a problem occurs during a particular delivery, the goods might be delayed or lost but the contract specifies whether the seller or the buyer is responsible. If the buyer or seller contracted this delivery to a logistics company, it can claim damages from this third party and should decide whether to continue using its services.

Buyers and sellers of electrical energy do not have a choice: they have to use the existing power system infrastructure to complete their transactions. Furthermore, this infrastructure operates on a continuous rather than batch basis and pools all transactions into a flow of power. A breakdown in this system affects indiscriminately all its users rather than individual transactions. As we will see later in this chapter, avoiding outages requires an awareness of the state of the entire system and the ability to act in a coordinated manner across the grid. Since this is not something that individual market participants have the ability to do, this responsibility is entrusted to an independent entity called the system operator. Despite their best efforts, occasional large blackouts and more frequent smaller outages do happen as a result of the unanticipated failure of a system component or a sudden large imbalance between load and generation.

System operators are typically not liable for the damages that these incidents might cause to the market participants. Consumers and generators of electrical energy therefore shoulder the socioeconomic costs of power outages. One of the roles of regulatory authorities is therefore to ensure that the monopoly transmission and distribution companies provide a satisfactory level of reliability, i.e. that the number, extent, and duration of power outages remain sufficiently low. Regulators usually do this by setting standards for the design of the networks, by obliging system operators to follow operating criteria, by periodically reviewing their reliability performance, and by conducting inquiries in the aftermath of major incidents. Power system reliability is a "public good" because all consumers and generators benefit from an uninterrupted service without subtracting from the reliability enjoyed by other users of the system.

6.1.2 The Value of Reliability

Since a higher level of reliability means a smaller probability of outages, the value of improving reliability should be measured in terms of a reduction in the expected cost of outages. Estimating this cost reduction is not a trivial matter. We must first estimate how the measures that are put in place to improve reliability affect the frequency of outages, the number of customers that are disconnected, and the duration of these outages. We should then assess the cost of each outage. Unfortunately, this depends on the type of customer, how long the outage lasts, and when it occurs. To get hold of this data, we have to survey different categories of customers and ask them what they would be willing to pay to avoid an outage of a certain duration. Once we have collected this data, we can calculate a quantity called the value of lost load (VoLL), which is defined as the amount of money that an *average* customer would be willing to pay for not being deprived of a kilowatt-hour of electrical energy without sufficient advance notice. Alternatively, VoLL can be estimated using a macroeconomic analysis. VoLL is meaningless for an individual customer because outages affect them in vastly different ways. For example, a momentary interruption has an almost negligible effect on most residential customers, while it can cost millions of dollars to a semiconductor manufacturer. On the other hand, VoLL is useful when studying reliability on a broad scale because large outages affect all types of customers indiscriminately. Table 6.1 summarizes the VoLL that are used or have been estimated in various countries or regions. The large range of values from similar countries suggests that VoLL provides only a rough estimate of the potential cost of outages. However, all these values are at least two orders of magnitudes larger than the average price of electricity. The socioeconomic cost to the consumers of an outage is thus considerably larger than the corresponding loss of revenue to the generators.

6.1.3 The Cost of Reliability

Reliability also has a cost. In Chapters 7 and 8, we will discuss how the need for reliability affects investments in new generation and transmission capacity. In this chapter, we focus

Table 6.1 Value of lost load (VoLL) used or estimated in various regions.

Country/Region	VoLL ($/MWh)	Source
United Kingdom	$22 000	London Economics
Countries of the European Union	$12 290–$29 050	European Commission
United States	$7 500	Brattle Group
MISO	$3 500	MISO
New Zealand	$41 269	London Economics
Victoria – Australia	$44 438	London Economics
Australia	$45 708	London Economics
Ireland	$9 538	London Economics
Northeast USA	$9 283–$13 925	London Economics

Where necessary, the values have been converted into US$ as of 09/19/2016.

on how it affects the cost of operating the power system using the existing assets. The operating cost of reliability stems from two types of activities that the system operator engages in to maintain the stability of the system and avoid the need to disconnect customers involuntarily:

- *Preventive measures:* Operators usually want to ensure that the system would remain stable if a contingency were to occur. Preventive measures represent the deviations from the pure economic optimum that are required to keep the system in such a state. Such measures have a cost because they imply a departure from the least cost solution. The differences between the economic dispatch and OPF solutions that we encountered in Chapter 5 are an example of preventive measures.
- *Corrective actions:* Unlike preventive measures that are in place at all times, operators take corrective actions only after a contingency has occurred. However, to be able to implement effective corrective actions, operators must secure the required resources before any contingency occurs. Examples of such resources include a generator's ability to increase rapidly its output or a large consumer's willingness to reduce its demand at short notice. Corrective actions therefore have a two-part cost: a procurement cost that is incurred any time the necessary resources are made available and a deployment cost that is paid only if the resource is used.

Implementing more extensive preventive measures and procuring more resources for corrective actions improve the system operators' ability to deal with unforeseen contingencies and thus the reliability of the system. However, there is a point at which the marginal cost of these measures and resources is equal to the value that they provide. In theory, this point should be used to determine the optimal amount of money that should be spent on operational reliability because one extra dollar spent on reliability measures will, on average over the long run, avoid $1 in outage costs. Unfortunately, a number of factors make the calculation of this equilibrium point very difficult in practice. First, while the number of possible contingencies is vast, each of them has a relatively small probability of occurrence. This probability is not known accurately because it depends on factors such as the weather and the condition of the affected components. The probability of cascading outages (i.e. the failure of one component causing the failure of another) is also non-negligible and depends in complex ways on the structure and state of the power system. Second, the most cost-effective way to deal with a particular contingency also depends on the state of the system. Third, because a particular measure or resource can often be used to handle several contingencies, its exact value is difficult to calculate.

Because of this uncertainty and complexity, system operators prefer that their responsibilities for reliability be defined using deterministic rules that do not take into account the cost of the measures taken for the sake of avoiding outages. Typically, these rules specify a set of contingencies that are deemed credible and require the operator to run the system in such a way that it would be able to withstand any of these contingencies without the need for involuntary load shedding. The set of credible contingencies usually encompasses the outage of all system components (branches, generators, and shunt elements) taken separately. The probability of two nearly simultaneous independent faults or failures is generally assumed to be so small that such events do not need to be considered. When regulatory authorities approve these criteria and hence allow operators to charge the users of the system for their implementation, they are, in essence, purchasing a certain level of reliability on behalf of consumers.

6.1.4 Procuring Reliability Resources

Operators do not simply react to sudden events and slower changes in the state of the system. Instead, they spend a considerable amount of time and effort pondering what dangerous contingencies might occur[1] and what needs to be done to avoid instabilities. The process of operating a power system must therefore start significantly ahead of real time to ensure that the preventive measures and the necessary corrective resources are in place. In a vertically integrated environment, all of the resources required to implement these actions are under the control of the utility. On the other hand, in a competitive environment, some of these resources belong to market participants, mostly on the supply side but increasingly also on the demand side. These resources are therefore no longer automatically and freely available to the system operator and have to be purchased on a commercial basis.

Procuring reliability resources well in advance of real time has two advantages. First, their cost is likely to be lower because more options are typically available. Second, it reassures system operators that they will have on hand what they need to deal with credible contingencies. On the other hand, uncertainty about the state of the system decreases as we approach real time. The amount of resources procured can then be tailored more closely to the actual needs. This trade-off is at the heart of the continuing debate about how best to integrate electrical energy markets with the need to maintain reliability. When electricity markets were first introduced, energy trading closed at least several hours ahead of real time because operators felt that they needed this lead time to decide what they needed to do to ensure the operational reliability of the system. Reliability resources were procured based on either long-term contracts for ancillary services[2] or on what generators offered at the closure of the energy market. As operators have gained confidence in their ability to maintain reliability in a competitive environment, their reliance on long-term contracts for some ancillary services has decreased and been replaced by balancing markets with trading periods and lead times as short as 5 min. Closing markets nearer to real time is particularly beneficial in systems with a significant proportion of stochastic renewable generation because the need to keep resources in reserve to deal with the uncertainty associated with these generators is reduced.

6.1.5 Outline of the Chapter

In the rest of this chapter, we will first analyze the different types of perturbations that affect power systems and the impact that these perturbations have on operational reliability. We will also describe the types of resources that are needed to deal with these perturbations. We will then discuss how to determine the required amount of each resource and explore the mechanisms that can be set up to procure them. Finally, we will take the perspective of a provider of reliability resources and investigate how they can be combined with transactions for electrical energy to maximize operating profits.

1 An experienced operator once described his job to us as "being paid to worry."

2 *Ancillary services* were given this name because they are auxiliary to the trading in the main commodity, i.e. electrical energy, and because they represent the potential to deliver energy (or another resource) upon request rather than the actual delivery of this resource.

6.2 Operational Issues

System operators must consider not only the current state of the system but also all that might happen over the next few hours. The current state of the system should be such that the system should be able to continue operating indefinitely if external conditions remain unchanged. This implies that no component should be operated outside its continuous rating. For example, as we discussed in the previous chapter, no transmission line should be loaded to such an extent that the temperature rise in the conductors due to ohmic losses causes the line to sag low enough to create a fault.

Assuming that external conditions do not change is unfortunately a very optimistic assumption. The demand for electricity changes continuously as a function of human activity, while the power produced by wind farms and solar generation depends on the weather and the time of day. Furthermore, in a system that consists of tens of thousands of components, the failure of a single component is not a rare event. This is particularly true if some of these components (such as transmission lines) are exposed to inclement weather conditions and others (such as generating plants) are subjected to repeated changes in operating temperature.

Example 6.1

Let us consider the two-generator system shown in Figure 6.1. If both generating units have a capacity of 100 MW, the maximum load that can be handled securely by this system is typically taken to be 100 MW and not 200 MW as one might have expected. The spare capacity is indeed needed when one of the generating units suddenly fails. A system with more generating units would obviously be able to operate with a much smaller capacity margin.

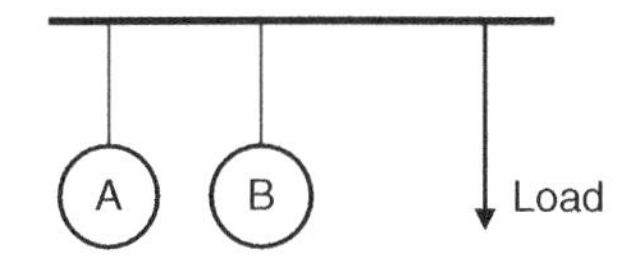

Figure 6.1 Two-generator power system illustrating the limitations that reliability places on operation.

We will first consider the operational reliability issues caused by a global imbalance between load and generation. Then, we will discuss the operational reliability problems that arise from the transmission network. This distinction is far from perfect, and on several occasions, we will have to highlight interactions between balancing and network issues.

6.2.1 Balancing Issues

When discussing the global balance between load and generation, we assume that all the loads and generators are connected to the same bus. In an interconnected system, this bus is also the terminal of all the tie lines with other regions or countries. At this level of abstraction, the only system variables are the generation, load, frequency, and net interchange with other systems. As long as the production is equal to the consumption, the frequency and the interchanges remain constant. However, the balance between load and generation is constantly perturbed by fluctuations in the load, by variations in the output of renewable generators, by deviations in the production of controllable generators, and occasionally by the sudden outage of a generating unit or of an interconnection. In an isolated system, a surplus of generation boosts the frequency while a deficit depresses it. The rate at which the frequency changes because of an imbalance is determined by the inertia of all the generators and rotating loads directly coupled to the system. Since larger wind turbines and photovoltaic generation are connected to the system through a power electronics interface, they do not contribute to the inertia of the system. As the proportion of these types of generators increases, the inertia of the system will therefore decrease, leading to concerns about frequency stability. See the papers by Wang et al. (2016) and O'Sullivan et al. (2014) for a detailed discussion of this issue.

Large frequency deviations can lead to a system collapse. Generating units are indeed designed to operate within a relatively narrow range of frequencies. If the frequency drops too low, protection devices disconnect generating units from the rest of the system to protect them from damage. Such disconnections exacerbate the imbalance between generation and load, causing a further drop in frequency and additional disconnections. There have also been instances where a system collapsed because protection relays tripped generating units that were exceeding their safe operating speed. The loss of these units caused a deficit of generation that led to a frequency collapse. Frequency deviations are less of a problem in an interconnected system because the total inertia increases with the size of the system. However, a large and sudden regional imbalance between load and generation in an interconnected system can overload the tie lines and trigger their disconnection, which might affect the stability of the neighboring networks. System operators must therefore always have available the resources they need to correct large imbalances as soon as they arise.

Minor imbalances between load and generation do not represent an immediate operational reliability issue because the resulting frequency deviations and inadvertent interchanges are small. However, these imbalances should be eliminated quickly because they weaken the system. A system that is operating below its nominal frequency or where the tie lines are inadvertently overloaded is indeed less able to withstand a possible further major incident.

The following example illustrates the imbalances that might be observed in an isolated power system.

Example 6.2

Figure 6.2a shows the net load profile observed in the Bordurian power system over five trading periods. The net load is defined as the power actually consumed minus the power produced from stochastic renewable energy sources such as wind and solar. This profile exhibits random fluctuations superimposed on a slower trend. The random fluctuations combine the ebb and flow in the load with the vagaries of renewable generation. Similarly, the trend in the net load amalgamates the cyclical variations in the demand driven by human activities with the changes in renewable generation driven by the weather and the time of day.

Like all other electricity markets, the Bordurian market makes the simplifying assumption that the demand is constant over each trading period. Figure 6.2a displays a staircase function that illustrates the energy that was traded on the market for each period by

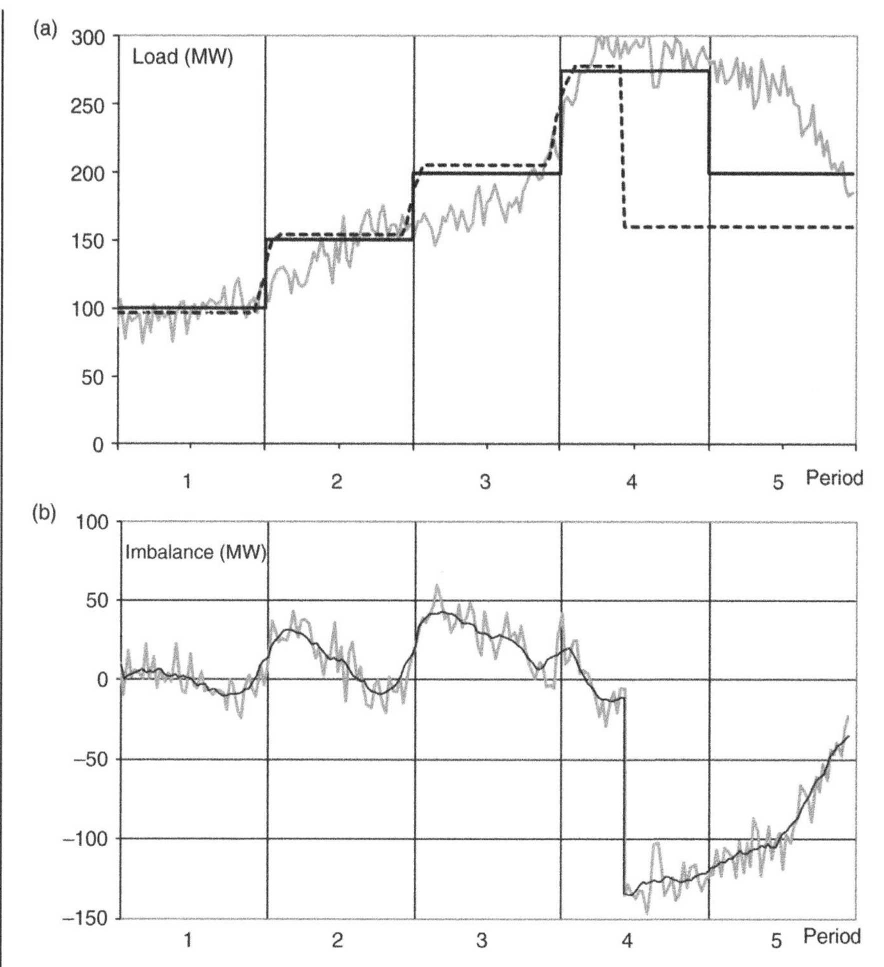

Figure 6.2 (a) Typical profiles of net load and generation over five market periods. (b) Imbalances resulting from the differences between these profiles.

conventional generators. This staircase function differs from the net load in two ways. First, it obviously cannot track the random changes in net load within each period. Second, if the market were able to predict the net load with perfect accuracy, the energy traded during each period would be equal to the integral of the instantaneous net load over the period. In practice, because markets operate on the basis of forecasts that are always inaccurate, the amount traded on the energy market is not an exact average of the net load. The staircase function also represents the expected total output of conventional generators. However, even conventional generators are not able to meet their commitments with perfect accuracy. The dashed line in Figure 6.2a represents the actual output of these conventional generating units. In addition to some minor discrepancies during each period, there are differences at the transitions between periods. Because of limits on the rate at which units can adjust their output, conventional generators are unable to achieve the idealized production profile that results from market trading. Figure 6.2b shows the difference between the actual production of the units scheduled through the energy market and the load. While these imbalances are relatively small during the first period, the net load is generally smaller than the scheduled conventional generation during periods 2 and 3. This could be due to an underforecast of renewable generation. A much more severe imbalance arises suddenly in the middle of period 4 because of the sudden outage of a large generating unit. The shape of the curve in Figure 6.2b suggests that imbalances have three main components with different time signatures: rapid random fluctuations, slower but larger deviations, and occasional large deficits. A smoothed version of the profile of imbalances has been added to the figure to highlight slower deviations.

6.2.1.1 Balancing Resources

As Example 6.2 shows, various phenomena create imbalances with different "time signatures." System operators need different resources to handle each type of imbalance. Note that the names used to describe these various resources differ from market to market. See Rebours et al. (2007a) and Rebours et al. (2007b) for a survey of the various types of resources.

Regulation resources are intended to handle rapid fluctuations in loads and small unintended changes in conventional and renewable generation. These resources help maintain the frequency of the system at or close to its nominal value and reduce unintended interchanges with other power systems. Generating units that can increase or decrease their output quickly will typically provide this service. These units must be connected to the grid and must be equipped with a governor. They usually operate under automatic generation control.

Generating units providing a *load following* resource handle slower fluctuations, in particular the intraperiod changes that the energy market does not take into account. These units obviously must be connected to the system and should have the ability to respond to these changes in load.

Regulation and load following require more or less continuous action from the generators providing these resources. However, regulation actions are relatively small and load following actions are fairly predictable. By keeping the imbalance close to zero and the frequency close to its nominal value, these resources provide preventive operational reliability measures. On the other hand, resources used for *contingency reserve* are designed to handle large and unpredictable power deficits that could threaten the stability of the system. These resources thus provide corrective actions. Obtaining reserve resources, however, can be considered a form of preventive operational reliability action.

Reserves are usually classified in two categories. Units that provide *spinning reserve* must start responding immediately to a change in frequency and the full amount of reserve capacity that they are supposed to contribute must be available very quickly. On the other hand, generating units providing *supplemental reserve services* do not have to start responding immediately. Depending on local rules, some forms of supplemental reserve services may be provided by units that are not synchronized to the grid but can be brought on-line quickly. In some cases, customers who agree to have their load disconnected during emergencies can also provide reserve services. Besides the speed and rate of response, the definition of reserve services must also specify the amount of time during which generating units must be able to sustain this response. All these parameters vary considerably from system to system depending on their size and the applicable reliability criteria. For example, preventing unacceptable frequency deviations in a small isolated system requires faster acting reserves than in a large interconnected system.

It would be nice if we could draw a clear distinction between balancing resources that are purchased as ancillary services and balancing energy that is traded on the spot energy market. Unfortunately, the wide variety of electricity markets designs makes an unambiguous classification impossible. In general, if the time that elapses between the closure of the open market and real time is short, the system operator is able to buy a substantial portion of its balancing needs on the spot energy market. On the other hand, if the market operates on a day-ahead basis, a complex mechanism is likely to be needed for the procurement of balancing services.

The rate at which the output of a generating unit can be adjusted is obviously the most important factor in determining its ability to provide balancing services. In some cases, however, its location may affect its ability to provide these services. A generating plant that is connected to the "main" part of the system through a transmission corridor that is often congested would not be a suitable candidate to provide these services. Its ability to increase its output could indeed be limited by these transmission constraints.

Example 6.3

Figure 6.3 illustrates the frequency response of a power system following a major generation outage and the response of the reserve services. This example is based on an actual incident. On August 15, 1995 at 12:25:30, 1220 MW of generation was suddenly disconnected from the power system of Great Britain. This system had at the time a total installed capacity of about 65 GW but does not have AC interconnections with any other system. It is therefore prone to significant frequency fluctuations. The two main categories of reserve services that have been defined for the operation of this system reflect this characteristic. *Primary response* must be fully available within 10 s and sustainable for a further 20 s. *Secondary response* must be fully available within 30 s of the incident and must be sustainable for a further 30 min. As can be seen from the figure, primary response succeeded in arresting the frequency drop before it reached the statutory limit of 49.5 Hz. Secondary response then helped bring the system frequency closer to its nominal value. Gas turbines were started at 12:29:20 to restore the frequency to its nominal value.

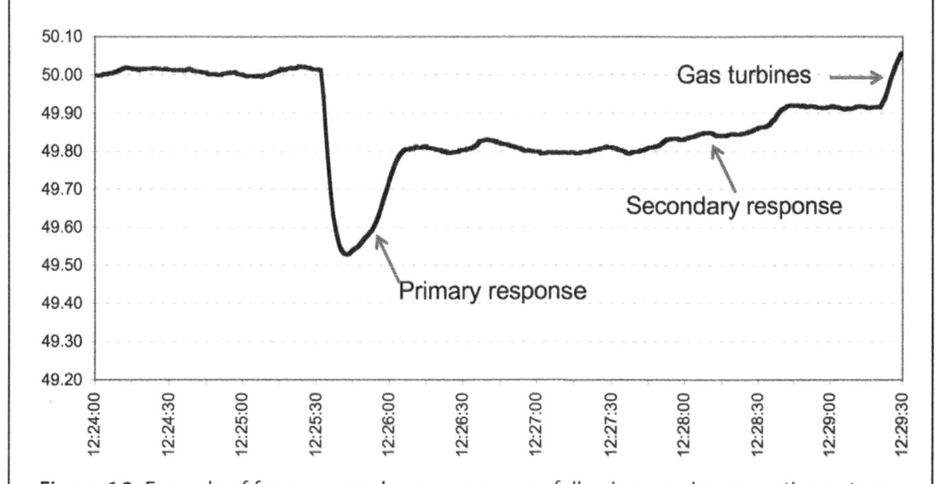

Figure 6.3 Example of frequency and reserve response following a major generation outage.

6.2.1.2 Effect of Generation from Stochastic Renewable Sources

As the proportion of generation from stochastic renewable sources such as wind and solar increases, operators must procure additional reserves to retain the ability to compensate for unpredicted changes in the renewable production, particularly during periods of high renewable production. The range of uncertainty on the wind generation increases significantly with the lead time of the forecast. For example, Table 6.2 shows the maximum range of change that should be expected if 26 GW of wind generation were installed in Great Britain. In particular, between 7200 and 9650 MW of additional reserve must be procured if this capacity is to be provided by combined-cycle gas turbines (CCGT) because these units require approximately 4 h to synchronize to the system.

Table 6.2 Estimates of wind forecast error and reserve requirements for 26 GW of installed wind generation capacity in Great Britain.

Lead time (h)	Maximum change observed (MW)
0.5	1090–1450
1	2100–2800
2	4050–5400
4	7200–9650

6.2.2 Network Issues

6.2.2.1 Limits on Power Transfers

In all but the smallest power systems, consumers and producers are connected by a network. As loads and generations vary, the flows in the branches and the voltages at the nodes of the network fluctuate. The system operator must therefore consider the effect of these changes on operational reliability. Besides continuously checking that no equipment is being operated outside its safe operating range, the operator periodically performs a computerized contingency analysis. This analysis takes as its starting point the current state of the power system and checks that no credible contingency would destabilize the system. Depending on the nature of the power system, such a destabilization could take several forms:

- Following the outage of a branch, the power that was carried by that branch reroutes itself through the network. In this postcontingency state, one or more other branches may be loaded beyond their thermal capacity. Unless the system operator can correct this situation quickly, overloaded lines will sag, cause a fault, and be disconnected. Similarly, overloaded transformers may be taken out of service to prevent heat-related damage. These additional outages further weaken the network and may lead to a system collapse as more and more branches become overloaded.
- The sudden outage of a generating unit or of a reactive compensation device can deprive the system of essential reactive support. Similarly, the outage of an important

branch can increase the reactive losses in the network beyond what the system can provide. The voltage in a region or even in the entire network may then collapse.

- A fault on a heavily loaded line may cause the rotor angle of some generators to increase so much that a portion of the network dynamically separates from the rest, causing one or both regions to collapse because generation and load are no longer balanced.

When the state of the system is such that a credible contingency would trigger any of these types of instabilities, operators must act by taking preventive actions.

Implementing some types of preventive actions involves a cost that is either very small or negligible. For example, operators can increase the margin to voltage collapse by adjusting transformer taps and the voltage set point of generators or by switching in or out banks of capacitors and reactors. They can also reduce the potential for postcontingency overloads by rerouting active power flows using phase-shifting transformers. While these low-cost preventive measures can be very effective, there is a limit to the contribution they can make to the operational reliability of the system. As the loading of the system increases, there comes a point where reliability can only be maintained by placing restrictions on the flow of active power on some branches. These restrictions constrain the amount of power that can be produced by generating units located upstream from the critical branches and prevent them from producing all the energy that they could sell on the market. As we saw in the previous chapter, limiting active power flows affects the market prices and often carries a very significant cost.

Example 6.4

Let us consider the two-bus power system shown in Figure 6.4. We want to quantify the amount of power that the generating unit located at bus A is able to sell to the load connected at bus B. Limitations imposed by the thermal capacity of the lines are the easiest ones to calculate. If each line is designed to be able to carry 200 MW continuously without overheating, the maximum amount of power that the load at bus B can obtain from unit A is limited to 200 MW. The spare 200 MW of transmission capacity must be kept in reserve in case a fault occurs and one of the lines must be disconnected. This very substantial reliability margin can be reduced if we consider the possibility of postcontingency corrective action. Let us suppose that either line can withstand a 10% overload for 20 min without sagging and causing another fault and without damage to the conductors. If the generating unit at bus B can guarantee that it will increase its output by 20 MW in 20 min if necessary, the maximum amount of power that can be transmitted from bus A to bus B can be raised from 200 to 220 MW.

In order to calculate the effect of transient stability on the maximum power that can be transmitted from A to B, we need more information about the system. To avoid

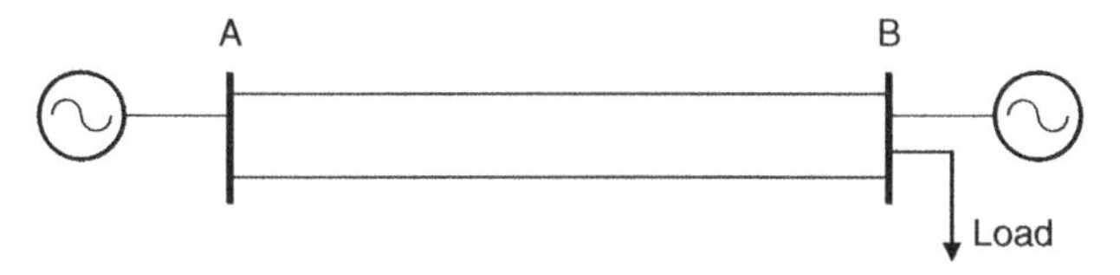

Figure 6.4 Two-bus power system used to illustrate the limitations that maintaining the reliability of the transmission network places on the operation of the system.

unnecessary complications, we will assume that bus B behaves like an infinite bus and that the generator at bus A has an inertia constant H of 2 s and can be modeled as a constant voltage behind a transient reactance X' of 0.9 p.u. The reactance of each line is equal to 0.3 p.u. The voltage at both buses is kept constant at 1.0 p.u. The worse contingency in this system is a fault on one of the lines close to bus A. We will assume that such a fault would be cleared in 100 ms by tripping of the faulted line. Using a transient stability program, it is easy to compute that, under these conditions, the maximum power that can be transmitted from A to B without endangering the transient stability of the system is 108 MW.

Let us now consider how voltage instability might limit the power transfer from A to B. Again, to avoid unnecessary complications, we will adopt a very simple system model and assume that the point of voltage collapse is reached when the power flow stops converging. This assumption gives us a good first approximation of the maximum flow that the system can handle. More complex analysis techniques have been developed if a more accurate measure of voltage stability is required.

The amount of reactive support available at bus B has a strong influence on the transfer capacity. Let us first consider the case where no voltage support is available because the generator at bus B has reached its upper MVAr limit. Using a power flow program, we can calculate that, when both lines are in service, 198 MW can be transmitted from A to B before the voltage at B drops below the usual 0.95 p.u. limit. However, if the power transfer exceeds 166 MW and one of the lines is disconnected, the voltage collapses. On the other hand, if 25 MVAr of reactive support is available at bus B, the power transfer can be increased up to 190 MW before a line outage would cause a voltage collapse.

In the previous example, transient stability places the most severe restriction on the maximum power transfer. In actual power systems, the limiting factor depends on their structure: networks with long transmission lines tend to be constrained by stability issues, while more densely meshed networks are limited by thermal or voltage considerations.

6.2.2.2 Voltage Control and Reactive Support

The previous example also shows how the operator can use reactive power resources to increase the amount of power that can be transferred from one part of the network to another. Some of these reactive resources and voltage control devices (e.g. mechanically switched capacitors and reactors, static VAR compensators, tap-changing transformers) are typically under the direct control of the network operator and can be used at will. Generating units, however, provide the best way to control voltage. *Voltage control services* therefore need to be defined to specify the conditions under which the system operator can make use of the resources owned by generating companies. Generators providing this service produce or absorb reactive power in conjunction with their active power production. It is also conceivable that businesses might be setup for the sole purpose of selling reactive support or voltage control.

The definition of a voltage control service must consider the operation of the system under normal conditions, but also the possibility of unpredictable outages. Under normal operating conditions, operators use reactive power resources to maintain the voltage at all buses within a relatively narrow range around the nominal voltage.

Typically, this range is:

$$0.95\,\text{p.u.} \leq V \leq 1.05\ \text{p.u.} \tag{6.1}$$

Keeping transmission voltages within this range is partially justified by the need to facilitate voltage regulation in the distribution network. It also makes the operation of the transmission system more secure. Maintaining the voltage at or below the upper limit reduces the likelihood of insulation failures. The lower limit is more arbitrary. In general, keeping voltages high under normal conditions makes it more likely that the system would avoid a voltage collapse if an unpredictable outage does occur. A good voltage profile, however, does not guarantee voltage stability. The outage of a heavily loaded transmission line increases the reactive losses in the remaining lines in a nonlinear manner. If these losses cannot be supplied, the voltage collapses. The amount of reactive power needed following an outage is therefore much larger than what is required during normal operation. Voltage control services must therefore be defined not only in terms of the ability to regulate the voltage during normal operation but also to provide reactive power during emergencies. The voltage control service is, in fact, often called *reactive support service*.

Example 6.5

Using a power flow program, we can explore the nature of the voltage control or reactive support service on a two-bus example similar to the one shown in Figure 6.4. Each of the transmission lines in this system is modeled using the π-equivalent circuit shown in Figure 6.5. The load at bus B has a power factor of unity. Let us first examine how the operator might control the voltage at bus B using the reactive capability of the generator at this bus. We assume that the voltage at bus A is kept constant at its nominal value.

Figure 6.6 shows that when the amount of power transferred from bus A to bus B is small, the reactive power produced by the equivalent shunt capacitances of the lines exceeds the reactive power consumed in the equivalent series reactances. The generator at bus B must absorb this excess to keep the voltage at the upper limit of the acceptable range. When the amount of power transferred is between 100 and 145 MW, the reactive power balance is such that the voltage stays naturally within the acceptable limits. A reactive injection is not needed at bus B for these conditions. When the power transfer exceeds 145 MW, the reactive losses in the lines must be compensated by a reactive injection at bus B to keep the voltage from dropping below its lower limit.

If the generator connected to bus B is disconnected or asks too high a price for regulating the voltage, the system operator could attempt to control it by adjusting the

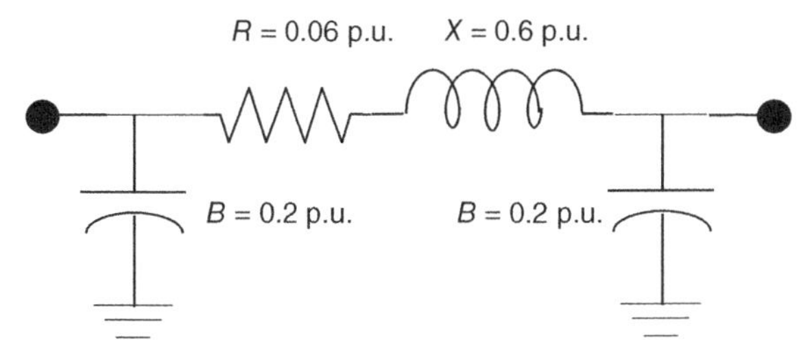

Figure 6.5 π-Model of the transmission lines of Example 6.5.

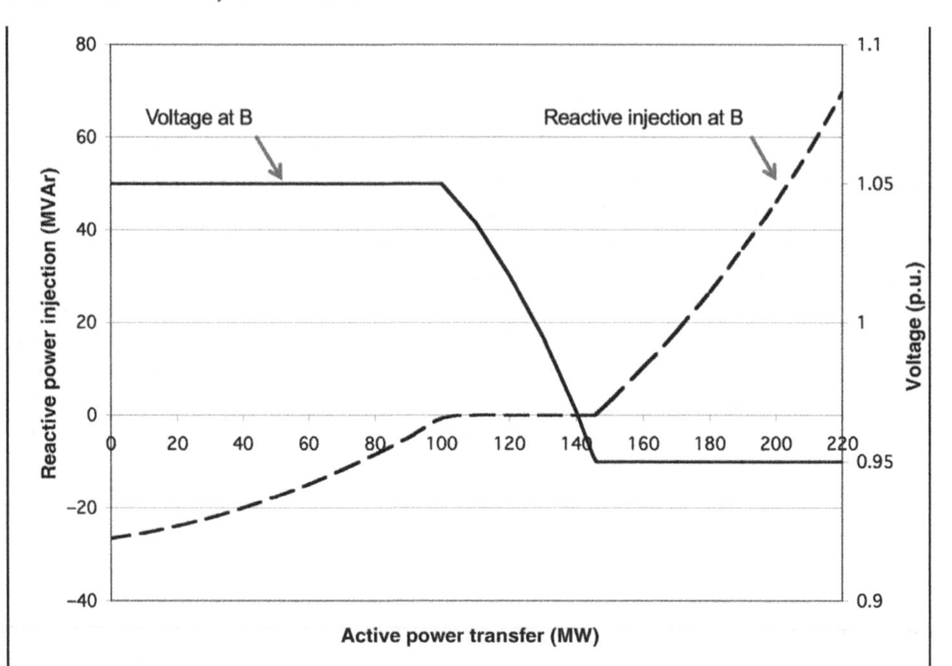

Figure 6.6 Variation in the reactive power injection and the voltage at bus B of the two-bus system of Example 6.5. The voltage (solid line) and the reactive injection needed to keep this voltage within normal limits (dashed line) are plotted as a function of the power transferred from bus A to bus B.

Table 6.3 Limits on the control of the voltage at bus B using the voltage set-point of the generator at bus A.

Power transfer (MW)	V_B (p.u.)	V_A (p.u.)	Q_A (MVAr)
49.0	1.05	0.95	−68.3
172.5	0.95	1.05	21.7

voltage set-point of the generator at bus A. When the amount of power transferred is small, the voltage at bus B is high. To keep it below its upper limit, the voltage set-point of the generator at bus A must be lowered. This implies that reactive power must be absorbed by this generator. Table 6.3 shows that when 49 MW is transferred, the voltage at B is at its upper limit and the voltage at A is at its lower limit. A lower power transfer could therefore not be accommodated. On the other hand, when the power transfer is high, the voltage set-point of generator A must be increased to keep the voltage at B above its lower limit. When this power transfer reaches 172.5 MW, the voltage at A is at its upper limit and the voltage at B is at its lower limit. A power transfer smaller than 49.0 MW or larger than 172.5 would therefore cause a violation of a voltage limit at either bus A or bus B. Further reactive power injections at bus A are pointless outside of this range of

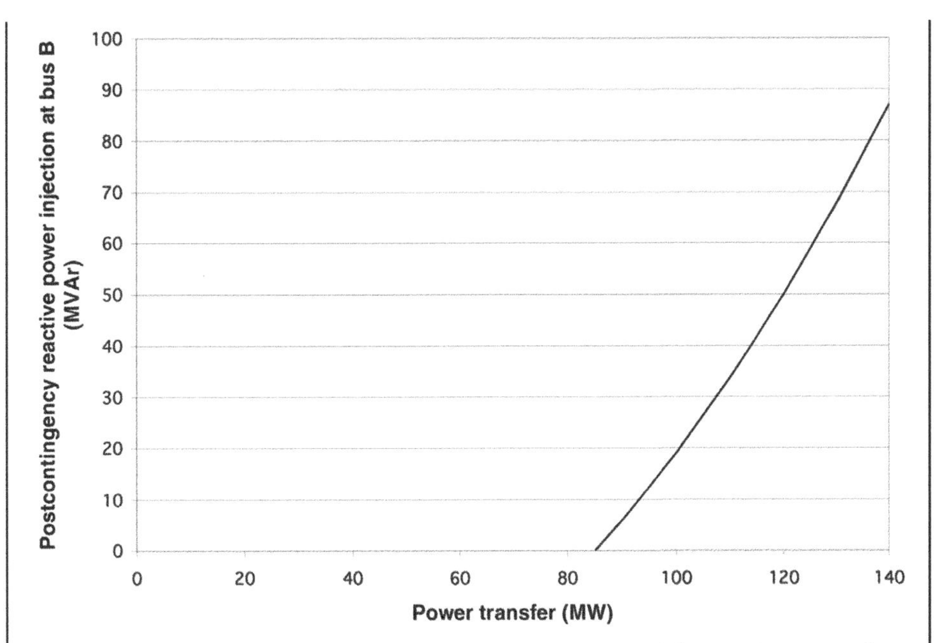

Figure 6.7 Postcontingency reactive support requirement at bus B following the outage of one of the lines connecting buses A and B.

power transfers. We can therefore conclude that local control of the voltage is much more effective than remote control, even under normal operating conditions.

As already discussed earlier, the true value of reactive support services does not reside in the actual production of VARs but instead in the ability to supply reactive power and prevent a voltage collapse following an outage. A power flow program can provide a rough estimate of the amount of reactive power that must be injected after an outage to prevent a voltage collapse. A more precise calculation of the need for reactive power reserves requires the consideration of dynamic effects. Figure 6.7 shows how much reactive power must be injected at bus B to prevent a voltage collapse following the outage of one of the two lines of our two-bus system. In the precontingency state, the voltage at bus A is kept at its nominal value by the generator connected to that bus. This graph shows that the system can withstand a line outage without reactive support at B when the power transfer is smaller than 85 MW. However, the postcontingency reactive support requirement increases rapidly when the power transfer exceeds this value.

Figure 6.8 illustrates the pre- and postcontingency reactive power balances for the case where 130 MW is transferred from A to B. The generator at B maintains the voltage at its nominal value before and after the outage. Under precontingency conditions, the lines produce about 25 MVAr that must be absorbed by the generator at bus A. The active power losses are about 6 MW. After the contingency, both generators must inject reactive power in the remaining line to prevent a voltage collapse. Instead of producing reactive power, the line now consumes 107 MVAr. On the other hand, the active power losses increase only to 15 MW.

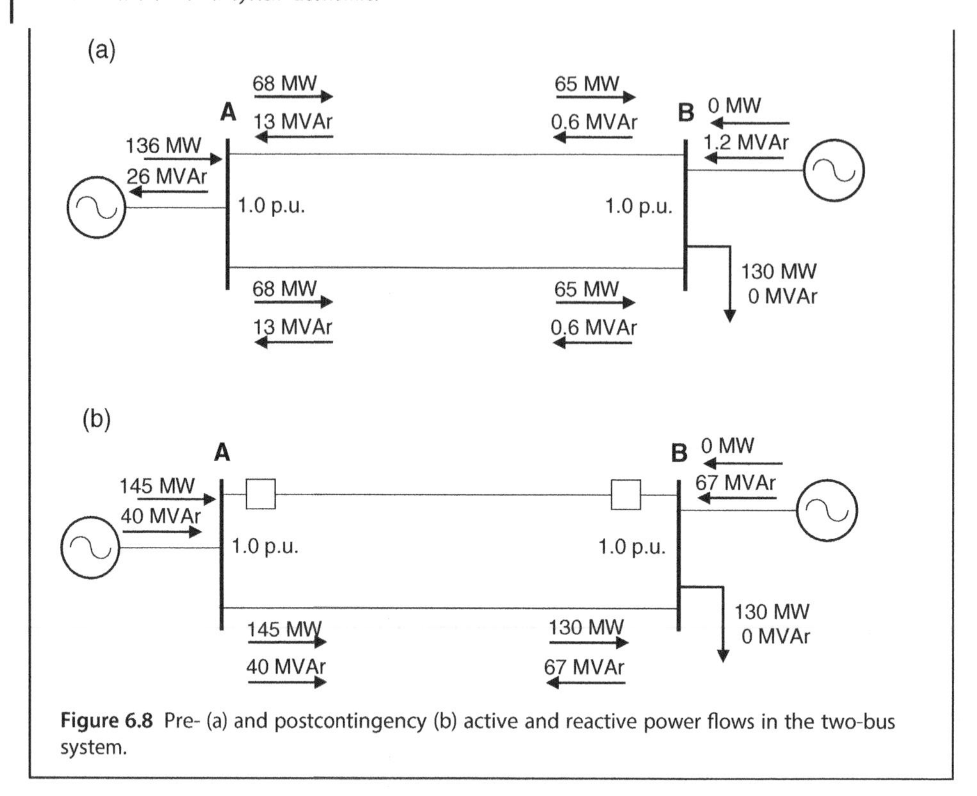

Figure 6.8 Pre- (a) and postcontingency (b) active and reactive power flows in the two-bus system.

6.2.2.3 Stability Services

Some system operators may also need to obtain other network stability services from generators. For example, *intertrip schemes* can mitigate transient stability problems. These schemes have no effect on the current state of the power system but, in the event of a fault, they automatically disconnect some generation and/or some load to maintain the stability of the system. Similarly, *power system stabilizers* make minute adjustments to the output of generators to dampen oscillations that might develop in the network. The action of these stabilizers increases the amount of power that can be transmitted.

6.2.3 System Restoration

Despite the best efforts of the system operator, a disturbance occasionally spirals out of control and the entire power system collapses. It is then the responsibility of the system operator to restore the system to a normal operating state as soon as possible. However, restarting large thermal generating plants requires a significant amount of electric power that is not available if the entire system has collapsed. Fortunately, some types of generators (e.g. hydro plants and small diesel generators) are able to restart either manually or using energy stored in batteries. The system operator must ensure that enough of these restoration resources are available to guarantee a prompt restoration of service at any time. This ancillary service is usually called *black-start capability*.

6.2.4 Market Models Vs Operational Models

An accurate determination of the limits that the transmission network imposes on the amount of power that various generators can produce requires computations using sophisticated system models that account for the current or expected operating conditions of the system. Such models are far too complex to be incorporated in the locational marginal price calculations described in the previous chapter. Instead, all stability limits are translated into approximate proxy limits on line flows that can be handled efficiently within the linear programming framework of Section 5.3.4.4. Such linear approximations are important because they ensure that market clearing programs terminate successfully in a reasonable amount of time. However, they are not sufficiently accurate to guarantee the operational reliability of the system. Operators therefore perform, in parallel with the market clearing software, a separate suite of analyses using more detailed and accurate power system models. Based on the results produced by these models, they modify the generation dispatch stemming from the market to make it compatible with the reliability criteria with which they have to comply.

6.3 Obtaining Reliability Resources

In the previous section, we saw that the system operator needs some resources to maintain the operational reliability of the system and that some of these resources must be obtained from other industry participants in the form of ancillary services. At this point, we need to examine the two mechanisms that can be used to ensure that the system operator has at its disposal the resources that are required. The first approach consists in making the provision of some resources compulsory. The second entails the creation of markets for different resources. As we will see, both approaches have advantages and disadvantages. The choice of one mechanism over the other is influenced not only by the type of resource but also by the nature of the power system and historical circumstances.

6.3.1 Compulsory Provision

In this approach, as a condition for being allowed to connect to the power system, a category of industry participants is required to provide some resources. For example, connection rules may require all generating units:

- To be equipped with a governor with a 4% droop coefficient. This requirement ensures that all units contribute equally to frequency regulation.
- To be capable of operating at a power factor ranging from 0.85 lead to 0.9 lag and be equipped with an automatic voltage regulator. This forces all units to participate in voltage regulation and contribute to voltage stability.

This approach represents the minimum deviation from the practice of vertically integrated utilities. It also guarantees that sufficient resources will be available to maintain the operational reliability of the system. While compulsion is apparently simple, it is not necessarily good economic policy and presents some implementation difficulties:

- These mandates may cause unnecessary investments and produce more resources than what is actually needed. For example, not all generating units need to take part in frequency control to maintain the stability of the system. Similarly, not all generating units need to be equipped with a power system stabilizer to dampen system oscillations.
- This approach does not leave room for technological or commercial innovation. New, more efficient ways of providing a service are unlikely to be developed by industry participants or sought by the system operator if traditional providers are compelled to offer this service.
- Compulsion tends to be unpopular among providers because they feel that they are forced to supply a service that adds to their costs without being remunerated. For example, generating companies complain that producing reactive power increases the losses in the synchronous machine and sometimes reduces the amount of active power that they are able to produce and sell.
- Some participants may be unable to provide some services or may be unable to provide them cost-effectively. Nuclear units, for example, are unable to provide services that demand rapid changes in active power output. Highly efficient conventional units or renewable generators should not be forced to operate at part-load so they can provide reserve. It is considerably cheaper to determine centrally how much reserve is needed and to schedule a few marginal or extramarginal units to provide this reserve. Compulsion is therefore not suitable for all services and even for those services where it seems appropriate, some participants may need exemptions. Such exemptions may be seen as distorting competition.

6.3.2 Market for Reliability Resources

Considering the economic disadvantages and the practical difficulties of compelling participants to provide reliability resources, it is often desirable to set up a market mechanism for the procurement of at least some resources. The preferred form of this mechanism depends on the nature of the service. Long-term contracts are suitable for services where the amount needed does not change or changes very little over time and for services where the availability is determined mostly by equipment characteristics. Black-start capability, intertrip schemes, power system stabilizers, and frequency regulation are typically procured under long-term contracts. On the other hand, a spot market is more appropriate for resources where the needs vary substantially over the course of the day and the offers change because of interactions with the energy market. For example, at least part of the necessary reserve is typically procured through a short-term market mechanism. However, the system operator will often seek to reduce the risk of not having enough reserve capacity or of having to pay too much for this capacity by arranging some long-term contracts for the provision of reserve. In a mature market, providers of reserve services may also prefer a mixture of short- and long-term contracts.

Markets provide a more flexible and hopefully more economically efficient procurement mechanism than compulsion. However, it is not clear at this point if a market-based approach can be applied to all resources. In some cases, the number of participants that are actually able to provide a certain resource is so small that the potential for abuse of market power precludes procurement on a competitive basis. For example, in some remote parts of a transmission network, there may be only one generating unit that can effectively support the voltage by providing reactive power in case of emergencies. A reactive power market would therefore need to be strictly regulated to avoid possible abuses.

6.3.3 System Balancing with a Significant Proportion of Variable Renewable Generation

As we discussed in Section 6.2.1.2, the integration of significant amounts of generation capacity from stochastic renewable sources requires an increase in the amount of reserve that the system operator must procure. Conventional generating units often do not have the flexibility needed to provide this additional reserve because of their large minimum stable generation, their limited ramp rate, and large minimum up- and downtimes. These limitations may force system operators to curtail generation from renewable sources. Furthermore, providing all this additional reserve from synchronized conventional power plants would negate at least some of the environmental benefits of renewable generation and could also significantly increase the cost of operating the system. The following example illustrates how procuring reserve from the demand side would reduce the need for reserve from conventional generation and the curtailments of renewable generation.

Example 6.6

Table 6.4 describes the characteristics of a system with a substantial amount of wind generation capacity as well as the demand, the expected wind generation, and the reserve that is required to ensure reliable operation. This reserve requirement is based on the data given in Section 6.2.1.2 and reflects the need to cover a potential drop in wind generation from 12 000 to 5500 MW within the 4 h lead time required to start CCGT units.

Table 6.5 summarizes the dispatch of the available generation resources for these conditions under three assumptions about the provision of reserve.

First, if we ignore the need to provide reserve, the 25 000 MW of load would be met by a combination of 12 000 MW of wind generation, 8400 MW of inflexible nuclear generation, and 4600 MW of generation from CCGT. Since each CCGT plant can provide up to 550 MW, 9 of these plants would need to be synchronized to the grid and each of them would

Table 6.4 System characteristics and snapshot of operating conditions.

Installed wind capacity (MW)	Must-run nuclear capacity (MW)	Each CCGT plant		Demand (MW)	Expected wind output (MW)	Reserve requirement (MW)
		P^{max} (MW)	P^{min} (MW)			
26 000	8 400	550	300	25 000	12 000	6 500

Table 6.5 Dispatch of the system of Example 6.6 considering three modes of reserve provision.

	No reserve	CCGT reserve only	CCGT + demand reserve
Nuclear (MW)	8 400	8 400	8 400
CCGT (MW)	4 600	7 800	5 400
Wind (MW)	12 000	12 000	12 000
Wind curtailment (MW)	0	3 200	800
CO_2 emission rate (g/kWh)	68.8	137.28	95.04
Total CO_2 emission (ton/30 min)	860	1 716	1 188

produce 511.1 MW on average. From an environmental perspective, no wind generation would need to be curtailed and the average CO_2 emission rate would be 68.8 g/kWh.[3]

Second, we consider the 6500 MW reserve requirement but assume that all this reserve will be provided by synchronized, part-loaded CCGT units. Because these units have a minimum stable generation (P^{min}) of 300 MW, each of them can only deliver 550 − 300 = 250 MW of reserve. Providing 6500 MW of reserve would therefore require the synchronization of 26 CCGT units operating at their minimum stable generation. These units would then also generate 26 × 300 = 7800 MW of power. Since nuclear generation is considered inflexible, balancing load and generation would require curtailing 3200 MW of wind generation, i.e. 26.7% of the available production capacity. The average CO_2 emission rate would be 137.3 g/kWh.

Third, we assess the value of obtaining 2000 MW of reserve from the demand side. In this case, only 18 CCGT units are required to provide the remaining 4500 MW of reserve. Operating at their minimum stable generation, these units produce 5400 MW of power. The remaining load is met by the 8400 MW of inflexible nuclear generation and 11 200 MW of wind generation. In this case, only 800 MW of wind power needs to be curtailed, i.e. 6.7% of the available production capacity. The average CO_2 emission rate would be 95.04 g/kWh, which is 30% less than when only synchronized conventional power plants provide reserve.

This example demonstrates the importance of alternative technologies (such as demand response or energy storage) in the provision of the system flexibility required to accommodate renewable generation.

6.3.4 Creating a Level-playing Field

Before the introduction of competition in the supply of electricity, generating units owned by vertically integrated utilities provided virtually all the resources needed to ensure reliability. Unfortunately, the definitions of ancillary services in many electricity

3 Based on the assumption that the emission rate of a CCGT plant is 368 g/kWh at full load and that the efficiency of such a plant decreases by 20% when running at P^{min}.

markets still reflect this practice. In a truly competitive environment, the system operator should have no obligation or incentive to favor generators in the procurement of reliability resources as long as other providers are able to deliver services of the same quality. Encouraging consumers and others (such as owners of energy storage systems) to offer reliability resources has several advantages. First, a larger number of providers should increase competition in these markets. Second, from a global economic perspective, participation by the demand side improves the utilization of the assets. For example, if interruptible loads or storage devices provide some of the reserve requirements, generation capacity does not have to be held in reserve. These generating units can then be used for producing electrical energy, which is what they were designed for. If the mix of generation technologies continues to evolve toward a combination of large inflexible units and renewable generation, resources for system control may have to come from the demand side. Finally, diversifying the provision of resources is likely to enhance the reliability of the system. The probability that a larger number of providers may simultaneously fail to deliver a critical service in time is indeed smaller than if this service is provided by a small number of large generators.

The demand side is probably most competitive in the provision of the different types of reserve services. Some consumers (for example, those who have large water pumping loads equipped with variable speed drives) might also be able to compete for the provision of regulation. The flexibility of pumped hydro plants makes them very competitive in the provision of regulation and reserve services. Battery energy storage systems are also increasingly deployed to provide these services. Being connected to the grid through a power electronics interface, they have the advantage of being able to respond extremely fast. On the other hand, their relatively small energy capacity limits their ability to respond to a generation deficit for a sustained period of time.

6.4 Buying Reliability Resources

We argued at the beginning of this chapter that the purpose of these resources is to maintain the operational reliability of the system in the face of unpredictable events. Because failure of one component can cascade through the network, operational reliability is a *system* concept that must be centrally managed. The system operator is thus responsible for purchasing operational reliability on behalf of all the users of the system. If we assume that a market mechanism has been adopted for the procurement of the necessary resources, operators have to pay the providers and then recover this cost from the users of the system. Since the amount of money involved is not negligible, these users are likely to scrutinize this purchasing process. They need to be convinced that the optimal amount of resources is purchased, that the right price is paid, and that each user pays its fair share of their cost.

6.4.1 Quantifying the Needs

Ideally, the level of operational reliability should be determined through a cost/benefit analysis. This analysis would set this level at the optimal point where the marginal cost of providing more reliability is equal to the marginal value of this reliability. While the marginal cost is relatively easy to calculate, the marginal value represents mostly the expected cost to consumers of load disconnections that are avoided through the provision of reliability. This cost is much harder to compute. Since performing a cost/benefit analysis in every case is not practical, reliability standards that approximate the optimal solution have been developed. These standards usually specify the contingencies that the system must be able to withstand. Sophisticated models and computational tools have been developed to help system operators manage the power system in accordance with these standards and to quantify the ancillary services that they need to achieve this goal. A discussion of these techniques is beyond the scope of this book. The interested reader is encouraged to consult Billinton and Allan (1996) for an exhaustive discussion of the techniques used for calculating reserve requirements. A method for determining and allocating the needs for reactive support is described by Pudjianto et al. (2002).

If the cost of running the system is simply passed on to the users, system operators may be tempted to purchase more reliability resources than is strictly needed. Having a bigger pool of resources to call upon in case of difficulties makes operating the system easier and less stressful. It is therefore desirable to develop an incentive scheme that encourages system operators not only to minimize the cost of purchasing reliability resources but also to limit the amount of resources purchased to what is truly necessary.

6.4.2 Co-optimization of Energy and Reserve in a Centralized Electricity Market

Setting the price for a reliability resource at the right level is not easy because the procurement of a particular resource often cannot be decoupled from the procurement of electrical energy or other related services. In the early years of competitive electricity markets, this issue was not fully understood. Energy and each type of reserve were traded in separate markets, which were cleared successively in a sequence determined by the speed of response of the resource. For example, the market for primary reserve would be cleared first, followed by the market for secondary reserve, and finally by the energy market. The idea was that resources that had not been successful in one market could then be offered in other markets where the performance requirements are not as demanding. Bids that were successful in one market would not be considered in the subsequent ones. Experience showed that this approach led to problems and it has since been abandoned. See Oren (2002) for more details on these problems.

There is now a wide consensus that energy and reserve should be offered in joint markets and that these markets should be cleared simultaneously to minimize the overall cost of providing electrical energy and reserve. This co-optimization is necessary because of the strong interaction between the supply of energy and the provision of reserve. To get a more intuitive understanding of this interaction, consider that to provide spinning reserve, generators must operate part-loaded. This mode of operation has several consequences:

- Part-loaded generators cannot sell as much energy as they might otherwise do.
- To meet the demand, other generators, which are generally more expensive, have to produce more energy.
- The efficiency of the generators that provide spinning reserve may be less than it would be if they were running at full load. These generators therefore may need to be paid more for the energy that they provide.

Meeting the reserve requirements therefore increases the price of electrical energy. In the rest of this section, we use simple examples to discuss how co-optimization in a centralized electricity market minimizes this additional cost while ensuring that no generator is disadvantaged when being asked to provide reserve rather than produce electrical energy. For a more detailed discussion of this topic, see Read et al. (1995).

Example 6.7

Let us consider a small electricity market where the demand varies between 300 and 720 MW. For the sake of simplicity, we assume that only one type of reserve is needed and that 250 MW of this reserve is required to avoid having to disconnect consumers for all loading conditions. Table 6.6 shows the relevant characteristics of the four generators that are connected to this system.

Table 6.6 Marginal cost, maximum output, and reserve capability of the generating units of Example 6.7.

Generating units	Marginal cost of energy ($/MWh)	P^{max} (MW)	R^{max} (MW)
1	2	250	0
2	17	230	160
3	20	240	190
4	28	250	0

These generators are assumed to have a constant marginal cost and are ranked in decreasing order of merit. While they have similar capacities, their ability to provide reserve is quite different. Units 1 and 4 cannot provide any reserve that meets the requirements set by the system operator. On the other hand, the amount of reserve that units 2 and 3 can provide is limited not only by their capacity but also by their speed of response. Figure 6.9 shows how much reserve they can provide as a function of their power output. For simplicity, we ignore all limitations and complications caused by the minimum stable generation of the units.

We assume that this market is centralized, that the generators' bids to produce electrical energy are equal to their marginal costs and that the market rules do not include separate bids for the provision of reserve. This last assumption is reasonable if the generators do not incur a direct cost when providing reserve. We will relax this assumption in the next example. To clear the market, the operator determines the dispatch that minimizes the cost of production (as measured by the bids) while respecting the operational constraints. Formally, this problem can be expressed as follows:

Find the power produced by each of the four generating units (P_1, P_2, P_3, and P_4) and the amount of reserve provided by these same units (R_1, R_2, R_3, and R_4) that minimize

$$2 \cdot P_1 + 17 \cdot P_2 + 20 \cdot P_3 + 28 \cdot P_4 \tag{6.2}$$

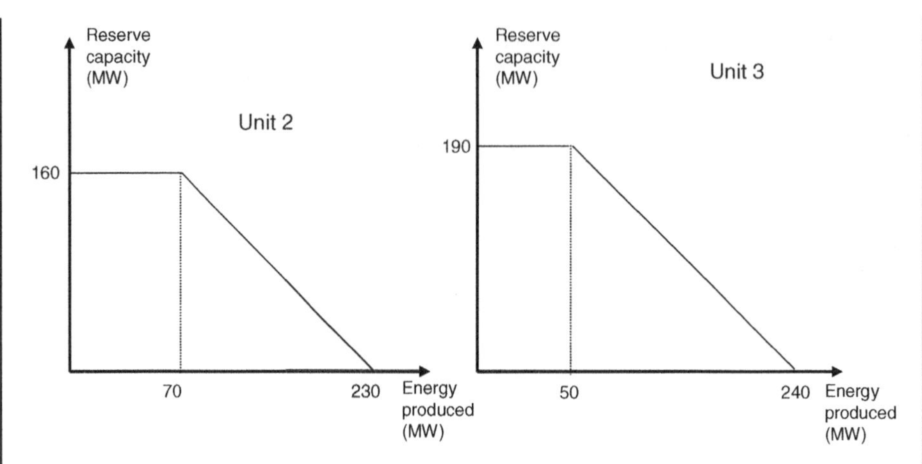

Figure 6.9 Amount of reserve that generating units 2 and 3 can provide as a function of the amount of electrical energy that they produce.

subject to the following constraints:

Balance between load and generation:

$$P_1 + P_2 + P_3 + P_4 = D \tag{6.3}$$

Minimum reserve requirement:

$$R_1 + R_2 + R_3 + R_4 \geq 250 \tag{6.4}$$

Limits on the output of the generating units:

$$\begin{aligned} 0 \leq P_1 &\leq 250 \\ 0 \leq P_2 &\leq 230 \\ 0 \leq P_3 &\leq 240 \\ 0 \leq P_4 &\leq 250 \end{aligned} \tag{6.5}$$

Limits on the reserve capabilities of the generating units:

$$\begin{aligned} &R_1 = 0 \\ &0 \leq R_2 \leq 160 \\ &0 \leq R_3 \leq 190 \\ &R_4 = 0 \end{aligned} \tag{6.6}$$

Limits on the capacity of the generating units:

$$\begin{aligned} P_1 + R_1 &\leq 250 \\ P_2 + R_2 &\leq 230 \\ P_3 + R_3 &\leq 240 \\ P_4 + R_4 &\leq 250 \end{aligned} \tag{6.7}$$

Any linear programming package can easily solve this problem. Table 6.7 shows the results for values of the demand D ranging from 300 to 720 MW. In addition to finding the optimal dispatch of energy and reserve, such a package calculates the dual variables or Lagrange multipliers associated with each constraint. The Lagrange multiplier associated

Table 6.7 Solution of the optimization problem of Example 6.7 for a range of values of the demand.

Demand (MW)	P_1 (MW)	R_1 (MW)	P_2 (MW)	R_2 (MW)	P_3 (MW)	R_3 (MW)	P_4 (MW)	R_4 (MW)
300–420	250	0	50–170	60	0	190	0	0
420–470	250	0	170	60	0–50	190	0	0
470–720	250	0	170	60	50	190	0–250	0

Each line in this table corresponds to a subrange where the output of only one of the generating units changes.

with the constraint on the production–demand balance gives the marginal cost of producing electrical energy. Similarly, the multiplier associated with the minimum reserve requirement constraint gives the marginal cost of providing reserve. In a centralized market, these marginal costs are deemed to be the market clearing prices for electrical energy and reserve, respectively.

For this simple example, we can easily check these solutions by hand and get a better understanding of the physical meaning of the price of reserve and its evolution as the demand changes.

Given that the minimum load we consider is 300 MW and that unit 1 has the lowest marginal operating cost and cannot provide reserve, we conclude immediately that this unit must produce its maximum output of 250 MW for all values of the demand. Since units 2 and 3 are the only ones that can provide reserve and since unit 2 can provide at most 160 MW, unit 3 has to provide at least 90 MW. Given that this unit has a capacity of 240 MW, its energy production must be less than 150 MW.

$$0 \leq P_3 \leq 150 \tag{6.8}$$

Similarly, since unit 3 can provide at most 190 MW of reserve, unit 2 has to provide at least 60 MW. It energy output is thus limited to 170 MW:

$$0 \leq P_2 \leq 170 \tag{6.9}$$

For a demand in the range between 300 and 420 MW, unit 2 is the marginal generator. It produces between 50 and 170 MW, i.e. the power that is not supplied by unit 1, which operates its maximum capacity of 250 MW. The marginal cost of unit 2 sets the price for energy at 17 $/MWh. In this range of demand, the inequality constraint for the minimum reserve requirement is not binding because units 2 and 3 provide more than enough reserve. The price of reserve is thus zero.

As the demand increases from 420 to 470 MW, the production of unit 2 is capped at 170 MW because it must provide at least 60 MW of reserve. Unit 3 becomes the marginal generator and progressively increases its output from 0 to 50 MW. The price of energy is thus set at the marginal cost of unit 3, which is 20 $/MWh. To determine the price of reserve, we must figure out where an additional megawatt of reserve would come from and how much it would cost. Figure 6.9 shows that over this range of output, unit 3 provides 190 MW of reserve, which is the maximum it is able to deliver under any circumstance. To get an additional megawatt of reserve beyond the basic 250 MW requirement, we would have to reduce the output of unit 2 by 1 MW. Instead of producing 170 MW, it would thus produce only 169 MW. To compensate for this reduction, we would have to increase the output of unit 3 by 1 MW. This extra megawatt from unit 3 would cost $20, while the reduction in the output of unit 2 would save $17. The net cost of getting an additional megawatt of reserve, and hence the price of reserve, is thus 20 − 17 = 3 $/MWh.

As the demand increases from 470 to 720 MW, the marginal producer is unit 4 and its production increases from 0 to 250 MW. The reserve constraint keeps the energy productions of units 2 and 3 at 170 and 50 MW, respectively. In this range, the price of energy is thus 28 $/MWh.

The marginal price for reserve increases to 11 $/MWh. This is because in order to make one more megawatt of reserve available, we need to reduce the energy output from unit 2 by 1 MW and increase the production of unit 4 by the same amount. The cost of this marginal redispatch sets the price of reserve at 28 − 17 = 11 $/MWh.

Figure 6.10 summarizes the prices of energy and reserve for the various ranges of demand.

Let us calculate the revenues collected by each generating unit, the costs that they incur and the profits that they achieve by producing energy and providing reserve. This analysis is not particularly interesting in the case of unit 1 because it always operates at full output and sells its energy at a price determined by the marginal cost of other generators. Since its own marginal cost is always lower than this price, it always makes a healthy operating profit.

In the range of demand between 300 and 420 MW, the market price for energy is equal to the bid price of generating unit 2. Since we assume that all generators bid at their marginal cost of production, this unit does not make an economic profit on the sale of energy. Given that the reserve price is zero, it does not make a profit on the provision of

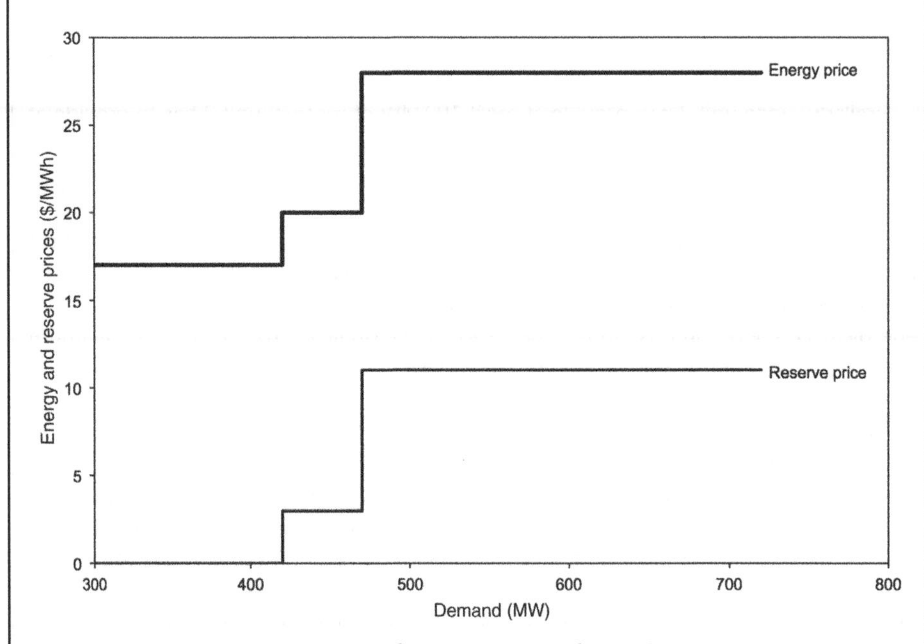

Figure 6.10 Energy and reserve prices for the conditions of Example 6.7.

reserve either. On the other hand, when the demand is in the range between 420 and 470 MW, even though unit 2 has a lower marginal cost than other units, its output is capped at 170 MW by the reserve requirement constraint. Unit 3 is then the marginal energy producer. The energy price jumps from 17 to 20 $/MWh, which means that unit 2 makes a 3 $/MWh profit on each megawatt-hour it produces. At first glance, one might think that the owner of unit 2 is unfairly treated because the reserve constraint prevents it from selling the additional 60 MWh of energy it could sell because it bid at a lower price than unit 3. Observe, however, that the price of reserve in this range of demand is 3 $/MWh and that unit 2 provides 60 MW of reserve. The revenue it collects for the reserve it provides is thus exactly equal to the opportunity cost of not selling energy. The owner of unit 2 is therefore indifferent to producing more electrical energy or providing reserve. In this same range of demand, unit 3 does not make an economic profit from the sale of energy because it is the marginal producer. On the other hand, it makes a profit of 3 $/MWh from the provision of reserve because the marginal provider of reserve is unit 2.

When the demand increases beyond 470 MW, unit 4 becomes the marginal producer and sets the energy price at 28 $/MWh. Unit 2 thus makes a profit of 11 $/MWh on each of the 170 MW that it is allowed to produce. Its owners do not mind the limitation that the reserve constraint puts on their unit's output because they also make an 11 $/MWh profit on every megawatt of reserve it provides. Unit 2 is still the marginal provider of reserve in this range of demand. On the other hand, unit 3 makes a profit of 8 $/MWh on its energy production and a profit of 3 $/MWh on the reserve it provides because it is marginal neither for energy nor for reserve.

Figure 6.11 summarizes the revenues that unit 2 derives from the energy and reserve markets as well as its costs and profits.

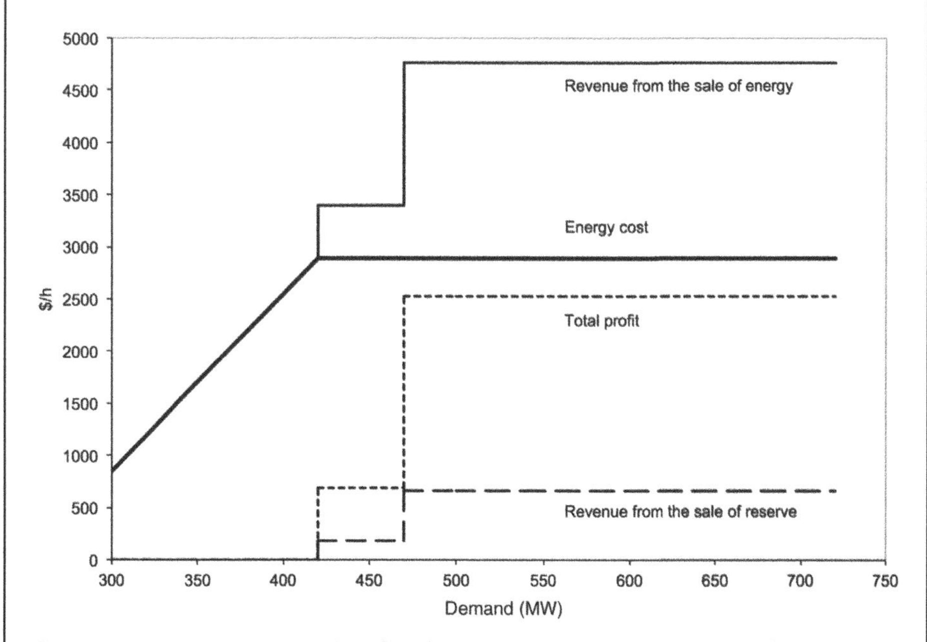

Figure 6.11 Revenues, costs, and profits of unit 2 of Example 6.7 for a range of demand.

Example 6.8

Let us assume that the rules of the market we considered in our previous example are changed to take into consideration the costs that generators must bear when they provide reserve. These costs may reflect the loss in efficiency of units that operate part-loaded or the additional maintenance costs that the provision of reserve may require. Under these rules, generators are allowed to submit separate bids in the reserve market. In a less than perfectly competitive market, these bids would not reflect the marginal cost of providing reserve, but the value that generators believe the market places on the reserve they provide. We also assume that unit 4 can now provide a maximum of 150 MW of reserve. Table 6.8 shows the bids that the generators have submitted as well as the relevant unit characteristics.

Table 6.8 Marginal energy and reserve costs, maximum output, and reserve capability of the generating units of Example 6.8.

Generating units	Marginal cost of energy ($/MWh)	Marginal cost of reserve ($/MWh)	P^{max} (MW)	R^{max} (MW)
1	2	0	250	0
2	17	0	230	160
3	20	5	240	190
4	28	7	250	150

Given that the generators now bid explicitly to provide reserve, the objective function of the optimization problem that the market operator has to solve becomes:

$$\min \; (2 \cdot P_1 + 17 \cdot P_2 + 20 \cdot P_3 + 28 \cdot P_4 + 0 \cdot R_1 + 0 \cdot R_2 + 5 \cdot R_3 + 7 \cdot R_4) \qquad (6.10)$$

The constraints remain the same as in Example 6.7, except for the constraint on the maximum reserve that unit 4 can provide:

$$0 \leq R_4 \leq 150 \qquad (6.11)$$

Table 6.9 summarizes the dispatch for the conditions of this example, while Figure 6.12 shows the evolution of the energy and reserve prices.

Let us analyze this solution. When the demand is in the range between 300 and 320 MW, unit 1 produces at its maximum capacity of 250 MW while unit 2 produces the rest of the demand and is thus the marginal generator. The price of energy is thus 17 $/MWh. Unit 2 provides the maximum amount of reserve that it can deliver (160 MW) because it is willing to provide this reserve at no cost. Unit 3 provides the remainder of the reserve requirement and is thus the marginal provider. The price of reserve is thus 5 $/MWh. Unit 2 makes a profit of $5 per MWh of reserve, while unit 3 just covers its cost of providing reserve.

In the range of demand extending from 320 to 470 MW, the output of unit 2 is kept at 70 MW so that it can provide 160 MW of reserve. Unit 3 is the marginal producer of energy

Table 6.9 Solution of the optimization problem of Example 6.8 for a range of values of the demand.

Demand (MW)	P_1 (MW)	R_1 (MW)	P_2 (MW)	R_2 (MW)	P_3 (MW)	R_3 (MW)	P_4 (MW)	R_4 (MW)
300–320	250	0	50–70	160	0	90	0	0
320–470	250	0	70	160	0–150	90	0	0
470–560	250	0	70	160	150–240	90–0	0	0–90
560–620	250	0	70–130	160–100	240	0	0	90–150
620–720	250	0	130	100	240	0	0–100	150

Each line in this table corresponds to a subrange where the output of only one of the generating units changes.

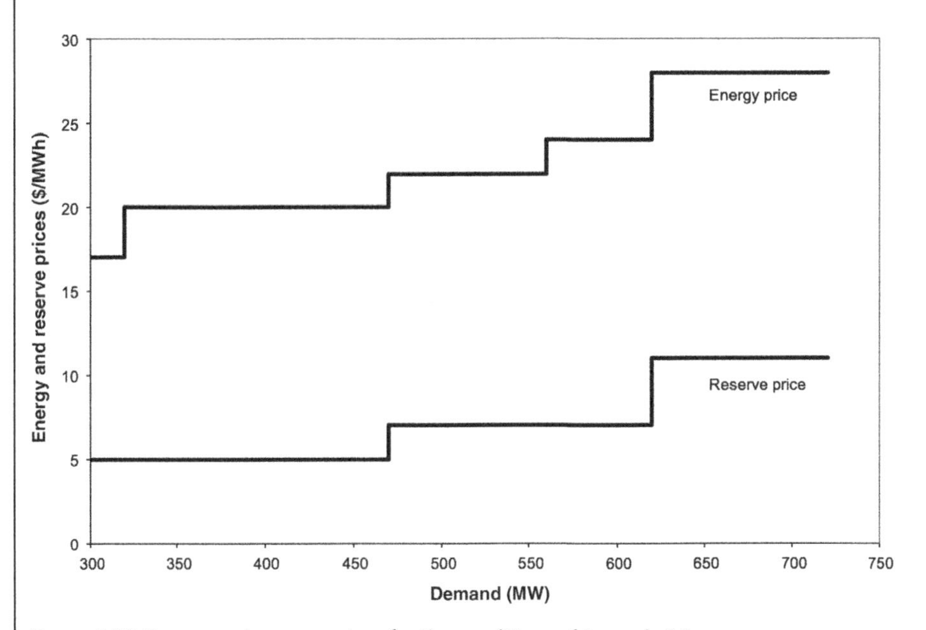

Figure 6.12 Energy and reserve prices for the conditions of Example 6.8.

and sets the price at 20 $/MWh. Unit 3 is also the marginal provider of reserve and the price of reserve thus stays at 5 $/MWh. Unit 2 makes a 3 $/MWh profit on the energy it sells and a profit of 5 $/MWh on the reserve it provides. It thus benefits from being limited in the amount it can sell in the energy market.

If the demand increases from 470 to 560 MW, unit 2 continues to produce 70 MW and to provide 160 MW of reserve. Unit 3 is the marginal producer of electrical energy. As its energy production increases, its contribution to reserve must go down. Unit 4 compensates for this reduction. In this case, the price of energy is not equal to the marginal cost of unit 3 because increasing the energy production of unit 3 has an impact on the allocation of reserve. Producing an additional megawatt with unit 3 costs 20 \$/MWh, but reduces its contribution to the reserve by the same amount, thereby saving 5 \$/MWh. This megawatt of reserve is then provided by unit 4 at a cost of 7 \$/MWh. The price of energy is thus $20 - 5 + 7 = 22$ \$/MWh, which is not equal to the marginal cost of production of any individual generator. Unit 3 thus earns 2 \$/MWh above its marginal cost for the energy it produces. The price of reserve is 7 \$/MWh because the marginal provider is unit 4. Observe that this solution is indeed optimum. Keeping the output of unit 3 at 150 MW so that it can provide 90 MW of reserve would require an increase in the energy production of unit 4. This approach would be more expensive because the extra energy produced would cost 8 \$/MWh while the saving in the provision of reserve would be only 2 \$/MWh.

For the range of demand between 560 and 620 MW, unit 3 produces its maximum capacity of 240 MW and therefore cannot provide any reserve. Interestingly, unit 2 becomes again the marginal producer of energy. However, the price of energy is not 17 \$/MWh but 24 \$/MWh. While producing an additional megawatt with unit 2 costs 17 \$/MWh, this reduces its contribution to the reserve by the same amount. No money is saved, however, because unit 2 is willing to provide reserve for free. Since the compensating megawatt of reserve is provided by unit 4 at a cost of 7 \$/MWh, the price of energy is $17 + 7 = 24$ \$/MWh. The price of reserve remains 7 \$/MWh because the marginal provider is unit 4.

Finally, for a demand greater than 620 MW but smaller than 720 MW, units 2 and 3 produce 130 and 240 MW, respectively. Unit 4 is the marginal producer of energy, while unit 2 is the marginal provider of reserve. The price of energy is thus 28 \$/MWh. The price of reserve is 11 \$/MWh because to get one additional megawatt of reserve we must reduce the production of unit 2 (thereby saving 17 \$/MWh) and increase the output of unit 4 (at a cost of 28 \$/MWh).

These two examples show that it is possible to clear the energy and reserve markets simultaneously in a way that minimizes the cost to consumers, meets the operational reliability requirements, and also ensures a fair treatment of all the providers of energy and reliability resources.

6.4.3 Allocation of Transmission Capacity Between Energy and Reserve

Operators must not only ensure that they have available a sufficient amount of reserve but also that this reserve capacity can actually be delivered across the network when the need arises. This issue is likely to become more important as the reserve requirements increase to cope with the larger uncertainty caused by generation from renewable sources. The available transmission capacity therefore must be optimally allocated between actual energy transfers and capacity set aside for the delivery of reserve. The following example illustrates this trade-off.

Example 6.9

Let us consider the slightly modified version of our Borduria/Syldavia example shown in Figure 6.13. Borduria still has cheaper generation and exports to Syldavia through an 800 MW interconnection. During the operating period that we consider, the loads are as

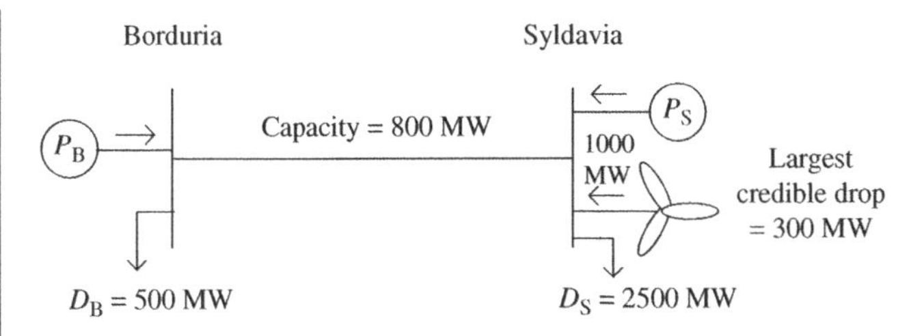

Figure 6.13 Modified Borduria–Syldavia system for Example 6.9.

indicated in this figure and the Syldavian wind farm produces 1000 MW. Among other things, operators of this interconnected system must guard against a credible drop of 300 MW of wind generation in Syldavia.

Both regions operate perfectly competitive markets for energy and reserve, where the total cost functions of Borduria $C_B(P_B, R_B)$ and Syldavia $C_S(P_S, R_S)$ are given by the following expressions:

$$C_B(P_B, R_B) = 10P_B + \frac{0.01}{2}P_B{}^2 + R_B + \frac{0.001}{2}R_B{}^2 + 0.001P_BR_B \tag{6.12}$$

$$C_S(P_S, R_S) = 13P_S + \frac{0.02}{2}P_S{}^2 + 5R_S + \frac{0.019}{2}R_S{}^2 + 0.001P_SR_S \tag{6.13}$$

where P_j and R_j denote the volumes of energy and reserve provided in each region (j = B for Borduria and j = S for Syldavia). Given Equations (6.12) and (6.13), the prices for energy π_j^E in Borduria and Syldavia are given by the partial derivatives of the cost functions with respect to P_j as follows:

$$\pi_B^E = 10 + 0.01P_B + 0.001R_B \tag{6.14}$$

$$\pi_S^E = 13 + 0.02P_S + 0.001R_S \tag{6.15}$$

While the prices for reserve π_j^R are given by the partial derivatives of the cost functions with respect to R_j:

$$\pi_B^R = 1 + 0.001R_B + 0.001P_B \tag{6.16}$$

$$\pi_S^R = 5 + 0.019R_S + 0.001P_S \tag{6.17}$$

A coupling factor of 0.001 between the energy and reserve prices is introduced because the same assets are used to supply both. Providing more reserve thus affects not only the price of reserve but also the price of energy and vice versa.

Comparing these supply curves shows that Bordurian generators are able to provide both energy and reserve at a lower cost than their counterparts in Syldavia. The interconnection therefore allows them to sell not only energy but also reserve that Syldavia may need to compensate for a potential decrease in the wind production. From a global economic perspective, this is more efficient than providing reserve in Syldavia to deal locally with the variability of the wind generation. However, accessing both lower cost energy and lower cost reserve from Borduria requires sharing the transmission network capacity between energy and reserve. An optimal allocation of this capacity between energy and reserve minimizes the sum of the costs of energy and reserve in both Borduria and Syldavia. Mathematically, this is expressed as the following optimization problem:

$$\text{Min}\{C_B(P_B, R_B) + C_S(P_S, R_S)\}$$

This minimization is subject to the following constraints:

Power balance in Borduria:

$$P_B - F^E = 500 \ \left(\pi_B^E\right) \tag{6.18}$$

Power balance in Syldavia:

$$P_S + F^E = 2500 - 1000 \ \left(\pi_S^E\right) \tag{6.19}$$

Reserve requirement in Borduria:

$$R_B - F^R = R_B^R \ \left(\pi_B^R\right) \tag{6.20}$$

Reserve requirement in Syldavia:

$$R_S + F^R = R_S^R \ \left(\pi_S^R\right) \tag{6.21}$$

Total transmission capacity limit:

$$F^E + F^R \leq 800 \ \left(\pi^T\right) \tag{6.22}$$

F^E and F^R denote, respectively, the capacities allocated for energy and reserve transfer. The reserve requirements in Borduria $\left(R_B^R\right)$ and Syldavia $\left(R_S^R\right)$ can be satisfied either locally (R_B, R_S) or from reserve located in the other country if enough transmission capacity has been allocated for reserve (F^R).

Dual variables π_x^y are associated with each constraint and shown in parentheses. Writing the Karush–Kuhn–Tucker (KKT) conditions of this optimization problem shows that these dual variables are equal to the energy and reserve prices defined in Equations (6.14)–(6.17).

Let us first consider the case where no transmission capacity is allocated for the sharing of reserve, and this reserve must then be provided locally. Let us also assume that reserve is required only for handling the largest credible drop in wind generation, i.e. 300 MW. In other words, we assume that the loads are forecast with perfect accuracy and that the conventional generators are 100% reliable. Mathematically, this means setting $F^R = 0$, $R_B^R = 0$, and $R_S^R = 300$ MW in Equations (6.20)–(6.22) and leads to the operating conditions illustrated in Figure 6.14, where the transmission capacity is fully used to access low marginal cost energy from Borduria.

Figure 6.15 illustrates the solution of this same optimization problem if we set $R_B^R = 0$ and $R_S^R = 300$ MW but do not impose $F^R = 0$. In this case, the transfer of energy is scaled back from 800 to 696 MW to allocate F^R=104 MW of network capacity to the sharing of reserve.

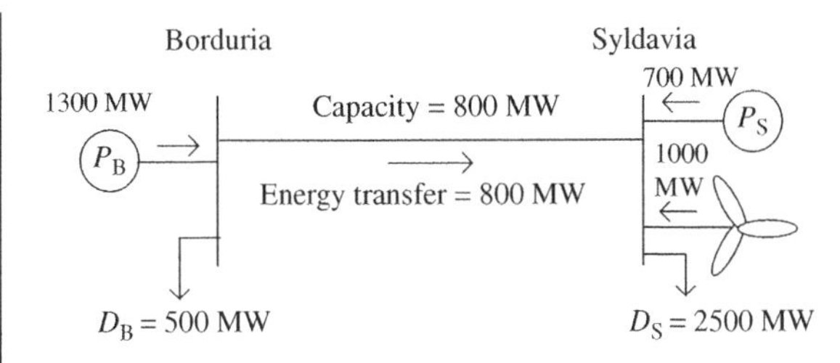

Figure 6.14 Optimal operation of the Borduria–Syldavia system when no transmission capacity is allocated to the sharing of reserve.

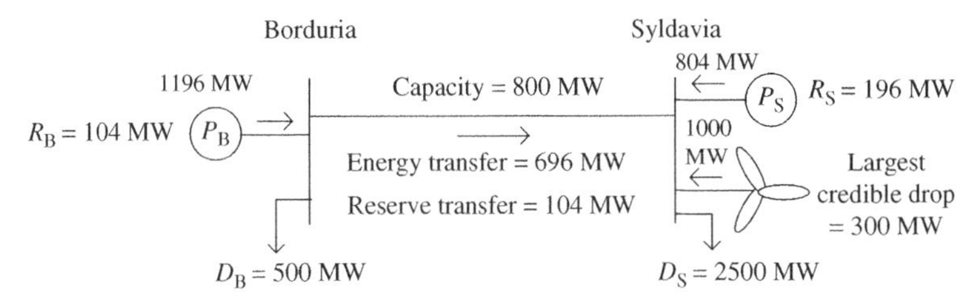

Figure 6.15 Optimal operation of the Borduria–Syldavia system when the transmission capacity is optimally allocated between the transfer or energy and the sharing of reserve.

Table 6.10 shows how the total cost of operating the system, as well as its energy and reserve components, changes as the transmission capacity allocated to energy increases. As this allocation increases from 0 to 500 MW, the cost of energy decreases because the system benefits from the cheaper generation in Borduria. Beyond 500 MW, the cost of reserve increases because more expensive reserve from Syldavia replaces cheaper reserve from Borduria. The total cost goes through a minimum at $F^E = 696$ MW.

Table 6.10 Evolution of the components of the total cost of operating the Borduria–Syldavia system as a function of the transmission capacity allocated to energy transfer.

Transmission capacity (MW)	0	100	200	300	400	500	600	700	800
Energy cost ($/h)	48 400	45 780	43 460	41 440	39 720	38 300	37 160	36 280	35 660
Reserve cost ($/h)	345	345	345	345	345	345	815	1 485	2 355
Total cost ($/h)	48 745	46 125	43 805	41 785	40 065	38 645	37 975	37 765	38 015

Table 6.11 summarizes the physical positions of all market participants at the point of minimum operating costs. Negative quantities denote demand for energy and reserve.

At the optimal point (denoted by *), the energy prices in Borduria and Syldavia are:

$$\pi_B^{*E} = 10 + 0.01P_B^* + 0.001R_B^* = 22.1\ \$/\text{MWh} \tag{6.23}$$

Table 6.11 Power generation, reserve capacity, demand, and power transfer between Borduria and Syldavia.

		Power (MW)	Reserve (MW)
Borduria	Conventional generation	1196	104
	Demand	−500	0
Syldavia	Conventional generation	804	196
	Wind generation	1000	−300
	Demand	−2500	0
	Power transfer	696	104

$$\pi_S^{*E} = 13 + 0.02P_S^* + 0.001R_S^* = 29.3\ \$/\text{MWh} \tag{6.24}$$

while the reserve prices in Borduria and Syldavia are:

$$\pi_B^{*R} = 1 + 0.001R_B^* + 0.001P_B^* = 2.3\ \$/\text{MWh} \tag{6.25}$$

$$\pi_S^{*R} = 5 + 0.019R_S^* + 0.001P_S^* = 9.5\ \$/\text{MWh} \tag{6.26}$$

The price differentials between energy and reserve are:

$$\pi_S^{*E} - \pi_B^{*E} = 7.2\ \$/\text{MWh} \tag{6.27}$$

$$\pi_S^{*R} - \pi_B^{*R} = 7.2\ \$/\text{MWh} \tag{6.28}$$

These differences are equal because, at the optimum, allocating one more megawatt of transmission capacity to the sharing of reserve or to the transfer of energy has the same value and should be charged at the same price. Using the KKT conditions of the optimization problem, it can be shown that this differential is also equal to the marginal value or price of transmission capacity π^T (in term of its absolute value), which is the dual variable associated with Equation (6.22). Given the results shown in Equations (6.23)–(6.28), Table 6.12 shows every market participant's remuneration or payment for energy and reserve services, as well as congestion surplus.

Table 6.12 Remunerations (positive values) and payments (negative values) for energy and reserve.

		Energy ($/h)	Reserve ($/h)	Total ($/h)
Borduria	Conventional generation	26 431	239	26 670
	Consumers	−11 050		−11 050
Syldavia	Conventional generation	23 557	1 862	25 419
	Wind generation	29 300	−2 850	26 450
	Consumers	−73 250		−73 250
	Congestion surplus	5 011	749	5 760

This congestion surplus is equal to 5760 \$/h, which is the net difference between what consumers pay and generators are paid for both energy and reserve services. This value can also be calculated by multiplying the marginal price of transmission $\pi^{*T} = 7.2$ \$/MWh (locational energy price difference or locational reserve price difference) by the total network capacity (800 MW, 696 MW of which is allocated for energy and 104 MW for reserve). In this example, we assume that the 2850 \$/h cost of reserve should be paid by the wind generators in Syldavia because this reserve is provided to mitigate fluctuations in their output. This amount is divided among the reserve providers in Borduria (239 \$/h), the reserve providers in Syldavia (1862 \$/h), and the reserve-related congestion surplus (749 \$/h).

6.4.4 Allocating the Costs

Not all consumers value reliability equally. For example, the cost of a service interruption is much larger for a semiconductor factory or a paper mill than it is for residential customers. Some consumers might therefore be willing to pay more for an improved level reliability, while others would accept a less reliable system in exchange for a reduction in the price they pay for their supply of electricity. Such reliability-based pricing would be economically efficient. Unfortunately, the current state of the technology does not enable system operators to deliver differentiated levels of reliability. The operational reliability standards that they apply must therefore aim to achieve an average level of reliability that is hopefully at least acceptable to all. Since all users get the same level of reliability, it seems logical to share the cost of the reliability resources among all users based on some measure of their use of the system. This measure is typically the energy consumed or produced.

There is, however, another aspect to this issue. The behavior of some users may cause a disproportionate amount of stress in the power system. Penalizing these users might encourage them to change their behavior or to implement mitigating measures. Ultimately, these adjustments should decrease the amount of resources needed and reduce the cost of achieving the desired level of reliability. Let us explore this concept using two examples.

6.4.4.1 Who Should Pay for Reserve?

Generation capacity is held in contingency reserve to avert a system collapse when a large imbalance develops between load and generation. In most cases, such imbalances originate from the sudden failure of a generating unit or the sudden disconnection of an interconnection with a neighboring system. If such a contingency occurs when the system does not carry enough reserve capacity, system operators must resort to shedding load to avoid a complete system collapse. Using historical data on the failure rate of generating units and interconnections, it is possible to calculate the amount of reserve required to reduce the probability of load disconnection to an acceptably low level (see, for example, Billinton and Allan 1996). These probabilistic calculations confirm that a system where generating units fail more frequently requires more reserve than a system where generators are more reliable. They also show that a system supplied by a few large generating units needs more reserve than one with many smaller generators. The unreliability of few large generating plants can therefore increase the need for operating reserve. Since our objective is to minimize the cost of reserve without reducing the level of operational reliability, we should give these generators an incentive to reduce their failure rates. If after some time they can demonstrate that they have succeeded in improving their performance, the system operator will be able to reduce the required amount of reserve. Strbac and Kirschen (2000) have argued that the fairest incentive involves charging the cost of the reserves to the generators in proportion to their contribution to the reserve requirement. Generators would obviously pass on this cost to their consumers under the form of higher prices for electrical energy. Smaller and more reliable generating units would then have a competitive advantage over larger and failure-prone units.

6.4.4.2 Who Should Pay for Regulation and Load Following?

Kirby and Hirst (2000) analyzed the requirements for load following and regulation in an actual and typical power system. They have also developed an equitable technique for allocating these requirements between industrial and nonindustrial consumers. For this particular power system, their analysis shows that industrial consumers account for 93% of the regulation and 58% of the load-following requirements even though they represent only 34% of the system load. Since the cost of these resources is charged to consumers on the basis of their energy consumption, the residential consumers are clearly subsidizing the industrial ones. It can be shown that there are also wide variations between the contributions of individual consumers within the industrial group. For example, aluminum smelters and paper mills have loads that are nearly time invariant and therefore do not contribute to the requirements for regulation and load following.

6.5 Selling Reliability Resources

Selling reliability resources represents another business opportunity for generating companies and other market participants. However, technical limitations and cost considerations inextricably link the sale of reserve and voltage control services to the sale of energy. For example, a generator cannot sell spinning reserve or reactive support if the unit is not running and producing at least its minimum power output. Conversely, a unit operating at maximum capacity cannot sell reserve capacity because it does not have any. If it decides to reduce its power output to be able to sell reserve, it forgoes an opportunity to sell energy. Since the cost of this opportunity can be significant, the generating company must optimize jointly the sale of energy and reserve.

Rather than attempting to develop a general formulation of this obviously complicated problem, let us explore these interactions using a simple example.

Example 6.10

Let us consider the operation of a generating unit that can sell both energy and spinning reserve in competitive markets. The precise characteristics of this spinning reserve service are not important for our analysis, and we will not consider the possibility of selling other reliability resources. We assume that the energy and reserve markets are sufficiently competitive that this unit can be treated as a price taker. This means that its bidding behavior has no effect on either the price of energy or the price of reserve and that it is able to sell any quantity it chooses in either market. We consider the operation of this unit over a single market period of 1 h and we assume that the unit is running at the beginning of the period. These assumptions allow us to neglect issues related to the startup cost of the unit, its minimum uptime, and its minimum downtime. In an actual application, the optimization would be carried out over a day or longer and all these issues would have to be taken into account.

We use the following notations:

π_1 price per MWh on the energy market

π_2 price per MWh of capacity on the spinning reserve market. A MWh corresponds to 1 MW of reserve capacity made available for 1 h. Since this reserve capacity may or may not be called, a MWh of reserve is not equivalent to a MWh of energy. For the sake of simplicity, we assume that the generator does not receive an additional exercise fee when the reserve it provides is actually called upon to provide energy. Considering such an exercise fee would not change the conclusions of this example

x_1 quantity bid by the generator in the energy market. Since this generator is a price taker, it is also the quantity of energy sold by the generator

x_2 quantity bid by the generator in the reserve market. Since this generator is a price taker, it is also the quantity of reserve sold by the generator

$P^{\min}$ minimum power output of the generating unit (minimum stable generation)

$P^{\max}$ maximum power output of the generating unit

$R^{\max}$ upper limit that the ramp rate of the unit and the definition of the reserve service place on the amount of reserve that the unit can deliver. For example, if the unit has a maximum ramp rate of 120 MW per hour and the reserve has to be delivered within 10 min, this unit cannot deliver more than 20 MW of reserve

$C_1(x_1)$ cost of producing the amount x_1 of energy. This function must be convex. It includes the costs of fuel and maintenance related to the production of energy but does not include any investment costs

$C_2(x_2)$ cost of providing the amount x_2 of reserve. This function must also be convex. It does not include the opportunity cost of selling energy or any investment costs. We assume that the generator can estimate the fraction of the reserve it offers that will be called upon to provide energy. The expected cost of producing this energy is included in this cost

Let us formulate this example as a constrained optimization problem. Since this generator is trying to maximize the profit it derives from the sale of energy and reserve, the objective function is the difference between the revenues and the costs from both energy and reserve:

$$f(x_1, x_2) = \pi_1 x_1 + \pi_2 x_2 - C_1(x_1) - C_2(x_2) \tag{6.29}$$

Several technical factors place constraints on the energy and reserve that can be provided by this unit. First, the sum of the bids for energy and reserve cannot exceed the maximum power output of the generating unit:

$$x_1 + x_2 \leq P^{\max} \tag{6.30}$$

Second, since the unit cannot operate below its minimum stable generation, the bid for energy should be at least equal to the minimum power output:

$$x_1 \geq P^{\min} \tag{6.31}$$

Third, the unit cannot bid for more reserve than it can deliver within the time allowed by the specification of the reserve service:

$$x_2 \leq R^{\max} \tag{6.32}$$

If $R^{\max} \geq P^{\max} - P^{\min}$, the amount of reserve that the unit can provide is not limited by the ramp rate and condition (6.32) is superfluous. We therefore assume that $R^{\max} < P^{\max} - P^{\min}$. This restriction implies that constraints (6.30) and (6.31) cannot be binding at the same time. We are not modeling explicitly the fact that reserve must be positive. Doing so would complicate our analysis without bringing additional insight. Some generators may obviously decide that, at least part of the time, providing reserve is not worthwhile.

Given this objective function and these constraints, we can form the Lagrangian function for this optimization problem:

$$\begin{aligned} \ell(x_1, x_2, \mu_1, \mu_2, \mu_3) &= \pi_1 x_1 + \pi_2 x_2 - C_1(x_1) - C_2(x_2) \\ &+ \mu_1(P^{\max} - x_1 - x_2) + \mu_2\left(x_1 - P^{\min}\right) + \mu_3(R^{\max} - x_2) \end{aligned} \tag{6.33}$$

Setting the partial derivatives of this Lagrangian with respect to the decision variables equal to zero, we obtain the conditions for optimality:

$$\frac{\partial \ell}{\partial x_1} \equiv \pi_1 - \frac{dC_1}{dx_1} - \mu_1 + \mu_2 = 0 \tag{6.34}$$

$$\frac{\partial \ell}{\partial x_2} \equiv \pi_2 - \frac{dC_2}{dx_2} - \mu_1 - \mu_3 = 0 \tag{6.35}$$

The solution must also satisfy the inequality constraints:

$$\frac{\partial \ell}{\partial \mu_1} \equiv P^{\max} - x_1 - x_2 \geq 0 \tag{6.36}$$

$$\frac{\partial \ell}{\partial \mu_2} \equiv x_1 - P^{\min} \geq 0 \tag{6.37}$$

$$\frac{\partial \ell}{\partial \mu_3} \equiv R^{\max} - x_2 \geq 0 \tag{6.38}$$

and the complementary slackness conditions:

$$\mu_1 \cdot (P^{\max} - x_1 - x_2) = 0 \tag{6.39}$$

$$\mu_2 \cdot \left(x_1 - P^{\min}\right) = 0 \tag{6.40}$$

$$\mu_3 \cdot (R^{\max} - x_2) = 0 \tag{6.41}$$

$$\mu_1 \geq 0; \; \mu_2 \geq 0; \; \mu_3 \geq 0 \tag{6.42}$$

The complementary slackness conditions assert the fact that an inequality constraint is either binding or nonbinding. If it is binding, it behaves like an equality constraint and it can be shown that the corresponding Lagrange multiplier μ_i is equal to the marginal or shadow cost of the constraint. Since a binding constraint always increases the cost of the optimal solution, the Lagrange multipliers of binding inequality constraints must be positive. On the other hand, since a nonbinding inequality constraint has no impact on the cost of the optimal solution, its Lagrange multiplier is equal to zero. Binding inequality constraints are thus associated with strictly positive Lagrange multipliers and vice versa. We will make repeated use of this observation in the discussion that follows.

Equations (6.34)–(6.42) form a set of necessary and sufficient optimality conditions for this problem. They are called the KKT conditions. Unfortunately, the KKT conditions do not tell us which inequality constraints are binding. Software packages for optimization try various combinations of binding constraints until they find one that satisfies the KKT conditions. We examine here all the possible combinations because each of them illustrates a different form of interaction between the energy and reserve markets. Since there are three inequality constraints in this problem, we have to consider eight possible combinations.

Case 1: $\mu_1 = 0; \quad \mu_2 = 0; \quad \mu_3 = 0$

Since all the Lagrange multipliers are equal to zero, none of the constraints is binding. Equations (6.34) and (6.35) simplify to:

$$\frac{dC_1}{dx_1} = \pi_1 \tag{6.43}$$

$$\frac{dC_2}{dx_2} = \pi_2 \tag{6.44}$$

These conditions mean that the generating unit should bid to provide energy and reserve up to the point where their respective marginal costs are equal to their price. Since there are no interactions between energy and reserve, this situation is similar to the sale of energy in a perfectly competitive market, as described in Chapter 4.

Case 2: $\mu_1 > 0; \mu_2 = 0; \mu_3 = 0$

The generation capacity of the unit is fully utilized by the provision of a combination of energy and reserve:

$$x_1 + x_2 = P^{\max} \tag{6.45}$$

Replacing the values of the Lagrange multipliers in Equations (6.34) and (6.35), we get:

$$\pi_1 - \frac{dC_1}{dx_1} = \pi_2 - \frac{dC_2}{dx_2} = \mu_1 \geq 0 \tag{6.46}$$

Equation (6.46) shows that the provision of energy and reserve are both profitable. Maximum profit is achieved when the unit is dispatched in such a way that the marginal profit on energy is equal to the marginal profit on reserve. The value of the Lagrange multiplier μ_1 indicates the additional marginal profit that would be achieved if the upper limit on the unit's output could be relaxed.

Case 3: $\mu_1 = 0; \mu_2 > 0; \mu_3 = 0$

The unit produces just enough energy to operate at its minimum stable generation:

$$x_1 = P^{\min} \tag{6.47}$$

Equations (6.34) and (6.35) give:

$$\frac{dC_1}{dx_1} - \pi_1 = \mu_2 \tag{6.48}$$

$$\frac{dC_2}{dx_2} = \pi_2 \tag{6.49}$$

In order to be able to provide spinning reserve, the unit must be running and operating at least at its minimum stable generation. Equation (6.49) shows that this unit should provide reserve up to the point where the marginal cost of providing reserve is equal to the market price for reserve. On the other hand, since the Lagrange multipliers of binding constraints are positive, Equation (6.48) indicates that the production of energy is marginally unprofitable. If it were possible, the generator would prefer to produce less energy.

Note that the KKT conditions determine the operating point that will maximize the profit or minimize the loss. They do not guarantee that the generator will actually make a profit. In this case, the loss on the sale of energy might exceed the profit on the sale of reserve. To check if an operating point is actually profitable, we must replace the values of x_1 and x_2 in the objective function given in Equation (6.29) and check the sign of the result. If an optimal operating point turns out to be unprofitable, the generator might decide to turn off the unit for that hour. However, when the operation of a unit is optimized over a number of periods (e.g. over a day), the overall optimal solution may include some unprofitable periods because of the startup costs and the minimum time constraints. The sale of reserve may reduce the loss that must be accepted during these unprofitable periods.

Case 4: $\mu_1 > 0; \mu_2 > 0; \mu_3 = 0$ and *Case 5:* $\mu_1 > 0; \mu_2 > 0; \mu_3 > 0$

Since we assume that the ramp rate limit on reserve is smaller than the operating range of the unit, these cases are not physical and we will not discuss them further.

Case 6: $\mu_1 = 0; \mu_2 = 0; \mu_3 > 0$

The only binding constraint in this case is that the reserve is limited by the ramping rate. We have:

$$x_2 = R^{\max} \tag{6.50}$$

$$\frac{dC_1}{dx_1} = \pi_1 \tag{6.51}$$

$$\pi_2 - \frac{dC_2}{dx_2} = \mu_3 \geq 0 \tag{6.52}$$

These equations show that, while the profit from the sale of energy is maximized, relaxing the ramp rate constraint would increase the profit from the sale of reserve.

Case 7: $\mu_1 > 0; \mu_2 = 0; \mu_3 > 0$

Both the maximum capacity and the ramp rate constraint are binding:

$$x_1 + x_2 = P^{\max} \tag{6.53}$$

$$x_2 = R^{\max} \tag{6.54}$$

We can rewrite Equation (6.53) as follows:

$$x_1 = P^{\max} - R^{\max} \tag{6.55}$$

The optimality conditions (6.34) and (6.35) give, respectively, the marginal profitability for energy and reserve:

$$\pi_1 - \frac{dC_1}{dx_1} = \mu_1 \tag{6.56}$$

$$\pi_2 - \frac{dC_2}{dx_2} = \mu_1 + \mu_3 \tag{6.57}$$

Since both μ_1 and μ_3 are positive, Equations (6.56) and (6.57) show that selling more energy and more reserve would be profitable. However, since the marginal profit on the sale of reserve is higher than on the sale of energy, the maximum profit is achieved by selling as much reserve as the ramp rate constraint allows.

Case 8: $\mu_1 = 0$; $\mu_2 > 0$; $\mu_3 > 0$

In this case, the values of both x_1 and x_2 are determined by the binding inequality constraints:

$$x_1 = P^{\min} \tag{6.58}$$

$$x_2 = R^{\max} \tag{6.59}$$

Once again, we can use the optimality conditions to determine the marginal profitability of both transactions:

$$\pi_1 - \frac{dC_1}{dx_1} = -\mu_2 \tag{6.60}$$

$$\pi_2 - \frac{dC_2}{dx_2} = \mu_3 \tag{6.61}$$

These equations indicate that the sale of reserve is profitable and would be even more so without the ramp rate constraint. On the other hand, the sale of energy is unprofitable and would be further reduced if it were not for the minimum stable generation constraint. Once again, the actual profitability of this operating point should be checked using the objective function.

6.6 Problems

6.1 A power system is supplied by three generating units that are rated at, respectively, 150, 200, and 250 MW. What is the maximum load that can be securely connected to this system if the simultaneous outage of two generating units is not considered to be a credible event?

6.2 Identify the documents describing the operational reliability criteria governing the operation of the power system in the region where you live or some other region of your choice. Summarize the main points of these operational reliability rules.

6.3 A small power system consists of two buses connected by three transmission lines. Assuming that this power system must be operated according to the $N-1$ operational reliability criterion and that its operation is constrained only by thermal limits on the transmission lines, calculate the maximum power transfer between these two buses for each of the following conditions:

a All three lines are in service and each line has a continuous thermal rating of 300 MW.

b Only two lines rated at 300 MW are in service.

c All three lines are in service. Two of them have a continuous thermal rating of 300 MW and the third is rated at 200 MW.

d All three lines are in service. All of them have a continuous thermal rating of 300 MW. However, during emergencies, they can sustain a 10% overload for 20 minutes. The generating units on the downstream bus can increase their output at the rate of 4 MW/min.

e Same conditions as in (d), except that the output of the downstream generators can only increase at the rate of 2 MW/min.

f Low temperatures and high winds improve the heat transfer between the conductors and the atmosphere. Assume that this dynamic thermal rating increases the continuous and emergency loadings of (d) by 15%.

6.4 A generator is connected to a large power system by a double-circuit transmission line. Each line has a negligible resistance and a reactance of 0.2 p.u. The transient reactance of the generator is 0.8 p.u. and its inertia constant is 3 s. The large power system can be modeled as an infinite bus and the voltages are kept at their nominal value. Assume that single circuit faults on the transmission line are cleared in 120 ms. Using a transient stability program, calculate the maximum power that this generator can produce without risking instability. Use a 100 MW base.

6.5 Repeat the calculations of Problem 6.4 for the case where the generator is connected to the power system by two identical double-circuit transmission lines.

6.6 Consider a power system with two buses and two lines. One of these lines has a reactance of 0.25 p.u. and the other a reactance of 0.40 p.u. The series resistances and shunt susceptances of the lines are negligible. A generator at one of the buses maintains its terminal voltage at nominal value and produces power that is consumed by a load connected to the other bus. Using a power flow program, calculate the maximum active power that can be transferred without causing a voltage collapse when one of the lines is suddenly taken out of service under the following conditions:

a The load has unity power factor and there is no reactive power injection at the receiving end.

b The load has unity power factor and a synchronous condenser injects 25 MVAr at the receiving end.

c The load has a 0.9 power factor lagging and there is no reactive power injection at the receiving end.

6.7 Consider the small power system shown in Figure P6.7. Each line of this system is modeled using a π equivalent circuit. The parameters of each line are given in this table.

Lines	*R* (p.u.)	*X* (p.u.)	*B* (p.u.)
A–B	0.08	0.8	0.3
A–C	0.04	0.4	0.15
C–B	0.04	0.4	0.15

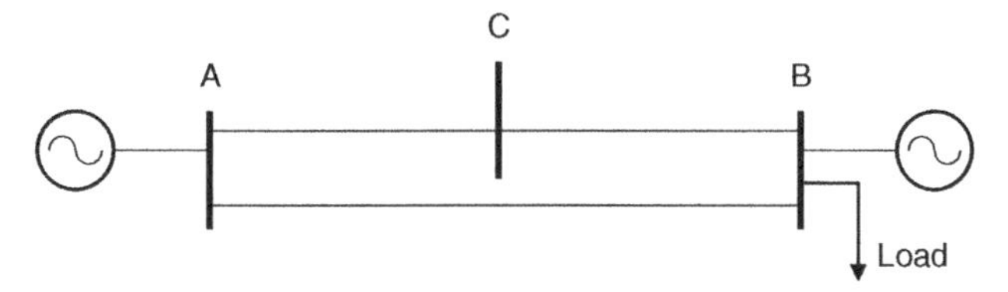

Figure P6.7 Schematic diagram of the power system for Problem 6.7.

Using a power flow program, study the reactive support requirements as a function of the amount of power transferred from bus A to bus B for both normal conditions and abnormal conditions (i.e. avoiding a voltage collapse following the sudden outage of a line). Consider both a unity power factor and a 0.9 power factor lagging load at bus B. Analyze and discuss the usefulness of a source of reactive power at bus C.

6.8 Identify the documents governing the provision of ancillary services in the region where you live or in another region of your choice. Determine the mechanism used to obtain each service. When services are compulsory, determine their parameters (e.g. minimum lead and lag power factor for generators). When services are procured on a competitive basis, describe the structure of the markets for ancillary services (duration of contracts, bid parameters). Pay particular attention to the definition of reserve services. Identify the mechanism used to pass on the cost of ancillary services to consumers.

6.9 Analyze prices and volumes in the markets for ancillary services in the region where you live or in another region for which you have access to the necessary data.

6.10 The owners of a generating unit would like to maximize their profit by selling both energy and balancing services. Write the objective function and the constraints for this optimization problem. Discuss the various cases that might arise depending on the price paid for energy and balancing. Neglect the constraint introduced by the ramp rate of the generating unit.

(*Hint:* The equations in Example 6.10 must be modified because providing balancing services might involve a reduction in the output of the generating unit. On the other hand, you may need to consider explicitly that the amount of balancing service sold must be positive, i.e. $x_2 \geq 0$.)

6.11 In the three-bus power system shown in Figure P6.11, each bus is connected to the other two buses by two circuits on the same set of towers. This is called a double-circuit line.

Assume the following:

- The DC power flow approximation is valid.
- All lines have the same reactance.
- Each double-circuit line has the same transmission capacity F.

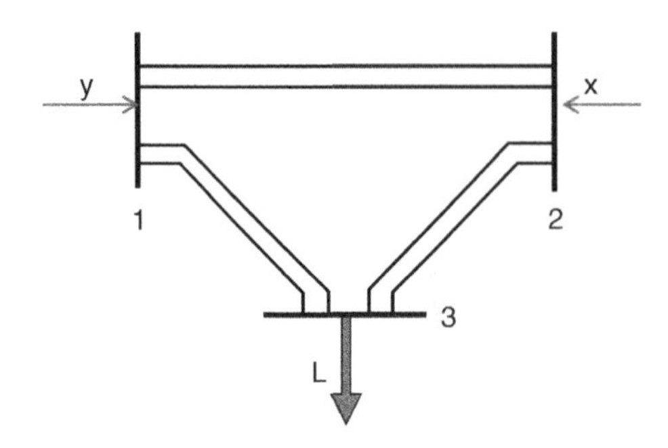

Figure P6.11 Three-bus power system of Problem 6.11.

- Applying the $N-1$ security criterion means considering the outage of one circuit at a time.

Write all the inequality constraints of the preventive security-constrained OPF problem as a function of the injections x and y, and the capacity of the double-circuit line F.

(*Hints:* (i) You do not need to write the DC power flow explicitly, but you need to express the flows in each line as a function of x and y. (ii) You must consider the impact of each outage on the reactance and the capacity of each line.)

References

Billinton, R. and Allan, R.N. (1996). *Reliability Evaluation of Power Systems*, 2e. Plenum Press.

Kirby B. and Hirst E. (2000). Customer-specific Metrics for the Regulation and Load-following Ancillary Services. Report ORNL/CON-474, Oak Ridge National Laboratory, January 2000. www.ornl.gov/psr.

Oren, S.S. (2002). Auction design for ancillary reserve products. IEEE Power Engineering Society Summer Meeting, Vol. 3, pp. 1238–1239.

O'Sullivan, J., Rogers, A., Flynn, D. et al. (2014). Studying the maximum instantaneous non-synchronous generation in an island system – frequency stability challenges in Ireland. *IEEE Trans. Power Syst.* 29 (6): 2943–2951.

Pudjianto, D., Ahmed, S., and Strbac, G. (July 2002). Allocation of VAr support using LP and NLP based optimal power flows. *IEE Proc. Gener. Transm. Distrib.* 149 (4).

Read, E.G., Drayton-Bright, G.R., and Ring, B.J. (1995). An integrated energy and reserve market for New Zealand. In: *Transmission Pricing and Access* (ed. R. Siddiqi and M. Einhorn), 183–206. Kluwer.

Rebours, Y., Kirschen, D.S., Trotignon, M., and Rossignol, S. (2007a). A survey of frequency and voltage control ancillary services – part 2: Economic features. *IEEE Trans. Power Syst.* 22 (1): 358–366.

Rebours, Y., Kirschen, D.S., Trotignon, M., and Rossignol, S. (2007b). A survey of frequency and voltage control ancillary services – part 1: Technical features. *IEEE Trans. Power Syst.* 22 (1): 350–357.

Strbac, G. and Kirschen, D.S. (2000). Who should pay for reserve? *Electr. J.* 13 (8).

Wang, Y., Silva, V., and Lopez-Botet-Zulueta, M. (2016). Impact of high penetration of variable renewable generation on frequency dynamics in the continental Europe interconnected system. *IET Renew. Power Gener.* 10 (1): 10–16.

Further Reading

The concepts of power system security (a term that has since been replaced by operational reliability) and the application of cost/benefit analysis to the determination of its optimal level are discussed by Kirschen (2002). Ejebe (2000) describes computationally efficient methods for calculating the available transmission capacity. Billinton and Allan (1996) wrote a standard reference on power system reliability that explains the techniques used to calculate the amount of reserve needed to achieve a certain level of reliability. A method for calculating the reactive support requirements is proposed by Pudjianto et al. (2002). The reader interested in the definition of the various ancillary services should consult the papers by Rebours et al. (2007). Useful information on the reliability resources used in a particular power system can often be found on the website of the system operator.

Oren (2002) compares sequential and simultaneous approaches to clearing energy and reserve markets. Read et al. (1995) discuss the co-optimization of energy and reserve and the implementation of this approach in the New Zealand electricity market. Alvey et al. (1998) give the detailed formulation of this approach. The allocation of the cost of ancillary services to the parties that are responsible for the need for these services is discussed by Strbac and Kirschen (2000) for the case of reserve and by Kirby and Hirst (2000) for the cases of frequency regulation and load following. The provision of ancillary services by the demand side is addressed by Kirby and Hirst (1999).

The report by Lawrence Berkeley National Laboratory details a rigorous methodology and provides data for calculating the cost of outages. The other reports discuss how the Value of Lost Load was calculated in various jurisdictions.

Alvey, T., Goodwin, D., Ma, X. et al. (1998). A security-constrained bid-clearing system for the New Zealand wholesale electricity market. *IEEE Trans. Power Syst.* 13 (2): 340–346.

The Brattle Group and Astrape Consulting (2013). Resource Adequacy Requirements: Reliability and Economic Implications (September 2013). http://www.brattle.com/system/news/pdfs/000/000/618/original/Resource_Adequacy_Requirements_-_Reliability_and_Economic_Requirements.pdf (accessed 14 March 2018).

Ejebe, G.C., Waight, J.G., Santos-Nieto, M., and Tinney, W.F. (2000). Fast calculation of linear available transfer capability. *IEEE Trans. Power Syst.* 15 (3): 1112–1116.

Ernest Orlando Lawrence Berkeley National Laboratory. (2009). Estimated Value of Service Reliability for Electric Utility Customers in the United States. Report LBNL-2132E (June 2009). http://eetd.lbl.gov/ea/EMS/EMS_pubs.html (accessed 14 March 2018).

European Commission (2016). Interim Report of the Sector Inquiry on Capacity Mechanisms (April 2016). http://ec.europa.eu/competition/sectors/energy/capacity_mechanisms_swd_en.pdf (accessed 14 March 2018).

Kirby, B. and Hirst, E. (1999). Load as a resource in providing ancillary services. American Power Conference, Chicago, Illinois (April 1999). www.ornl.gov/psr (accessed 14 March 2018).

Kirschen, D.S. (2002). Power system security. *IEE Power Eng. J.* 16 (5): 241–248.
London Economics (2013a). The Value of Lost Load (VoLL) for Electricity in Great Britain, Final report for OFGEM and DECC (June 2013). https://londoneconomics.co.uk/blog/publication/the-value-of-lost-load-voll-for-electricity-in-great-britain/ (accessed 14 March 2018).
London Economics (2013b). Estimating the Value of Lost Load – Briefing paper prepared for the Electric Reliability Council of Texas, Inc. (June 2013). http://www.ercot.com/content/gridinfo/resource/2015/mktanalysis/ERCOT_ValueofLostLoad_LiteratureReviewandMacroeconomic.pdf (accessed 14 March 2018).
MISO Market Subcommittee (2016). Evaluating Energy Offer Cap Policy (May 2016). https://www.misoenergy.org/_layouts/MISO/ECM/Redirect.aspx?ID=223536 (accessed 14 March 2018).

7

Investing in Generation

7.1 Introduction

In the previous chapters, we studied the economics of operating a power system using a given set of generating plants. In this chapter, we consider the possibility of adding or removing generation capacity. In the first part of the chapter, we consider each generating plant independently. Taking the perspective of a potential investor, we examine the factors that affect the decision to build a new generating plant. We also consider the retirement of existing plants when their profitability becomes insufficient. For the sake of simplicity, we assume that all the revenues produced by these plants are derived from the sale of electrical energy and we neglect the additional revenues that a plant could obtain from the sale of ancillary services or reliability resources. We also assume that generating plants are not remunerated for providing capacity, i.e. simply for being available in case their output is needed.

In the second part of the chapter, we discuss the provision of generation capacity from the perspective of the consumers. Electricity is so central to economic activity and personal well-being that consumers expect their supply to remain available and affordable not only when the demand fluctuates but also when some generators are unable to produce because of technical difficulties. We must therefore consider whether profits from the sale of electrical energy result in a total generation capacity that is and remains sufficient to meet the consumers' expectations. Since in many electricity markets it has been decided that the answer to this question is negative, we discuss the additional incentives that can entice generating companies to provide the extra capacity needed to ensure reliability.

7.2 Generation Capacity from an Investor's Perspective

7.2.1 Building New Generation Capacity

Investor will finance a production facility if they believe that the plant will earn a satisfactory profit over its lifetime. More specifically, the revenues produced by the plant should not only exceed the cost of building and operating the plant but also the profits that it generates should be larger than what these investors could realize from any other venture with a similar level of risk. To make such an investment decision, investors

Fundamentals of Power System Economics, Second Edition. Daniel S. Kirschen and Goran Strbac.

compute the long-run marginal cost of the plant (including the expected rate of return) and forecast the price at which the output of this plant might be sold. Building a plant is a rational decision as long as the forecasted price exceeds the long-run marginal cost of the plant. In a competitive electricity market, this reasoning is applicable to investments in generation capacity. Relying on this type of investment decision leads to what is called "merchant generation expansion."

In practice, deciding to invest in a new generating plant is considerably more complex than this simplified theory would suggest. Both sides of the equation are indeed affected by a considerable amount of uncertainty. Construction delays and fluctuations in the price of fuel can affect the long-run marginal cost. On the other hand, the evolution of wholesale electricity prices over a long period is notoriously difficult to forecast because demand might change, competitors might enter the market or new, more efficient generation technologies might be developed. The development of merchant plants is often possible only when backed by upstream and downstream contracts. Upstream contracts guarantee a supply of fuel at a fixed price. Downstream contracts ensure that the electrical energy produced by the plant is sold at a price that is also fixed. Such arrangements eliminate the price risks over which the plant owner usually has very little control. With such contracts in place, the owners of the plant thus carry only the risk associated with operating the plant, i.e. the risk that a failure might prevent it from producing electrical energy and honoring its contracts.

A generating plant, like any other machine, is designed to operate satisfactorily for a certain number of years. Investors who decide to build a generating plant base their decision on this estimated lifetime. For generating plants, this lifetime typically ranges from 20 to 40 years. Some hydro power plants, however, have considerably longer lifetimes.

Example 7.1

Bruce, a young consulting engineer, has been asked by Borduria Power to help reach a preliminary decision on whether a new 500 MW coal-fired power plant should be built. Bruce begins by collecting some typical values for the essential parameters of a plant of this type. The table below shows the values he has gathered:

Investment cost	1021 $/kW
Expected plant life	30 years
Heat rate at rated output	9419 Btu/kWh
Expected fuel cost	1.25 $/MBtu

Adapted from DOE data cited by Stoft (2002).

Since he is only asked for a rough estimate, Bruce neglects the costs associated with starting up and maintaining the plant. Borduria Power has asked Bruce to use the Internal Rate of Return (IRR) method to estimate the profitability of the plant. This method, which is also called the discounted cash flow method, measures the internal earning rate of an investment. To implement this method, Bruce must determine the net cash flow for each year of this plant's lifetime. Bruce starts by calculating the cost of building the plant:

Investment cost:

$$1021\,\$/\text{kW} \times 500\,\text{MW} = \$510\,500\,000$$

Next, Bruce needs to estimate the annual production of this plant. Ideally, the plant should operate at full capacity at all times. In practice, this is not possible because the plant has to be shut down periodically for maintenance and because there will inevitably be failures resulting in unplanned outages. Based on his experience with plants of this type, Bruce postulates a utilization factor of 80%. Under these conditions, the estimated annual production of this plant would be:

Annual production:

$$0.8 \times 500\ \text{MW} \times 8760\ \text{h/year} = 3\,504\,000\,\text{MWh}$$

Bruce can then calculate the annual cost of producing this energy:

Annual production cost:

$$3\,504\,000\,\text{MWh} \times 9419\ \text{Btu/kWh} \times 1.25\ \$/\text{MBtu} = \$41\,255\,220$$

Finally, to estimate the revenue, Bruce assumes that this generating plant will be able to sell the energy it produces at 32 $/MWh. The annual revenue is thus:

Annual revenue:

$$3\,504\,000\,\text{MWh} \times 32\,\$/\text{MWh} = \$112\,128\,000$$

At this point, Bruce's spreadsheet looks like the following:

Year	Investment	Production	Production cost	Revenue	Net cash flow
0	$510 500 000	0	0	0	−$510 500 000
1	0	3 504 000	$41 255 220	$112 128 000	$70 872 780
2	0	3 504 000	$41 255 220	$112 128 000	$70 872 780
3	0	3 504 000	$41 255 220	$112 128 000	$70 872 780
...	0	...	...	...	...
30	0	3 504 000	$41 255 220	$112 128 000	$70 872 780

Bruce has thus assumed that all the investment costs are incurred during the year immediately before the plant starts generating and that the production, revenue, production cost, and net cash flow remain constant over the 30-year productive life of the plant. Using one of the functions provided by the spreadsheet software, Bruce then calculates the IRR for this stream of net cash flow. (See Sullivan et al. (2003) for a detailed explanation of the calculation of the IRR. Most spreadsheet programs provide a function for calculating this quantity.) He obtains a value of 13.58%. Borduria Power must then compare this value with their "minimum acceptable rate of return" (MARR) before they make their decision.

Before committing itself, Borduria Power will also want to consider the risks associated with this project. Two issues are of particular concern in this case: what happens if the price of electrical energy does not meet expectations or if the plant cannot achieve the targeted utilization factor? Using his spreadsheet, Bruce can easily recalculate the IRR for a

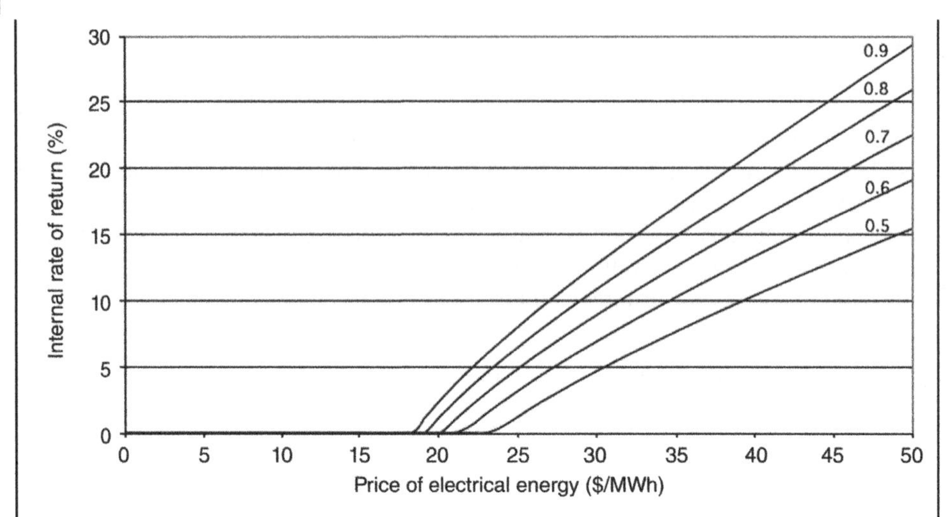

Figure 7.1 Internal Rate of Return for the coal unit of Example 7.1 as a function of the expected price of electrical energy for various values of the utilization factor.

range of prices and utilization factors and produce the graph shown in Figure 7.1. Assuming that the plant achieves a utilization factor of 80%, the average price at which it sells electricity cannot drop below about 30 $/MWh if the plant is to achieve an MARR of 12%. On the other hand, the average selling price would have to increase considerably if the utilization factor drops much below 80%.

Example 7.2

After considering the results shown in the previous section, the board of Borduria Power is concerned with the size of the initial investment and the risk associated with this project. It therefore asks Bruce to perform a similar analysis for a combined-cycle gas turbine (CCGT) plant. As the following table shows, the economics of this technology are quite different from those of a coal plant. The initial investment is much smaller and the energy conversion efficiency is much higher (because the heat rate is lower). On the other hand, a CCGT burns gas, a fuel that is much more expensive than coal in Borduria.[1]

Investment cost	533 $/kW
Expected plant life	30 years
Heat rate at rated output	6927 Btu/kWh
Expected fuel cost	3.00 $/MBtu

Adapted from DOE data cited by Stoft (2002).

1 When we developed this example for the first edition of this book, coal was indeed cheaper than natural gas in many countries. The development of fracking technology has significantly increased the supply of gas and reduced its price. This reversal illustrates the risks associated with making long-term investment decisions.

Assuming the same utilization factor (80%) and the same electricity price (32 \$/MWh), the annual production and the annual revenue from a CCGT plant of the same capacity would clearly be the same as for a coal plant:

Annual production:

$$0.8 \times 500 \text{ MW} \times 8760 \text{ h/year} = 3\,504\,000 \text{MWh}$$

Annual revenue:

$$3\,504\,000 \text{MWh} \times 32 \text{ \$/MWh} = \$112\,128\,000$$

On the other hand, the investment cost and the annual production cost would be different:

Investment cost:

$$533 \text{ \$/kW} \times 500 \text{ MW} = \$266\,500\,000$$

Annual production cost:

$$3\,504\,000 \text{MWh} \times 6927 \text{ Btu/kWh} \times 3.00 \text{ \$/MBtu} = \$72\,816\,624$$

Using his spreadsheet, Bruce again analyzes how the IRR varies with the projected price of electrical energy and the utilization factor. The results of this analysis, which are shown in Figure 7.2, suggest that a CCGT plant might yield a higher rate of return than a coal plant.

A decision between mutually exclusive investment alternatives, however, should not be based simply on a comparison of their respective rates of return. If the smaller investment (in this case the CCGT plant) produces an acceptable rate of return, the larger investment (the coal plant) should be treated as an incremental investment. Bruce therefore calculates the incremental net cash flow derived from this additional investment. This part of his spreadsheet looks like the following:

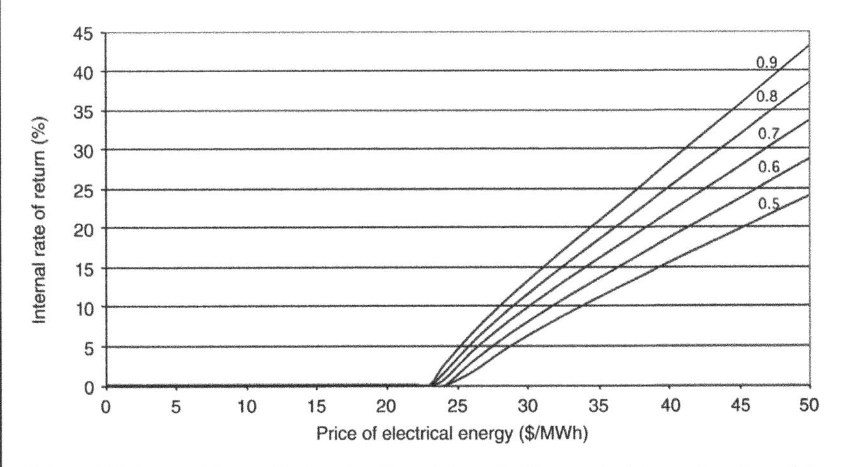

Figure 7.2 Internal Rate of Return for the CCGT unit of Example 7.2 as a function of the expected price of electrical energy for various values of the utilization factor.

Year	CCGT plant investment (A)	Coal plant investment (B)	CCGT plant production cost (C)	Coal plant production cost (D)	Incremental net cash flow (A) − (B) + (C) − (D)
0	$266 500 000	$510 500 000	0	0	−$244 000 000
1	0	0	$72 816 624	$41 255 220	$31 561 404
2	0	0	$72 816 624	$41 255 220	$31 561 404
3	0	0	$72 816 624	$41 255 220	$31 561 404
...	...	...	...	...	...
30	0	0	$72 816 624	$41 255 220	$31 561 404

The columns showing the annual production and the annual revenue are not shown because they are identical for both technologies. Bruce then calculates the IRR corresponding to the cash flow stream shown in the last column and gets a value of 12.56%. If the MARR of Borduria Power is set at 12%, building a coal plant rather than a CCGT plant would be justified, at least for this value of the utilization factor. In his report, Bruce includes the graph shown in Figure 7.3 and points out to the board of Borduria Power that this incremental IRR drops below 12% if the plant does not achieve an 80% utilization factor.

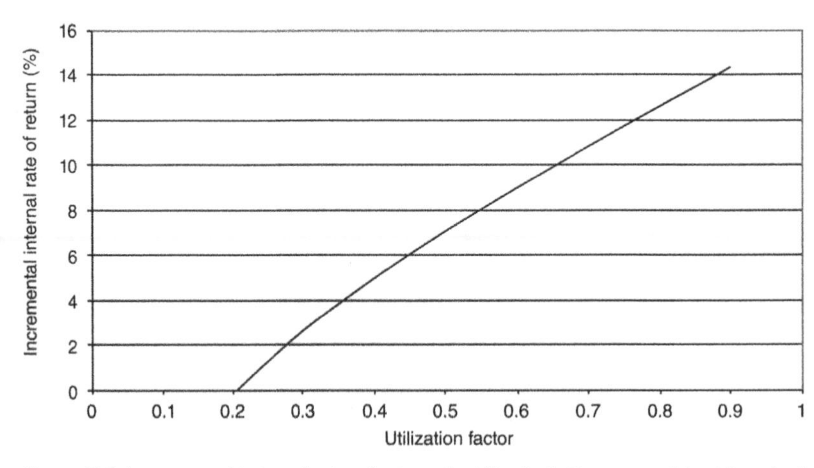

Figure 7.3 Incremental internal rate of return that Borduria Power would achieve by investing in a coal plant rather than a CCGT plant.

Example 7.3

While Borduria Power is considering building a plant that burns fossil fuel, Nick, the managing director of Syldavian Wind Power Ltd., has identified a promising site for the development of a 100 MW wind farm. The following table shows the plant parameters that Nick considers in his preliminary profitability calculation.

Investment cost	919 $/kW
Expected plant life	30 years
Heat rate at nominal output	0
Expected fuel cost	0

Adapted from DOE data cited by Stoft (2002).

The initial investment cost is thus: 919 $/kW × 100 MW = $91 900 000.

Since the wind is free and the maintenance and operation costs are neglected in this first approximation, Nick does not need to consider an annual production cost. At 32 $/MWh, his best estimate of the average price of electricity during the lifetime of the plant happens to be identical to the one used by Borduria Power. Even though the site that Nick is considering has an excellent wind regime, the utilization factor of a wind farm is unlikely to exceed 35%.

Annual production:

$$0.35 \times 100 \text{ MW} \times 8760 \text{ h/year} = 306\,600\,\text{MWh}$$

Annual revenue:

$$306\,600\,\text{MWh} \times 32\ \$/\text{MWh} = \$9\,811\,200$$

Nick's spreadsheet thus looks like the following one:

Year	Investment	Production	Production cost	Revenue	Net cash flow
0	$91 900 000	0	0	0	−$91 900 000
1	0	306 600	0	$9 811 200	$9 811 200
2	0	306 600	0	$9 811 200	$9 811 200
3	0	306 600	0	$9 811 200	$9 811 200
...	0	...	...	...	...
30	0	306 600	0	$9 811 200	$9 811 200

Over the 30-year expected lifespan of the wind farm, the stream of net cash flow shown in the last column yields an IRR of 10.08%. This is less than the 12% return that Borduria Power considers acceptable but it meets the less exacting 10% MARR used by Syldavian Wind Power Ltd.

7.2.2 Retiring Generation Capacity

Once a generating plant goes into operation, its designed lifetime becomes a theoretical reference point around which the actual lifetime can deviate significantly. Market conditions may indeed change so much that the revenues generated by the plant no longer cover its operating costs. Unless there are reasons to believe that market conditions will improve, the plant must then be retired. It is worth emphasizing that, in a competitive environment, such a decision is based solely on the future revenue and cost prospects for the plant and does not take into account the technical fitness of the plant or sunk costs. On the other hand, recoverable costs (such as the value of the land on which the plant is built) are taken into consideration in such a decision because they represent revenues that become available.

Example 7.4

On the basis of Bruce's report, the board of Borduria Power decided to build the coal plant discussed in Example 7.1. Unfortunately, after only 15 years of operation, this plant has run into trouble. Because of increased demand, the price of the low-sulfur coal burnt by the plant has climbed to 2.35 \$/MBtu. Moreover, the government of Borduria has imposed an environmental tax of 1.00 \$/MWh on the output of fossil-fuel plants. Under these conditions, the marginal cost of production of the plant has risen to:

$$2.35\ \$/\text{MBtu} \times 9419\ \text{Btu/kWh} + 1\ \$/\text{MWh} = 23.135\ \$/\text{MWh}$$

At the same time, competitors have put in service more efficient CCGT plants that have depressed the average price of electrical energy to 23.00 \$/MWh. Assuming a utilization factor of 80%, the plant makes an annual loss of:

$$(23.135 - 23.00)\ \$/\text{MWh} \times 0.8 \times 500\ \text{MW} \times 8760\ \text{h/year} = \$473040$$

A market analysis commissioned by Borduria Power suggests that this situation is unlikely to change because more high efficiency plants are expected to come on-line over the next few years. These changes in the generation mix are expected to result in a drop in the price of electrical energy to 22.00 \$/MWh. Furthermore, a further increase in the price of low-sulfur coal is predicted. On this basis, the plant should be decommissioned immediately to recover the value of the land, which is estimated at \$10 000 000.

Before making a final decision on the decommissioning of the plant, Borduria Power investigates another possibility. Instead of burning low-sulfur coal, it could switch to high-sulfur coal, which costs only 1.67 \$/MBtu. This change would require an investment of \$50 000 000 for the installation of flue gas desulfurization (FGD) equipment. This installation would take a year and would have a detrimental effect on the heat rate of the plant, which would increase to about 11 500 Btu/kWh. Keeping the other economic assumptions unchanged, the effect of this plant refurbishment over the remaining 15 years of the plant's life is summarized in the following spreadsheet:

Year	Investment	Production	Production cost (incl. tax)	Revenue	Net cash flow
0	\$50 000 000	0	0	0	−\$50 000 000
1	0	3 504 000	\$70 798 320	\$77 088 000	\$6 289 680
2	0	3 504 000	\$70 798 320	\$77 088 000	\$6 289 680
3	0	3 504 000	\$70 798 320	\$77 088 000	\$6 289 680
...	0	...	...	...	...
15	0	3 504 000	\$70 798 320	\$87 088 000	\$16 289 680

The revenue for year 15 includes the estimated value of the land. While this investment would create a positive cash flow, the Net Present Value of this cash flow stream is equal to −\$4 763 285. Investing in FGD would therefore not be profitable and the plant should definitely be retired.

7.2.3 Effect of a Cyclical Demand

If the demand for electricity increases without a compensating increase in generation capacity, or if the available capacity decreases because generating units are decommissioned, the market price for electrical energy will rise. Such a price increase provides an incentive for generating companies to invest in new plants. As we discussed in Chapter 2, production capacity will expand up to the point where the market price is equal to the long-run marginal cost of producing electrical energy. On a superficial level, investments in electrical generation capacity are governed by the principles that apply to the production of any commodity. We must, however, take into consideration the cyclical nature of the demand for electrical energy and the significant effect that the weather has on this demand. While electrical energy is by no means the only commodity that exhibits such fluctuations in demand, it is the only one that cannot easily be stored. Its production must therefore match the consumption, not only over a period of days or weeks but on a second-by-second basis. When the instantaneous demand decreases, some generating units have to operate below their rated capacity or be temporarily shut down. To properly assess investments in generation capacity, we therefore need to know for how many hours each year the load is less than a given value. This information is encapsulated in what is called a *load-duration curve*. Figure 7.4 shows this curve for the Pennsylvania–Jersey–Maryland (PJM) system during the year 1999. From this curve, we can observe that the load in this system was never less than 17 500 MW or greater than 51 700 MW during that year. We can also see that it exceeded 40 000 MW for only about 8% of the 8760 h that made up 1999.

Since the shape of this curve is typical, we can conclude that the installed generation capacity in a power system must be substantially larger than the demand averaged over a

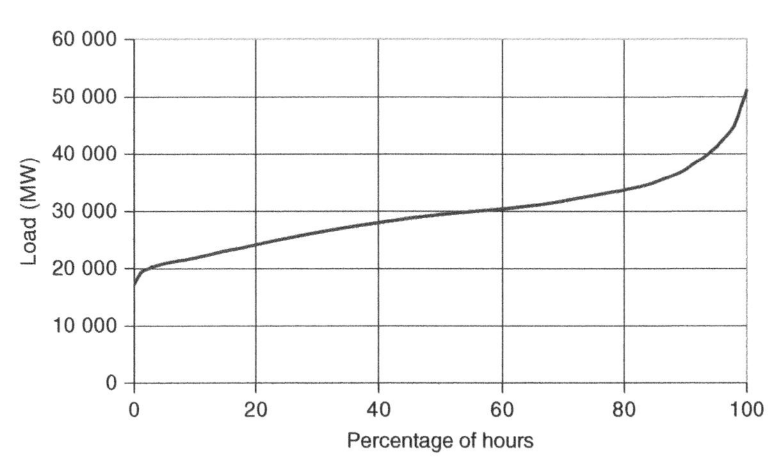

Figure 7.4 Load-duration curve for the PJM system during the year 1999. Source: www.pjm.com.

whole year. This means that not all generators can expect to have a utilization factor close to unity. In an efficient competitive market, a generator with a lower marginal production cost will usually have the opportunity to produce before a generator with a higher marginal cost. Cheap generators therefore achieve higher utilization factors than less efficient ones. As one would expect, prices are thus lower during periods of low demand than during periods of high demand. Competition during periods of low demand should also be more intense than during periods of high demand. During periods of high demand, most generators are indeed fully loaded and are not actively competing. Competition is then limited to a smaller number of expensive generating units. On the other hand, during periods of low demand, even efficient generators may have to compete to remain on-line and avoid incurring startup costs.

These conjectures are confirmed by the *price-duration curve* of the PJM system for 1999 shown in Figure 7.5. This curve shows the fraction of the number of hours of the year during which the price was less than a given value. The shape of this curve is similar to the shape of the load-duration curve, but its extremities are distorted by the variations in the intensity of competition.

As we discussed in previous chapters, the marginal generator sets the market price. In an efficient competitive market, this generator has no incentive to bid higher or lower than its marginal cost of production. The market price is therefore equal to the cost of *producing* the last megawatt hours. While the marginal generator will not lose money on the sale of the electrical energy it produces, it will not collect any economic profit either. On the other hand, the inframarginal generators collect an economic profit. Despite its name, all of this economic profit cannot be passed on to the shareholders to remunerate their investment. Part of this money must be used to cover the fixed costs of the plant. These fixed costs include the maintenance costs that do not vary with production, the personnel costs, taxes on the value or capacity of the plant, and the opportunity cost on the recoverable value of the plant.

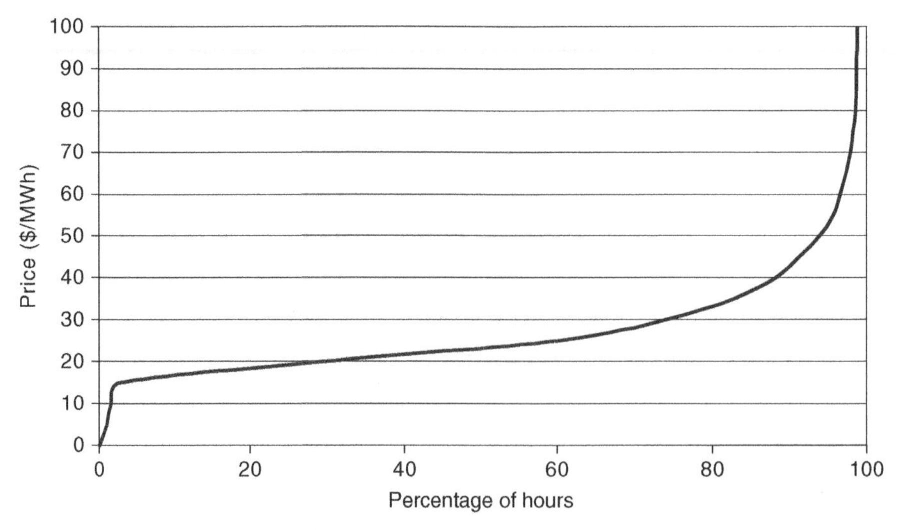

Figure 7.5 Price-duration curve for the PJM system during the year 1999. To enhance readability, the price axis has been limited to 100 $/MWh. In fact, the curve peaks at 1000 $/MWh for 100% of hours. The price shown is the average locational marginal price in the system. Source: www.pjm.com.

But what happens to the least efficient and therefore most marginal of generators? This generator is needed only when the load reaches its maximum on a very hot summer day or a very cold and dark winter evening. By definition, it is never inframarginal and thus never collects economic profits, while other generators set the market price. If this generator bids purely on the basis of its short-run marginal cost of production, it will never collect money to pay its fixed costs. If this generator is to stay in business, it must factor its fixed costs in its bids.

Example 7.5

Faced with the introduction of a competitive electricity market, Harry, the vice-president for operations of Syldavia Electric, must decide what to do with the Skunk River plant, an aging oil-fired 50 MW generating station. This plant has a heat rate of 12 000 Btu/kWh and burns a fuel that costs 3.00 $/MBtu. Since it is among its least efficient, Syldavia Electric has used the Skunk River plant in recent years only during periods of extremely high load. Harry first calculates the fixed costs associated with this plant and gets a figure of about $280 000 per year. He then tries to estimate the minimum bid that the Skunk River plant should submit to recover all of its costs. This means that the revenue produced by the plant should be equal to its total operating cost, which can be expressed as:

$$\text{Production (MWh)} \times \text{bid (\$/MWh)} = \text{fixed cost (\$)} + \text{production (MWh)} \times \text{heat rate (Btu/kWh)} \times \text{fuel cost (\$/MBtu)}$$

Since Harry does not really know what the production will be, he decides to calculate the minimum bid for a range of values. To simplify the calculations, he assumes that, if it runs, the unit will produce at maximum capacity. Harry can then calculate the minimum bid price as a function of the number of hours or of the utilization factor. Figure 7.6 summarizes his results and shows that the minimum bid increases beyond 1000 $/MWh if the unit runs only 5 h per year. For comparison, the marginal cost of production of the

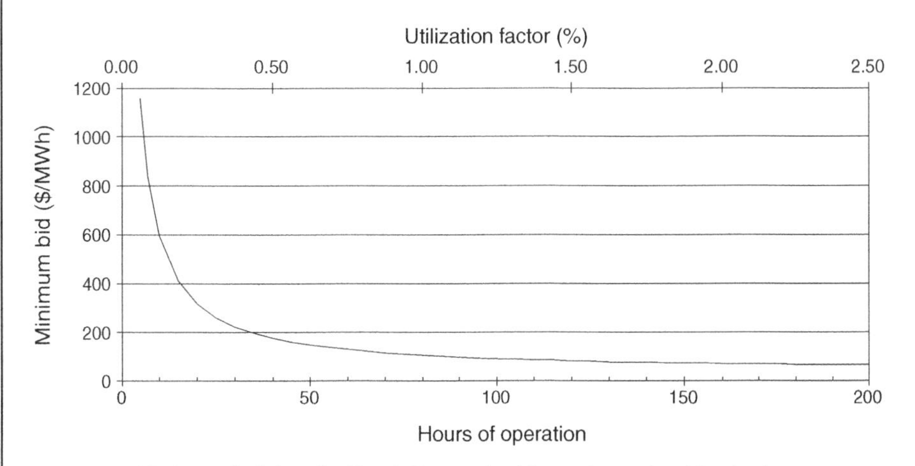

Figure 7.6 Minimum bid that the Skunk River unit of Example 7.5 should submit to recover both its fixed and variable costs, as a function of the number of hours of operation at full load that this unit expects to achieve.

Skunk River plant is 36 $/MWh. Such prices might seem totally unreasonable, but from Harry's perspective they are entirely justified. Moreover, he is also very likely to obtain whatever bids he submits during these few hours a year because the only competition he is likely to face would be from plants that are in the same situation as his. While consumers might balk at these prices, the alternative is not to consume at all during these periods because the Skunk River plant could be described as the plant of last resort.

Owners of very marginal plants who use this approach to set their prices need to estimate the number of hours that the plant is likely to run each year. This is not an easy task because this quantity is affected by several random factors. An average or expected value can be predicted based on historical data and predictions for load growth and the retirement of other generating plants. The actual value, however, may deviate significantly from this average. For example, during a warm winter or a cool summer, the demand may reach critical values less frequently and for fewer hours than expected. Similarly, insufficient precipitations may reduce the availability of hydro energy and increase the need for thermal generation. Depending on the conditions, very marginal plants therefore might be called frequently or not at all. While their revenues may be acceptable if averaged over a number of years, the possibility of losing money for one or more years might be more than a risk-averse owner would tolerate. Such plants are therefore prime candidates for retirement.

7.3 Generation Capacity from the Customers' Perspective

In the first part of this chapter, we took the perspective of a potential investor who tries to decide whether to build a new generating plant. We also considered the decision process of the owner of a plant who is trying to decide whether the time has come to shut it down. In this section, we consider the provision of generation capacity from the consumers' perspective. In a completely deregulated environment, there is no obligation on any company to build power plants. The total generation capacity that is available for supplying the demand therefore arises from individual decisions based on perceptions of profit opportunities.

We will first discuss whether the decision to build generation capacity can be driven entirely by the profits obtainable in the markets for electrical energy. If this is not satisfactory, market-based expansion needs to be supplemented by a centralized mechanism designed to ensure or encourage the availability of a certain amount of capacity. We will discuss three forms that this mechanism can take.

At this point, we must note that consumers have some reliability expectations when they purchase electrical energy. This means that the energy should be delivered when consumers demand it and not at some other time. Since generating units are occasionally unavailable because of breakdowns or the need to perform maintenance, the power system must have available more generation capacity than is necessary to meet the peak demand. Increasing this generation capacity margin improves the reliability of the system. The mechanisms described below should therefore be judged on not only their ability to provide enough generation capacity to deliver the electrical energy demanded by consumers but also their ability to meet their reliability expectations.

7.3.1 Expansion Driven by the Market for Electrical Energy

Some power system economists (see, for example, the comprehensive exposition of this point of view by Stoft (2002)) insist that electrical energy should be treated like any other commodity. They argue that:

- If electrical energy is traded on a free market, there is no need for a centralized mechanism for controlling or encouraging investments in generating plants.
- If left alone, markets will determine the optimal level of production capacity called for by the demand.
- Interfering with the market distorts prices and incentives.
- Centralized planning and subsidies lead to overinvestment or underinvestment, both of which are economically inefficient.

As we saw in Chapter 2, if the demand for a commodity increases, or its supply decreases, the resulting upswing in market price encourages additional investments in production capacity and a new long-run equilibrium is ultimately reached. Because of the cyclical nature of the demand for electricity and its lack of elasticity, price increases on electricity markets are usually not smooth and gradual. Instead, we are likely to observe price spikes (i.e. very large increases in price over short periods of time) when the demand approaches the total installed generation capacity. Figure 7.7 illustrates this phenomenon. A typical supply function is represented by a stylized, three-segment, piecewise linear curve. The first, moderately sloped, segment represents the bulk of the generating units in a reasonably competitive market. The second segment, which has a much steeper slope, represents the peaking units that are called infrequently. The third segment is vertical and represents the supply function when all the existing generation capacity is in use. An almost vertical line represents the low-elasticity demand function. This demand function moves horizontally as the demand fluctuates over time. Two curves are shown: one representing the minimum demand period and the other the peak demand period. The intersections of these curves with the supply function determine the minimum and the maximum prices. When the generation capacity is tight but sufficient to meet the load (Figure 7.7a), the price rises sharply during periods of peak demand because the market price is determined by the bids of generating units that operate very infrequently. These price spikes are much higher when all the generation capacity is in use under peak load conditions (Figure 7.7b). Such a situation could happen because the installed generation capacity has not kept up with the load growth, because some generation capacity has been retired or because some generation capacity is unavailable (for example, because a drought has reduced the amount of available hydro energy). Under these conditions, the only factor that would limit the price increase is elasticity of the demand. Note that this reasoning assumes that the demand does respond to prices. If this is not the case, load must be shed to prevent a collapse of the system.

In practice, these price spikes are significantly higher than what Figure 7.7 suggests, and they are sufficient to substantially increase the average price of electricity even if they occur only a few times a year. Price spikes therefore provide a vivid signal that there is not enough capacity to meet the demand and the "extra" revenue that they produce is essential to give generating companies the incentive that they need to invest in new generation capacity or keep older units available.

(a)

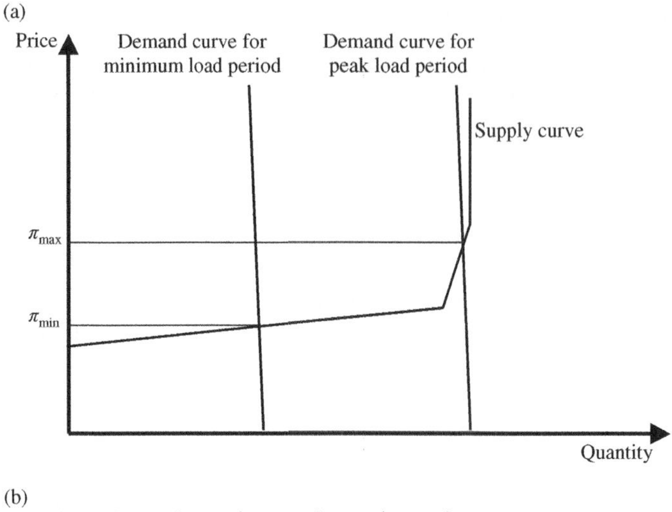

(b)

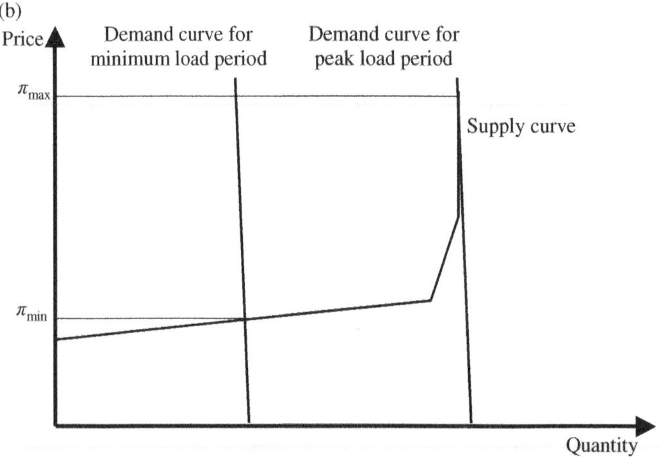

Figure 7.7 Illustration of the mechanism leading to price spikes in markets for electrical energy. (a) Sufficient generation capacity. (b) Insufficient generation capacity.

These price spikes are obviously very expensive (one might say painful) for the consumers. They should thus encourage them to become more responsive to price signals. As the price elasticity of the demand increases, the magnitude of the spikes decreases, even if the balance between peak load and generation capacity does not improve. Price spikes also give consumers a strong incentive to enter into contracts that encourage generators to invest in generation capacity.

According to this theory, which is supported by quite sophisticated mathematical models (see the classical work of Schweppe et al. (1988)), an equilibrium should eventually be reached. At this equilibrium, the balance between investments in generation capacity and investments in load control equipment is optimal and the global welfare is maximum. However, several practical and behavioral problems and their political consequences may prevent this equilibrium from being reached. First, the technology required to make a sufficient portion of the demand responsive to short-term price signals has not yet been deployed on a large scale. Until such technology becomes widely available and accepted, it may still be necessary to implement quantity rationing rather than price rationing when demand exceeds supply. In other words, the system operator may have to disconnect loads to keep the system in balance during periods of peak demand. Widespread load disconnections are extremely unpopular and often have disastrous social consequences, such as accidents, and vandalism. They are also economically very inefficient. Their impact can be estimated using the Value of Lost Load (VoLL), which is several orders of magnitude larger than the cost of the energy not supplied. Consumers are not used to such disruptions, and it is unlikely that their political representatives would tolerate them for any length of time.

When consumers are exposed to spot prices and are expected to adjust their demand accordingly, price spikes are very unpopular. Since the origin of these spikes is rather hard to explain and justify to nonspecialists, consumers often believe that they are being ripped off. Price spikes also have socially unacceptable consequences such as forcing poor and vulnerable people to cut back on their consumption of electrical energy for essential needs such as heating, cooking, and air conditioning. To be politically acceptable, many electricity markets therefore incorporate a price cap designed to prevent large price spikes. Such a price cap obviously removes a good part of the incentive for building or keeping generation capacity.

An electricity market that relies on spikes in the price of electrical energy to encourage the development of generation capacity is not necessarily good for investors either. Price spikes may not materialize and the average price of electricity may be substantially lower if the weather is more temperate or if higher-than-average precipitations make hydro energy more abundant. Basing investment decisions on such signals represents a significant risk for investors. This risk may deter them from committing to the construction of new plants.

Finally, simulation models developed by Ford (1999, 2001) suggest that the time it takes to obtain planning permission for a power plant and to build this plant can create instability in the market. Instead of increasing smoothly in response to load growth, the generation capacity might go through a series of boom and bust cycles. A lack of generation capacity produces very high electricity prices and triggers a boom in power plant construction. This boom results in a glut of capacity that depresses prices and discourages construction until the overcapacity has been resorbed. Such boom-and-bust cycles are not in the long-term interest of either producers or consumers.

In conclusion, it appears that relying solely on the market for electrical energy and its price spikes to bring about enough generation capacity is unlikely to give satisfactory results. This approach presumes that consumers are only buying electrical energy and that this transaction can be treated as the purchase of a commodity. In practice, consumers do not purchase only electrical energy but a service that can be defined as the provision of electrical energy with a certain level of reliability.

7.3.2 Capacity Payments

The risk associated with leaving generation investments to the invisible hand of the electrical energy market has often been judged to be too great. Market designers in several countries and regions have decided that, rather than occasionally paying generators large amounts of money because of shortage-induced price spikes, it was preferable to pay them a smaller amount on a regular basis. Payments would be proportional to the amount of capacity made available by each generator. These capacity payments form a stream of revenue that is separate from the money that generators derive from the market for electrical energy. They should cover at least part of the capital cost of new generating units and encourage generating companies to keep available units that are rarely called upon to produce energy. By increasing the total available capacity, these payments reduce, but do not eliminate, the likelihood of shortages. More production capacity also enhances competition and restrains prices in the market for electrical energy.

Capacity payments thus reduce the risks described in the previous section and spread them among all consumers, irrespective of the timing of their demand for electrical energy. At least in the short term, this socialization of the cost of peaking energy benefits the risk-averse market participants, whether they are consumers or producers. In the long term, this approach reduces the incentive for economically efficient behavior: too much capital may be invested in generation capacity and too little on devices that consumers could use to control their demand.

There are also practical difficulties. First, there is no clear way to determine either the total amount of money to be spent on capacity payments or the rate to be paid per MW of installed capacity. Second, such a system can also lead to endless debates about how much should be paid to each generator. For example, it can be argued that thermal and hydro plants do not make the same contribution to reliability because a drought may limit the output of hydro units. Finally, because capacity payments are not tied to any performance criteria, it is not obvious that they actually do enhance reliability.

In an attempt to get around these difficulties, the Electricity Pool of England and Wales adopted an alternative approach. The centrally determined price of electrical energy during each period t was increased by a capacity element equal to:

$$\mathrm{CE}_t = \mathrm{VoLL} \times \mathrm{LoLP}_t \tag{7.1}$$

where VoLL is the Value of Lost Load (determined through a customer survey and updated annually to take inflation into account) and LoLP_t is the Loss of Load Probability during period t. Since this probability depends on the margin between the load and the available capacity and on the outage rates of the units, this capacity element fluctuated from one period to the next and occasionally caused significant price spikes. In exchange for this capacity element, the energy price was capped at the VoLL. The money collected during each half-hourly period through this capacity element was divided among all the generating units that submitted bids to supply energy, regardless of whether or not they had been scheduled to produce energy. The capacity element was intended to send a short-term signal to consumers, while the associated capacity payments were designed to provide a long-term incentive to producers. While the capacity payments procured significant revenues to the generators and helped maintain a substantial generation capacity, their dependence on LoLP, which is a short-term variable, made them easy to manipulate by the large generating companies. These payments were abandoned when the New Electricity Trading Arrangements were introduced.

7.3.3 Capacity Market

Rather than fixing the total amount or the rate of capacity payments, some regulatory authorities set a generation adequacy target and determine the amount of generation capacity required to achieve this target. All energy retailers and large consumers (i.e. all entities that buy energy) are then obligated to buy their share of this requirement on an organized capacity market. While the amount of capacity to be purchased on this market is determined administratively, its price depends on the capacity on offer and may be quite volatile.

Implementing a capacity market that achieves its purpose is not a simple matter. Several important issues must be considered carefully. The first, and probably most fundamental, of these issues is the length of the market periods, i.e. the period over which each retailer's capacity obligations are calculated. Retailers prefer a shorter period (say a month or less) because it reduces the amount of capacity that they have to purchase during periods of light load. A shorter period also increases the liquidity of the capacity market. On the other hand, a longer period (e.g. a season or a year) favors generators and encourages the building of new capacity. In an interconnected system, it discourages existing generators from selling their capacity in a neighboring market. A longer period also matches more closely the frequency at which the regulatory authorities evaluate the reliability of the system.

The installed generation capacity must exceed the peak demand because generators can fail at any time. Unreliable generators therefore increase the size of the required generation capacity margin and impose a cost on the entire system. Choosing an appropriate method to evaluate and reward the performance of generators is thus the second major issue in the design of a capacity market. This method should track as closely as possible the reliability of the system. It should reward reliable plants and encourage the retirement of unreliable units. For example, in the Pennsylvania-Jersey Maryland market, the amount of capacity that generators are allowed to offer in the capacity market is derated by their historical forced outage rate. Generators thus have an incentive to maintain or improve the availability of their units. Ideally, these performance criteria should be refined to encourage generators not only to build or retain capacity but also to operate in such a way that it is available during critical periods.

An energy buyer who does not purchase its share of the target generation capacity benefits from the installed capacity margin paid for by the other market participants. It also has a cost advantage in the energy market. A deficiency payment or penalty must therefore be imposed on any entity that does not meet its obligations. The level of this payment and the rules for its imposition must be set in a way that encourages proper behavior and discourages free riders.

Some electricity markets have started to include the demand side in their capacity markets. In these markets, large consumers or entities who aggregate a sufficient number of small consumers bid their ability to reduce their demand when requested by the system operator. See, for example, PJM Interconnection (2017) for a detailed discussion of the operation of such a market.

7.3.4 Reliability Contracts

Ideally, every consumer should decide freely and independently how much it is willing to pay for reliability. In a mature electricity market, it would then be able to enter into a long-term contract with a generator that would guarantee the delivery of energy with this level of reliability. Such long-term contracts would give generators the incentive to build the amount of capacity required to achieve the desired level of reliability.

Until electricity markets achieve the level of maturity where this approach becomes possible, a central authority (for example, the regulator or the system operator) could purchase reliability on behalf of consumers. Instead of setting a target for installed capacity as happens in capacity markets, this central authority could auction reliability contracts as proposed by Vazquez et al. (2002). Such contracts consist essentially of long-term call options with a substantial penalty for nondelivery. The central authority uses reliability criteria to determine the total amount Q of contracts to be purchased and sets the strike price s of these contracts, typically at 25% above the variable cost of the most expensive generator that is expected to be called. It also sets the duration of the contracts. Bids for these contracts are ranked in terms of the premium fee asked by the generators. The premium fee P that clears the quantity Q is paid for all contracts.

Let us consider a generator that has sold an option for q MW at a premium P. This generator receives a premium fee of $P \cdot q$ for every period of the duration of the contract. For each period during which the spot price of electrical energy π exceeds the strike price s, this generator must reimburse $(\pi - s) \cdot q$ to the consumers. If this generator is only producing g MW during this period, it must pay an additional penalty of $\text{pen} \cdot (q - g)$.

Reliability contracts have a number of desirable consequences:

- They reduce the risks faced by marginal generators because the highly volatile and uncertain revenues derived from price spikes are replaced by a steady income from the option fees.
- The central authority can set the amount of contracts to be auctioned at a level that is likely to achieve the desired level of reliability.
- Generators have an incentive to maintain or increase the availability of their generating units because periods of high prices caused by shortages of capacity are less profitable. The penalty for nondelivery during periods of high prices discourages generators from bidding for contracts with less reliable units.
- In exchange for the money they pay above the cost of electrical energy, consumers get a hedge against very high prices. This is in direct contrast with capacity payments and capacity markets where the benefit for consumers is not tangible. Consumers also get the reassurance that the option fees are determined through a competitive auction.
- Finally, because the strike price is set significantly above competitive prices, the options become active only when the system is close to rationing. Interferences with the normal energy market are thus minimized.

7.4 Generation Capacity from Renewable Sources

7.4.1 The Investors' Perspective

As we illustrated in Example 7.3, generating units that rely on a renewable energy source have a very low operating cost. However, this does not mean that they are automatically profitable because their investment cost per MW of installed capacity are usually significantly higher than that for conventional generating plants. Their profitability thus depends on being able to generate as much energy as possible. As we discussed in Section 4.5.3.2, many governments encourage the development of renewable energy sources to combat climate change and have put in place mechanisms to reduce the financial risks associated with renewable generation.

7.4.2 The Consumers' Perspective

When system operators and regulators assess the adequacy of the installed generation to meet the expected demand, they always derate this capacity by a certain fraction to reflect the fact that no generating unit is available 100% of the time because of planned or forced outages. Because renewable generators rely on a source of energy that is often not controllable, their contribution to the capacity needed to ensure adequacy must be derated further because these generators may not be available during periods of peak demand. Estimating the actual contribution of renewable generators to adequacy can be controversial if the energy source is seasonal and the system load peaks at various times of the year. For example, solar generation can generally help meet the summer peak demand, but its contribution to a winter peak is more doubtful.

As the proportion of wind and solar generation capacity increases, their intermittency and stochasticity increase the need for resources able to compensate rapidly for substantial changes in the load/generation balance. Ensuring that enough generation capacity is available to meet reliably the peak demand is no longer sufficient. Generators and other resources (such as demand response and storage) must be sufficiently flexible to respond to these changes. Flexible generating units have large ramp-up and ramp-down rates, low minimum stable generation as well as short minimum up- and down-times. Storage typically has a fast response time but must have a large enough energy capacity to sustain this response. Demand-side resources must demonstrate their dependability. Taking the need for flexibility into account when assessing whether a portfolio of generation and other resources will meet future needs for electricity cannot be done on the basis of a projected load-duration curve because such curves do not reflect the time-domain variations of the load. To ensure that a set of resources will be able to meet a system's operating constraints, their operation must be simulated on a set of demand profiles that reflect a sufficiently wide range of anticipated system conditions. For a more detailed discussion of this issue, see Ma et al. (2013) or Ulbig and Andersson (2015).

7.5 Problems

Most of these problems require the use of a spreadsheet.

7.1 Calculate the IRR for an investment in a 400 MW power plant with an expected life of 30 years. This plant costs 1200 $/kW to build and has a heat rate of 9800 Btu/kWh. It burns a fuel that costs 1.10 $/MBtu. On average, it is expected to operate at maximum capacity for 7446 h per year and sell its output at an average price of 31 $/MWh. What should be the average price of electrical energy if this investment is to achieve an MARR of 13%?

7.2 What would be the IRR of the unit of Problem 7.1 if the utilization rate drops by 15% after 10 years and by another 15% after 20 years?

7.3 What would be the IRR of the unit of Problem 7.1 if the price of electrical energy was 35 $/MWh during the first 10 years of the expected life of the plant before dropping to 31 $/MWh? What would be the value to the IRR if this price was 31 $/MWh during the first 20 years and $35 $/MWh during the last 10 years? Compare these results with the IRR calculated in Problem 7.1 and explain the differences.

7.4 In an effort to meet its obligation under the Kyoto agreement, the government of Syldavia has decided to encourage the construction of renewable generation by guaranteeing to buy their output at a fixed price of 35 $/MWh. Greener Syldavia Power Company is considering taking advantage of this program by building a 200 MW wind farm. This wind farm has an expected life of 30 years and its building cost amounts to 850 $/kW. Based on an analysis of the wind regime at the proposed location, the engineers of Greener Syldavia Power Company estimate that the output of the plant will be as shown in the table below:

Output as a fraction of capacity (%)	Hours per year
100	1700
75	1200
50	850
25	400
0	4610

Given that the Greener Syldavia Power Company has set itself an MARR of 12%, should it take the government's offer and build this wind farm?

7.5 Syldavia Energy is exploring the possibility of building a new 600 MW power plant. Given the parameters shown in the table below, which technology should it adopt for this plant, assuming that the plant would have a utilization factor of 0.80 and would be able to sell its output at an average price of 30 $/MWh? Syldavia Energy uses an MARR of 12%.

	Technology A	Technology B
Investment cost	1100 $/kW	650 $/kW
Expected plant life	30 years	30 years
Heat rate at rated output	7500 Btu/kWh	6500 Btu/kWh
Expected fuel cost	1.15 $/MBtu	2.75 $/MBtu

7.6 Borduria Power has built a plant with the following characteristics:

Investment cost	1000 $/kW
Capacity	400 MW
Expected plant life	30 years

Heat rate at rated output	9800 Btu/kWh
Expected fuel cost	1.10 $/MBtu
Expected utilization factor	0.85
Expected average selling price	31 $/MWh

After 5 years of operation, market conditions change dramatically. The fuel price increases to 1.50 $/MBtu, the utilization factor drops to 0.45 and the average price at which Borduria Power can sell the energy produced by this plant drops to 25 $/MWh.

What should Borduria Power do with this plant? What should Borduria Power have done if it had known about this change in market conditions? Assume that Borduria Power uses an MARR of 12% and ignore the recoverable cost of the plant.

7.7 Assume that Borduria Power decides to continue operating the plant of Problem 7.6 and that the market conditions do not improve. Five years later, the plant suffers a major breakdown that would cost $120 000 000 to repair. It is expected that this repair would allow the plant to continue operating for the rest of its design life. What should Borduria Power do? What should it do if this breakdown occurs 15 years after the plant was built?

7.8 An old 100 MW power plant has a heat rate of 13 000 Btu/kWh and burns a fuel that costs 2.90 $/MBtu. The owner of the plant estimates the fixed cost of keeping the plant available at $360 000 per year. What is the minimum price that would justify keeping this plant available if it has a 1% utilization rate? Compare this price with the average production cost of the plant.

7.9 The investment analysis illustrated by Examples 7.1 and 7.2 is quite simplified. Discuss what factors should be considered in a more detailed analysis.

7.10 Plot the load-duration curves and the price-duration curve for the power system in the region where you live or another system for which you have access to the necessary data. Compare the peak demand to the installed generation capacity in the system.

7.11 Repeat the previous problem for several years. Try to explain any significant differences that you may observe in terms of weather conditions, commissioning of new generating plants, retirement of old plants, and other relevant factors.

7.12 Determine if there is a mechanism to encourage the provision of generation capacity in the region where you live (or in another region for which you have access to sufficient information).

References

Ford, A. (1999). Cycles in competitive electricity markets: a simulation study of the western United States. *Energy Policy* 27: 637–658.

Ford, A. (2001). Waiting for the boom: a simulation study of power plant construction in California. *Energy Policy* 29: 847–869.

Ma, J., Silva, V., Belhomme, R. et al. (2013). Evaluating and planning flexibility in sustainable power systems. *IEEE Trans. Sustainable Energy* 4 (1): 200–209.

PJM Interconnection (2017). Demand response strategy. http://www.pjm.com/~/media/library/reports-notices/demand-response/20170628-pjm-demand-response-strategy.ashx (accessed 3 March 2018).

Schweppe, F.C., Caramanis, M.C., Tabors, R.D., and Bohn, R.E. (1988). *Spot Pricing of Electricity*. Kluwer Academic Publishers.

Stoft, S. (2002). *Power System Economics*. Wiley.

Sullivan, W.G., Wicks, E.M., and Luxhoj, J.T. (2003). *Engineering Economy*, 12e. Prentice Hall.

Ulbig, A. and Andersson, G. (2015). Analyzing operational flexibility of electric power systems. *Int. J. Electr. Power Energy Syst.* 72: 155–164.

Vazquez, C., Rivier, M., and Perez-Arriaga, I.J. (2002). A market approach to long-term security of supply. *IEEE Trans. Power Systems* 17 (2): 349–357.

Further Reading

Sullivan et al. (2003) have written an easy-to-read introduction to the techniques used to make investment decisions. Stoft (2002) discusses in considerable details the perceived advantages of electricity markets where decisions to invest in generation capacity are driven entirely by prices for electrical energy. Schweppe et al. (1988) is the classic reference on spot pricing. Ford (1999, 2001) presents some very interesting simulations that explore the factors that might lead to boom-and-bust cycles in the building of generation capacity. De Vries and Hakvoort (2003) discuss the advantages and disadvantages of the various methods used to encourage investments in generation capacity. Vazquez et al. (2002) introduce the concept of reliability contracts. Billinton and Allan (1996) explain in great detail the relation between generation capacity and system reliability.

Billinton, R. and Allan, R.N. (1996). *Reliability Evaluation of Power Systems*, 2e. Plenum Press.

de Vries, L.J. and Hakvoort, R.A. (2003). The question of generation adequacy in liberalized electricity Markets. *Proceedings of the 26th IAEE Annual Conference* (June 2003).

8

Investing in Transmission

8.1 Introduction

In Chapter 5, we studied the effect that an existing transmission network has on electricity markets. Expanding this transmission network through the construction of new lines or the upgrade of existing facilities increases the amount of power that can be traded reliably and the number of generators and consumers that can take part in this market. Transmission expansion thus enhances the competitiveness of the market. On the other hand, investments in new transmission equipment are costly and should be undertaken only if they can be justified economically. In order to deliver maximum economic welfare to society, the electricity supply industry should therefore follow the path of least cost long-term development. This implies that generation and transmission investments should be jointly optimized. However, the creation of open markets required the separation of generation and transmission activities to avoid the potential for discrimination between market participants. Unfortunately, this separation means that investments in generation and transmission are no longer coordinated by a vertically integrated utility but are handled separately by organizations with different structures and motivations. On the one hand, generating companies aggressively seek investment opportunities that will maximize their profits and will build new plants only if they have access to enough transmission capacity to be able to bring their production to the market. On the other hand, most transmission facilities are owned by monopoly companies. Before these companies are allowed to build new lines or upgrade the transmission network, they must convince the regulators that these investments are in the public interest. Regulators and legislators around the world are working on developing frameworks that encourage all stakeholders to work together to produce optimal expansion plans.

In keeping with the underlying philosophy of this book, we will not discuss these evolving frameworks. Instead, we concentrate on the fundamental principles underlying any investment in transmission. After a brief review of the essential characteristics of the transmission business, we discuss the traditional approach to transmission investments where investors are remunerated on the basis of the cost of the installed transmission equipment. We then turn our attention to the various streams of value that transmission creates and how these benefits can justify investments in transmission. The most obvious source of value stems from transmission's innate ability to perform locational arbitrage, i.e. reduce price disparities between regions. However, the transmission network can also be used for sharing reserve, balancing capacity, and capacity

Fundamentals of Power System Economics, Second Edition. Daniel S. Kirschen and Goran Strbac.

margin. We show how these other sources of value can be quantified to provide additional justifications for building new lines or expanding existing transmission capacity.

8.2 The Nature of the Transmission Business

In liberalized electricity markets, transmission is usually separated from the other components of a traditional vertically integrated utility. It is therefore useful to begin our discussion of transmission investments by considering some of the characteristics of transmission as a standalone business.

Rationale for a Transmission Business

The transmission business exists only because generators and loads that use the network are in the wrong place. Transmission's value increases with the distance that separates producers and consumers. If a reliable and environmentally friendly generation technology became cost–effective for domestic and commercial installations, the transmission business would probably disappear.

Transmission Is a Natural Monopoly

It is currently almost inconceivable that a group of investors would decide to build a completely new transmission network designed to operate in competition with an existing one. Because of their visual impact on the environment, it is also most unlikely that the construction of competing transmission lines along similar routes would be allowed. Furthermore, the minimum efficient size of a transmission network is such that electricity transmission is a good example of natural monopoly.

Like all monopolies that provide an essential service, electricity transmission must be regulated to ensure that it delivers an economically optimal combination of quality of service and price. Such an objective is not easy to achieve. Even though the consumers and generators pay for using the transmission network, the regulator in essence "buys" transmission capacity on their behalf. Its best judgment about how much capacity is optimal thus replaces the multitude of independent purchasing decisions that make up the demand curve in a competitive market.

In exchange for being granted a regional monopoly, a transmission company must accept that the regulatory authorities will determine its revenues. These revenues are usually set in such a way that investors get a relatively modest return on their capital. However, compared to other stock market investments, transmission companies are relatively safe because they do not face competition. In fact, the biggest risk that these companies face is regulatory, i.e. the risk that a change in regulatory principles or practices could decrease their allowed revenues.

Transmission Is a Capital-intensive Business

Transmitting electric power reliably and efficiently over long distances requires large amounts of expensive equipment. While the most visible items of equipment are obviously aerial transmission lines, the cost of transformers, switchgear, and reactive compensation devices is also very high. Maintaining the operational reliability of the system while operating it close to its physical limits requires ubiquitous protection devices, an extensive communication network, as well as sophisticated control centers. The cost of these investments is high compared to the recurring cost of operating the system. Making good investment decisions is thus the most important aspect of running a transmission company.

Transmission Assets have a Long Life

Transmission equipment is usually designed for an expected life ranging from 20 to 40 years or even longer. Conditions can change drastically over such a long period. Generating plants that were expected to produce the bulk of the demand for electrical energy may become prematurely obsolete because of changes in the cost of fuels or because of the emergence of a better technology. At the same time, uneven economic growth may shift the geographical distribution of the demand. A transmission line built on the basis of erroneous forecasts may therefore be used at only a fraction of its rating.

Transmission Investments Are Irreversible

Once a transmission line has been built, it cannot be redeployed in another location where it could be used more profitably. Other types of transmission equipment may be easier to move, but the cost of doing so is often prohibitive. The resale value of installed assets is very low. Owners of transmission networks therefore have to live with the consequences of their investment decisions for a very long time. A large investment that is not used as much as was initially expected is called a *stranded investment*. Investors must therefore analyze what the performance of an asset might be under a wide variety of scenarios. In a regulated environment, they usually get some form of assurance that they will be able to recover the value of their investment even if it becomes stranded because of unforeseen changes in the demand for transmission.

Transmission Investments Are Lumpy

Manufacturers sell transmission equipment in only a small number of standardized voltage and MVA ratings. It is therefore often not possible to build a transmission facility with a rating that exactly matches the need. Furthermore, transmission investments also tend to have a large fixed cost that is independent of the rating of the new facility. While it is occasionally possible to upgrade facilities as demand increases, these factors, combined with the low resale value of installed equipment, often make this process impractical and economically difficult to justify. Investments in transmission facilities thus occur infrequently and in large blocks. Early in its life, the capacity of a transmission facility tends to exceed the demand. Later on, it is likely to be utilized much more intensively, at least if the situation evolves as forecast.

Economies of Scale

Ideally, investments should be proportional to the capacity they provide. For transmission lines, this is clearly not the case. The cost of building the line itself is primarily proportional to its length because of the need to acquire the right of way, adapt the terrain, and erect the towers. The rating of the line affects the cost only through the size of the conductors and the height that the towers must have to accommodate higher voltages. In addition, new substations must be built at both ends or existing ones must be expanded. This cost is significant and almost independent of the amount of active power that the line can transport. Because of these fixed costs, the average cost of transmitting electricity decreases with the amount transported. Transmission networks thus involve important economies of scale.

8.3 Cost-Based Transmission Expansion

Under the traditional regulatory compact, regulated transmission companies collect enough revenues to cover the cost of their investment plus a rate of return sufficient to attract investors seeking a relatively safe investment. While this approach is conceptually simple, we need to explore two important questions:

- How much transmission capacity should be built?
- How should the cost of transmission be allocated among the users of the transmission network?

8.3.1 Setting the Level of Investment in Transmission Capacity

Under the traditional model, investments in transmission facilities are carried out according to a process that typically works as follows:

- Using demographic and economic projections, the transmission company forecasts the needs for transmission capacity.
- Based on this forecast, it prepares an expansion plan and submits it to the regulatory authorities.
- The regulatory authorities review this plan and decide which facilities may be built or upgraded.
- The transmission company builds these new facilities using capital provided by shareholders and bondholders.
- Once the new facilities are commissioned, the transmission company begins recovering the cost of these investments through charges that users of the network have to pay.

The price that consumers pay for electricity is clearly a function of the capacity of the transmission network. If the regulator allows the transmission company to build too much transmission capacity, the users pay for capacity that is not used. On the other hand, if too little capacity is available, congestion in the network reduces trading opportunities, increases prices in some areas, and depresses them in others. Users pay less in transmission charges but do not reap the benefits of an efficient transmission network. Too little or too much transmission capacity thus causes a loss of global welfare. Because of the inevitable uncertainty on the evolution of demand and generation, achieving this optimum is not easy. In practice, one might argue that, from an economic perspective, it is better to err on the side of too much transmission capacity. Transmission indeed accounts for only about 10% of the total cost of electricity to consumers. While the cost of overinvesting is not small, the potential cost of underinvesting is much higher because even a small deficit in transmission capacity can have a very large effect on the price of electrical energy, which represents about 60% of the total cost to consumers.

On the other hand, remunerating transmission companies on a rate-of-return basis encourages them to overstate the need for transmission capacity because building more facilities increases the revenue that they are allowed to collect from the users of the network. Regulatory authorities rarely have the resources and technical expertise required to assess accurately the expansion plans prepared by the transmission company.

In conclusion, remunerating transmission investments on the basis of their cost keeps the transmission company in business, which is usually in the best interest of all parties involved. This approach also ensures some predictability in the cost of transmitting electricity. On the other hand, it does not guarantee that the level of investment in transmission capacity is economically optimal.

8.3.2 Allocating the Cost of Transmission

Once the regulator has determined the revenue that the transmission company can collect to recover its investment, this embedded cost must be divided between the producers and consumers that use the transmission network. In the following paragraphs, we discuss briefly the principles of the main allocation methods that have been proposed. Readers interested in the details of these embedded cost methods should consult Marangon Lima (1996).

8.3.2.1 Postage Stamp Method

Under this method, all users must pay a "use of system charge" to gain access to the network of their local transmission company. This charge usually depends on the MW rating of the generating units for a producer or the peak demand for a consumer. It can also factor in the annual energy produced or consumed (in MWh). It is also often a function of the voltage level at which a consumer is connected to reflect the fact that a user connected directly to the transmission network does not make use of the sub-transmission and distribution networks. However, like a postage stamp, this charge usually does not depend on where the energy is coming from or going to, as long as it is within the local system.

The charge that each user pays thus reflects an average usage of the entire network rather than the use of specific transmission facilities. Charges are adjusted proportionally to ensure that the transmission company recovers all the revenue that it is entitled to collect.

Because of its simplicity, this method is the most common charging mechanism for the utilization of the local transmission network. Its main disadvantage is that the charges paid by each user do not reflect the actual utilization that they make of the network or the value they derive from being connected. In many cases, some users cross-subsidize others. This is not economically desirable because it distorts competition. For example, generators connected close to the main load centers could argue that they should not pay the same charges as remote generators because the energy they produce does not need to transit through long and expensive transmission lines to reach the consumers.

Another problem with the postage stamp approach is that it only covers the cost of using the local transmission network. If a producer wants to sell energy in a neighboring system, it may have to buy an additional postage stamp to get access to the neighboring network. If two trading partners are not located in adjacent networks, each intermediate transmission company may require the purchase of a separate postage stamp. Like in a stack of pancakes, the cost of each stamp may not be very high, but the overall expense may be substantial. This phenomenon is dubbed the "pancaking of rates." It is usually undesirable because the overall charge overestimates the actual cost of transmitting the energy and may make economically justified transactions unprofitable.

8.3.2.2 Contract Path Method

The contract path method finds its origins in the days when the electricity supply industry consisted mostly of vertically integrated utilities and energy transactions were infrequent. When a consumer wanted to buy energy from a producer other than its local utility, it was still making use of the network of this utility and therefore had to bear a proportionate share of the embedded costs of this network. A wheeling contract had to be set up to formalize this arrangement. In this method, the contract specifies an electrically continuous path (the contract path) along which the power is assumed to flow from the generator to the point of delivery. The producer and the consumer agree to pay for the duration of the contract a wheeling charge proportional to the amount of power transmitted. This wheeling charge provides part of the revenue that the utility needs to recover the cost of the transmission assets included in the contract path.

The producer and consumer thus pay only for the usage of specific network facilities and not a fraction of the average cost of the entire network. This method is somewhat more cost reflective than the postage stamp approach and remains relatively simple. However, the power traded does not follow the contract path but usually flows through a multitude of paths, as dictated by Kirchhoff's laws. Whether the contract path method is truly more cost reflective is thus questionable.

8.3.2.3 MW-mile Method

With the MW-mile method, power flow calculations are used to determine the actual paths that the power follows through the network. The amount of MW-mile of flow that each transaction causes is calculated. This amount is then multiplied by an agreed per-unit cost of transmission capacity to get the wheeling charge. The method can be refined to account for the fact that some transactions reduce the flow on some lines. If transmission networks were linear systems, this approach would be rigorous. Unfortunately, they are not. The base case from which transactions are evaluated and the order in which they are considered have a significant effect on the results. This nonuniqueness is undesirable.

8.3.2.4 Discussion

All the methods discussed in the previous section have been criticized because they lack a credible foundation in economic theory. In particular, they produce charges that are proportional to the average rather than the marginal cost of the network. This means that they do not provide correct economic signals. Nevertheless, because of their simplicity and ease of implementation, they have been used extensively.

8.4 The Arbitrage Value of Transmission

In a competitive market for electrical energy, transmission can be viewed as being in competition with generation. The transmission network indeed allows remote generators to compete with local ones. We can thus quantify the value of transmission based on the differences in the marginal cost or price of generation across the network. This value provides a sound basis for setting the price that producers and consumers should pay to use the network.

Example 8.1

Let us consider the two-bus, one-line system shown in Figure 8.1. For the sake of simplicity, we neglect the losses and ignore operational reliability considerations. We also assume that the capacity of both generators is such that each of them can supply the 1000 MW load on its own. Finally, we assume that the capacity of the transmission line is sufficient to support any power flow that may be required.

The consumers at bus B can either buy energy at 45 \$/MWh from the local generator G_2 or buy energy at 20 \$/MWh from the remote generator G_1 and pay for the transmission of this energy. If the cost of this transmission is less than 25 \$/MWh, consumers will choose to buy their energy from generator G_1 because the overall cost would be less than the 45 \$/MWh they would have to pay to buy energy from generator G_2.

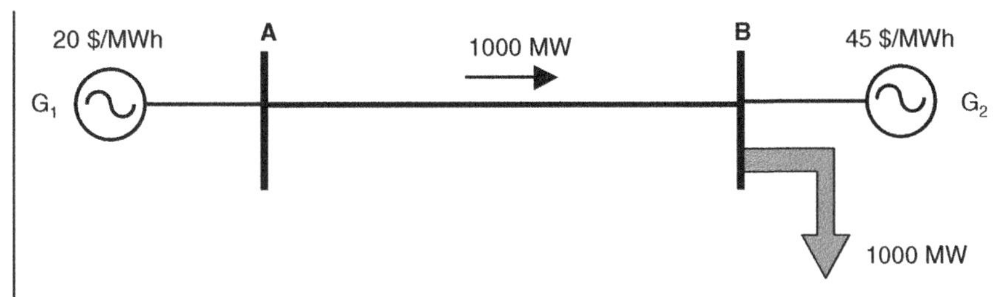

Figure 8.1 Simple example illustrating the value of transmission.

It is thus not in the best interest of the owner of the transmission line to charge more than 25 $/MWh because such a charge would discourage consumers from making use of the transmission system. In this example, the value of the transmission service is 25 $/MWh because at that price, consumers are indifferent between using and not using transmission. The value of transmission is thus a function of the short-run marginal cost (SRMC) of generation. In this case, this function is very simple because there is no limit to the substitution between transmission and local generation.

We can also look at the problem from an investment perspective. This transmission line should be built only if its amortized cost amounts to less than 25 $/MWh.

If the maximum output of the local generators is less than 1000 MW, the transmission line must be used to supply the load. The value of transmission is then no longer determined by the price of local generation but by the consumers' willingness to pay for electrical energy. In the short term, this could be significantly higher than 25 $/MWh. Limitations on local generation place the transmission provider in a monopoly position because the consumers have a choice between using transmission and giving up consumption. This monopoly position may not be sustainable in the long run because it would encourage the development of local generation.

Example 8.2

Let us revisit the Borduria/Syldavia example that we introduced in Chapter 5. In that chapter, we studied the effect that the operation of this interconnection has on market prices. We now want to determine the optimal capacity of this interconnection.

Our model for this interconnected system is the same as the one we used in Chapter 5 and is shown in Figure 8.2. The only difference is that the capacity of the interconnection is not fixed. Our starting point is the economic characteristics of the two markets when they operate independently. The supply functions for the electricity markets in Borduria and Syldavia are as follows:

$$\pi_B = MC_B = 10 + 0.01P_B \,\$/\text{MWh} \tag{8.1}$$

$$\pi_S = MC_S = 13 + 0.02P_S \,\$/\text{MWh} \tag{8.2}$$

The demands in Borduria and Syldavia are, respectively, 500 and 1500 MW. We continue to assume that these demands do not vary with time and are perfectly inelastic.

In the absence of an interconnection, the two national electricity markets operate independently and the prices in Borduria and Syldavia are, respectively, 15 and 43 $/MWh.

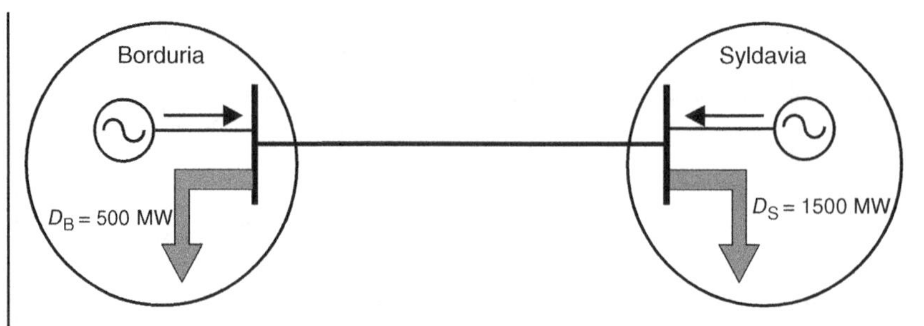

Figure 8.2 Model of the Borduria–Syldavia interconnection.

The value of transmitting the first megawatt-hour from Borduria to Syldavia is thus equal to the difference in price between the two countries, i.e. 28 \$/MWh.

We saw in Chapter 5 that when the flow through the interconnection is 400 MW, generators in Borduria produce 900 MW, of which 500 MW is consumed locally load while the remaining 400 MW is sold to consumers in Syldavia. The remaining 1100 MW of Syldavian load is produced locally. Under these conditions, the prices in Borduria and Syldavia are 19 and 35 \$/MWh, respectively. The value of transporting one additional megawatt-hour from Borduria to Syldavia is thus only 16 \$/MWh. This is also the maximum price that consumers in Syldavia would agree to pay for the transport of a megawatt-hour that they have bought in Borduria for 19 \$/MWh. If the price of transmission were any higher, they would prefer to buy this megawatt-hour from local generators.

When the flow on the interconnection reaches 933.3 MW, the prices in Borduria and Syldavia are equal:

$$\pi = \pi_B = \pi_S = 24.30\ \$/\text{MWh} \tag{8.3}$$

At that point, the marginal value of transmission is zero because Syldavian consumers can buy one extra megawatt-hour from local generators at the same price they would pay for a megawatt-hour purchased on the Bordurian market. They would therefore not be willing to pay anything for the transmission of this incremental energy. There is also no reason for any further increase in the amount of power transferred between the two countries because that would make the marginal value of transmission negative. Transmitting more power would require an increase in production in Borduria and would make the price of energy on that market higher than the price in Syldavia. The interconnection would then be transmitting energy from a higher priced location to a lower priced location. This would obviously be wasteful and economically inefficient. We can thus conclude that the marginal value of transmission is a function of the magnitude of the flow, which, in turn, depends on the energy prices and the capacity of the transmission network.

8.4.1 The Transmission Demand Function

We will now formalize the observations that we made in the examples above by introducing a demand function for transmission. This function gives the value of transmission in terms of the amount of the power F transmitted between Borduria and Syldavia:

$$\pi_T(F) = \pi_S(F) - \pi_B(F) \tag{8.4}$$

where $\pi_T(F)$ is the value of the transmission. The prices of electrical energy in Syldavia and Borduria, $\pi_S(F)$ and $\pi_B(F)$, are expressed in terms of the power transmitted. Substituting (8.1) and (8.2) into (8.4), we get the following expression:

$$\begin{aligned}\pi_T(F) &= [13 + 0.02P_S(F)] - [10 + 0.01P_B(F)]\\ &= 3 + 0.02P_S(F) - 0.01P_B(F)\end{aligned} \tag{8.5}$$

The production of the generators in Borduria and Syldavia can be expressed in terms of the flow on the interconnection and the local demands as follows:

$$P_B(F) = D_B + F \tag{8.6}$$

$$P_S(F) = D_S - F \tag{8.7}$$

Equation (8.5) then becomes:

$$\pi_T(F) = 3 + 0.02\ (D_S - F) - 0.01\ (F + D_B) \tag{8.8}$$

Substituting the known values for the demands, we get:

$$\pi_T(F) = 28 - 0.03F \tag{8.9}$$

Using this expression, we can check the results that we obtained above in an ad hoc manner. In particular, we see that when the flow is equal to zero, the price of transmission is 28 \$/MWh. Conversely, the transmission price drops to zero when the flow reaches 933.3 MW, which is the value of the flow for which the prices of generation in Borduria and Syldavia are equal.

Equation (8.9) can be inverted to get the demand for transmission as a function of its price:

$$F(\pi_T) = 933.3 - 33.3\pi_T \tag{8.10}$$

As Figure 8.3 shows, and as one would expect from any demand function, the demand for transmission increases when the price decreases.

It is interesting to examine the revenue that the owner of the transmission line would receive as a function of the capacity that it makes available. This revenue is equal to the price times the transmission capacity:

$$R(F) = \pi_T.F = (28 - 0.03F).F \tag{8.11}$$

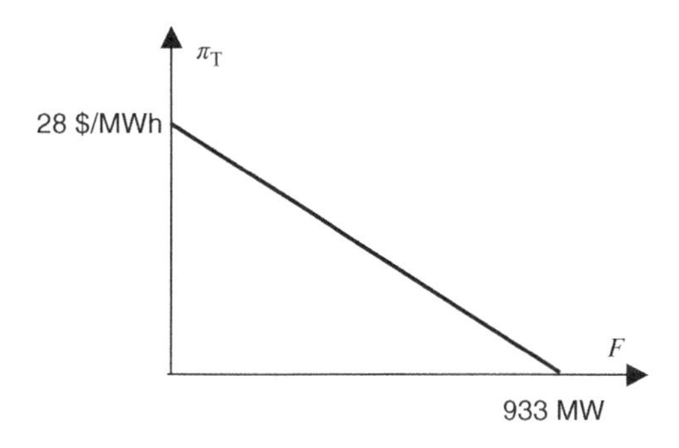

Figure 8.3 Transmission demand function for the interconnection between Borduria and Syldavia.

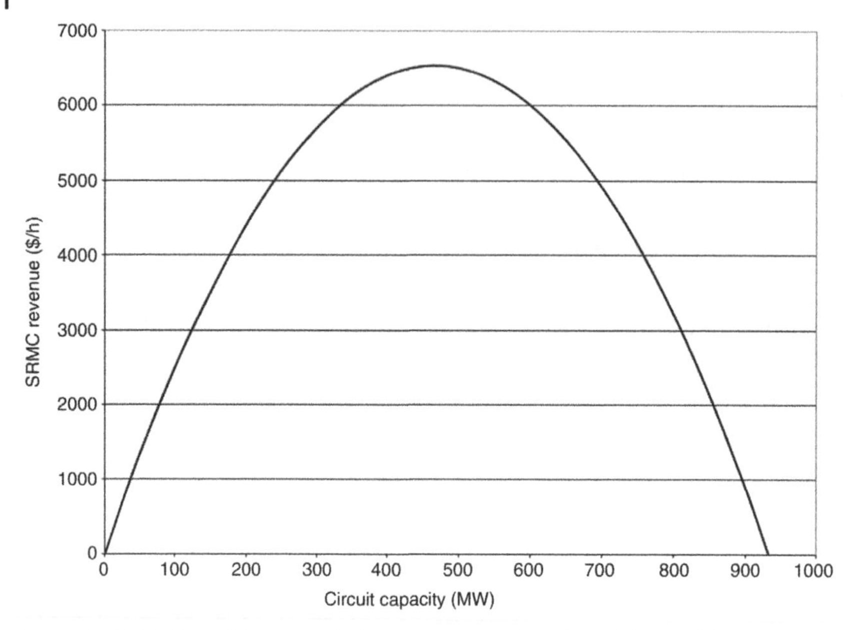

Figure 8.4 Variation of the transmission revenue as a function of the available capacity for the Borduria–Syldavia interconnection.

As illustrated in Figure 8.4, this revenue is a quadratic function of the amount of power transmitted. If no capacity is made available, the transmission owner obviously collects no revenue as no power is being transmitted. On the other hand, for a transmission capacity of 933 MW, the flow through the circuit is maximum and the nodal prices at both ends of the line are identical. We thus have $\pi_T = 0$ and the revenue is also zero. The revenue is maximum for a transmission capacity of 466 MW.

8.4.2 The Transmission Supply Function

Let us now look at the other side of the "market" for transmission and construct a supply function for transmission. The annuitized cost of building a transmission line consists of a variable cost component, which does depend on the capacity of the line T and a fixed cost component, which does not depend on this capacity:

$$C_T(T) = C_F + C_V(T) \tag{8.12}$$

For the sake of simplicity, we will assume that the variable component is a linear function of the capacity:

$$C_V(T) = k \cdot l \cdot T \tag{8.13}$$

If l is the length of the line in kilometers, k is the annuitized marginal cost of building 1 km of transmission line and its dimensions are \$/(MW × km × year). The annuitized marginal cost of transmission capacity is:

$$\frac{dC_T}{dT} = k \cdot l \tag{8.14}$$

This quantity is called the long-run marginal costs (LRMC) because it relates to the cost of investments in transmission. Dividing it by the number of hours in a year ($\tau_0 = 8760$ h), we get the hourly LRMC, which, as we need for the transmission supply function, is expressed in \$/MWh:

$$c_T(T) = \frac{k \cdot l}{\tau_0} \tag{8.15}$$

Because of the simplifying assumptions that we made in Equation (8.13), the marginal cost of transmission is a constant that does not depend on the capacity of the line.

If we assume that the line is 1000 km long and that:

$$k = 35\ \$/(\text{MW km year}) \tag{8.16}$$

The hourly LRMC of transmission is then:

$$c_T = 4.00\ \$/\text{MWh} \tag{8.17}$$

8.4.3 Optimal Transmission Capacity

When the interconnection capacity is optimal, the supply and demand for transmission are in equilibrium, i.e. the price that transmission users are willing to pay is equal to the marginal cost of providing this capacity. At this optimum, we have:

$$\pi_T = c_T = 4.00\ \$/\text{MWh} \tag{8.18}$$

Combining Equations (8.10) and (8.18), we get the optimal capacity:

$$T^{\text{OPT}} = 800\ \text{MW} \tag{8.19}$$

Figure 8.5 (which is identical to Figure 5.10) illustrates this optimization. It shows the nodal prices in Borduria and Syldavia as a function of the production in each country.

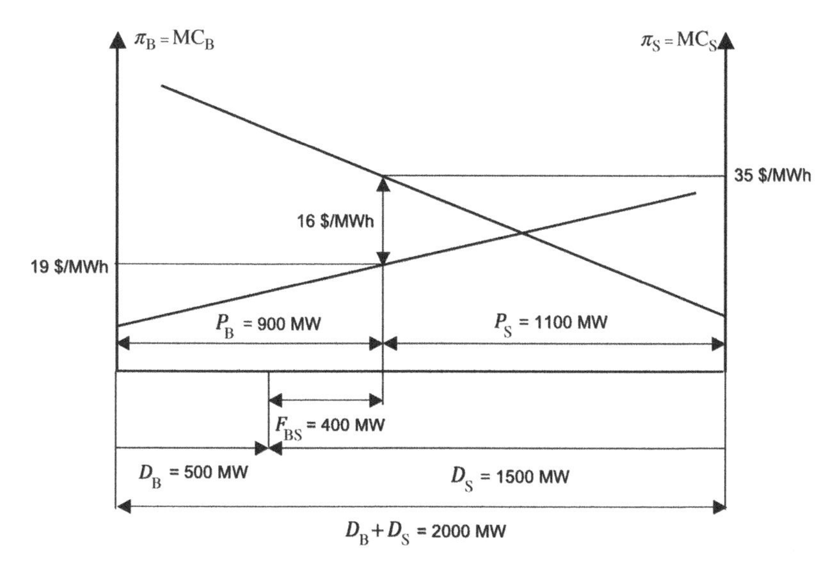

Figure 8.5 Relation between the capacity of the interconnection and the difference in nodal prices between Borduria and Syldavia.

Since we assume that the demands are constant, it also shows these nodal prices as a function of the flow on the interconnection. If this flow is limited by the capacity of the interconnection, the vertical distance separating the two curves gives the difference in nodal prices that arises between the two markets. We could call this difference the SRMC of not having more transmission capacity. If this interconnection has a transmission capacity of 800 MW, the flow from Borduria to Syldavia is equal to 800 MW ($F = T$). The SRMC is then 4.00 \$/MWh. This means that the SRMC is exactly equal to the LRMC of the interconnection. If the owner of the interconnection collected the difference in nodal prices between the two markets (or charged a transmission price equal to this difference), it would collect exactly enough revenue to pay for the construction of the line.

If the transmission capacity is larger than 800 MW, the operating point moves to the right in Figure 8.5 and the nodal price difference (SRMC) would be lower. Since the LRMC is constant, the value of the interconnection would be less than its cost. If the revenues of the transmission owner are proportional to the nodal price difference, it would not collect enough revenue to cover its investment costs. In other words, it would have overinvested.

On the other hand, if the transmission capacity were less than 800 MW, the operating point would move to the left in Figure 8.5. The difference in nodal prices would then be larger than the LRMC. This underinvestment is good for the owner of the interconnection because it can charge a higher price for the use of the transmission line. From a global perspective, this underinvestment is not good because it limits trading opportunities to a suboptimal level.

8.4.4 Balancing the Cost of Constraints and the Cost of Investments

Integrating the expressions for the marginal costs of generation in Borduria and Syldavia given in Equations (8.1) and (8.2), respectively, we get the variable generation costs in each country:

$$C_B = 10P_B + \frac{1}{2}0.01P_B^2 \text{ (\$/h)} \tag{8.20}$$

$$C_S = 13P_S + \frac{1}{2}0.02P_S^2 \text{ (\$/h)} \tag{8.21}$$

In Chapter 5 we determined that the productions that minimize the total generation cost when operation is not constrained by the transmission network are:

$$P_B = 1433.3 \text{ MW} \tag{8.22}$$

$$P_S = 566.7 \text{ MW} \tag{8.23}$$

The unconstrained flow in the interconnection is then:

$$F = 933.33 \text{ MW} \tag{8.24}$$

The corresponding generation costs in each country and in the whole system are:

$$C_B = 24\ 605 \text{ \$/h} \tag{8.25}$$

$$C_S = 10\,578 \text{ \$/h} \tag{8.26}$$

$$C^U = C_B + C_S = 35\,183 \text{ \$/h} \tag{8.27}$$

This unconstrained dispatch and the associated costs are often called, respectively, the *merit order dispatch* and *the merit order costs.*

If the transmission capacity (and hence the flow on the interconnection) is 800 MW, the generations and the corresponding costs are:

$$P_B = 1300\text{ MW}, \quad C_B = 21450\,\$/\text{h} \tag{8.28}$$

$$P_S = 700\text{ MW}, \quad C_S = 14000\,\$/\text{h} \tag{8.29}$$

The total cost of supplying the load for this constrained condition is:

$$C^C = 35450\,\$/\text{h} \tag{8.30}$$

The difference in cost between the constrained and unconstrained conditions is called the *cost of constraints* or the *out-of-merit generation cost*:

$$\Delta C = C^C - C^U = 267\,\$/\text{h} \tag{8.31}$$

The total cost of transmission is the sum of the cost of building the transmission system and the cost of constraints. As Figure 8.6 shows, the cost of building the transmission system increases with the transmission capacity while the cost of constraints decreases because the transmission network puts fewer limitations on the generation dispatch. Minimizing the total cost of transmission is thus the objective of the network development task. From Figure 8.6, we see that this optimum is achieved for a transmission capacity of 800 MW, which is consistent with the result that we obtained in Equation (8.19).

8.4.5 Effect of Load Fluctuations

So far, we have made the very convenient assumption that the load remains constant over time. This is obviously not realistic and we must analyze how the natural fluctuations of the load with the cycle of human activities affect the value of transmission.

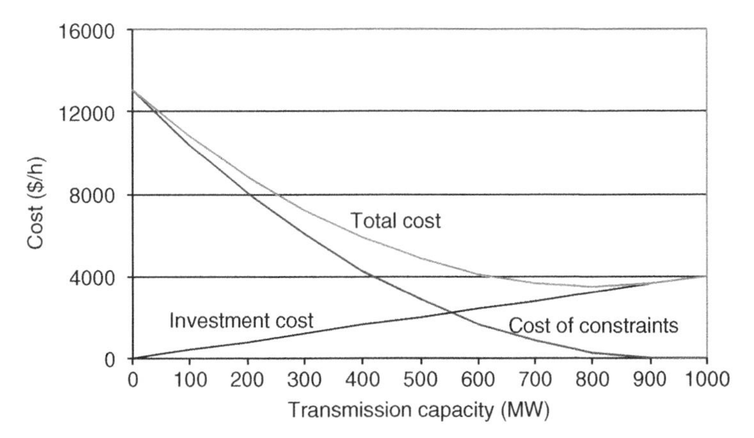

Figure 8.6 Evolution of the cost of constraints, the investment cost, and the total transmission cost for the Borduria–Syldavia interconnection.

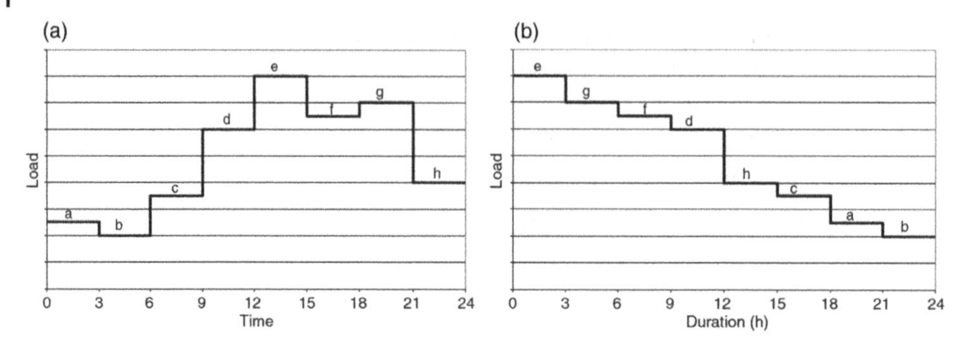

Figure 8.7 Chronological load profile (a) and load-duration curve (b).

8.4.5.1 Load-duration Curve

If we assume that the load fluctuations in the whole system follow similar patterns, we do not need to concern ourselves with the time at which the load achieves a particular value. What is important is the duration of each load level. Chronological load profiles, such as the one in Figure 8.7a, show how the load varies over the course of a day. This period is divided into a number of intervals during which the load is assumed to be constant. In this case, the day has been divided into eight intervals of 3 h labeled a to h. On the graph of Figure 8.7b, these intervals have been sorted in decreasing order of load. This graph thus shows the number of hours over the course of a day during which the load exceeded a certain value. This process can be applied over a longer period (for example, a year) and with shorter intervals (for example, 1 h). The resulting load-duration curve then shows the number of hours over the course of a year during which the load exceeded a certain value. We have already encountered such a curve in Chapter 7.

Since handling a load-duration curve with up to 8760 hourly intervals is not practical, some aggregation is usually performed. For example, Figure 8.8 shows how the load-duration curve of Figure 8.7 has been simplified by grouping the values of the load into four groups.

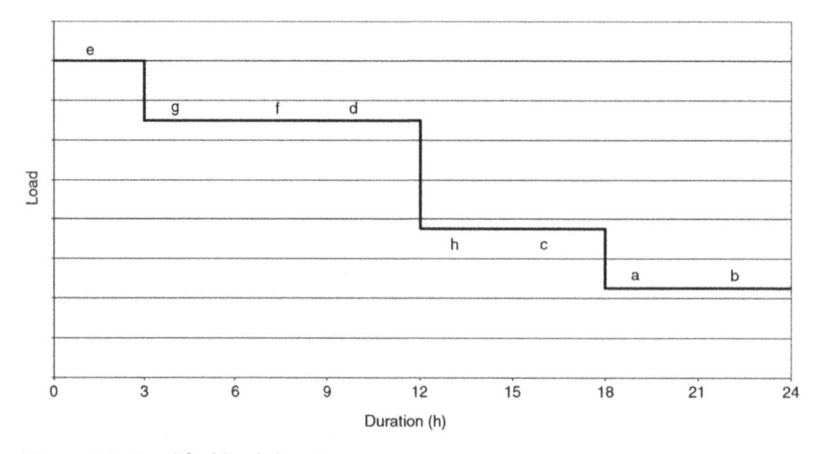

Figure 8.8 Simplified load-duration curve.

Example 8.3

Let us add the very simple load-duration curves shown in Figure 8.9 to our Borduria–Syldavia example. In this case, the load in each country has been divided into a peak level and an off-peak level. The peak period has a duration of 3889 h, while the off-peak period lasts 4871 h. For the sake of simplicity, we will assume that the on-peak and off-peak periods in the two countries coincide.

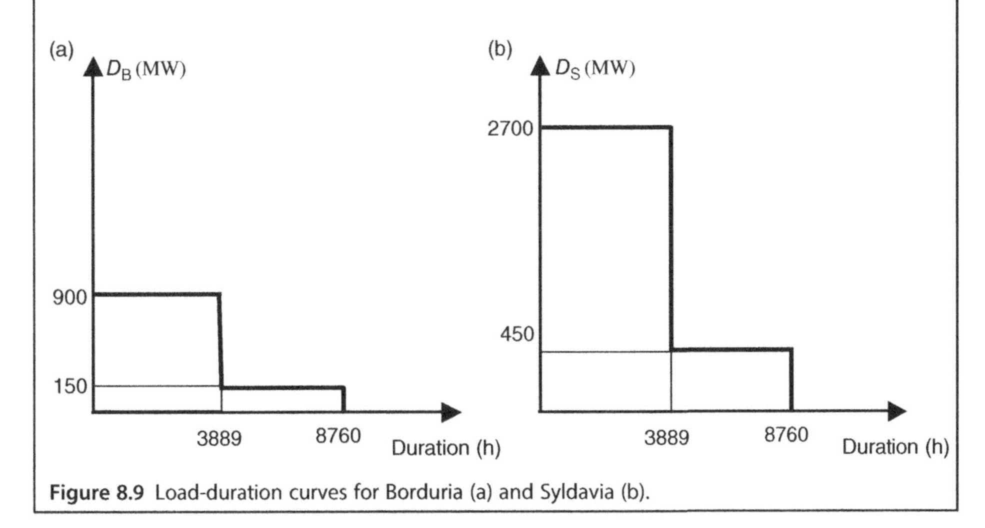

Figure 8.9 Load-duration curves for Borduria (a) and Syldavia (b).

As discussed, in order to determine the optimal transmission capacity, we must balance the annual savings in energy costs against the annuitized cost of transmission. While we could carry out this calculation analytically, we will instead compute the components of the cost for a range of transmission capacities and find the value that minimizes the total cost.

To calculate the hourly cost of constraints, we need to know the unconstrained cost of generation. Table 8.1 shows the unconstrained economic dispatch for the peak and off-peak loads and the corresponding generation costs as calculated using Equations (8.20) and (8.21). Tables 8.2 and 8.3 show the hourly generation costs for the off-peak and on-peak loads.

Given that the durations of the off-peak and on-peak periods are 4871 and 3889 h, respectively, we can compute the total annual cost of constraints for the values of the interconnection capacity in the previous two tables. In this example, we assume that the

Table 8.1 Unconstrained economic dispatch for the peak and off-peak load conditions in the Borduria–Syldavia system.

Total load in Borduria and Syldavia (MW)	Generation in Borduria (MW)	Generation in Syldavia (MW)	Total hourly generation cost ($/h)
600	500	100	7 650
3 600	2 500	1 100	82 650

Table 8.2 Hourly generations, total hourly generation costs, and hourly constraint cost for the Borduria–Syldavia system for the off-peak load as a function of the interconnection capacity.

Interconnection capacity (MW)	Generation in Borduria (MW)	Generation in Syldavia (MW)	Total hourly generation cost ($/h)	Hourly constraint cost ($/h)
0	150	450	9488	1838
100	250	350	8588	938
200	350	250	7988	338
300	450	150	7688	38
350	500	100	7650	0
400	500	100	7650	0
450	500	100	7650	0
500	500	100	7650	0
600	500	100	7650	0
700	500	100	7650	0
800	500	100	7650	0
900	500	100	7650	0

Table 8.3 Hourly generations, total hourly generation costs, and hourly constraint cost for the Borduria–Syldavia system for the on-peak load as a function of the interconnection capacity.

Interconnection capacity (MW)	Generation in Borduria (MW)	Generation in Syldavia (MW)	Total hourly generation cost ($/h)	Hourly constraint cost ($/h)
0	900	2 700	121 050	38 400
100	1 000	2 600	116 400	33 750
200	1 100	2 500	112 050	29 400
300	1 200	2 400	108 000	25 350
350	1 250	2 350	106 088	23 438
400	1 300	2 300	104 250	21 600
450	1 350	2 250	102 488	19 838
500	1 400	2 200	100 800	18 150
600	1 500	2 100	97 650	15 000
700	1 600	2 000	94 800	12 150
800	1 700	1 900	92 250	9 600
900	1 800	1 800	90 000	7 350

Table 8.4 Annual cost of constraints, annuitized cost of transmission investments, and total annual cost of transmission as a function of the transmission capacity of the Borduria–Syldavia interconnection.

Interconnection capacity (MW)	Annual constraint cost (k$/year)	Annuitized investment cost (k$/year)	Total annual transmission cost (k$/year)
0	158 304	0	158 304
100	135 835	14 000	149 835
200	115 993	28 000	143 993
300	98 780	42 000	140 780
350	91 159	49 000	140 159
400	**84 012**	**56 000**	**140 012**
450	77 157	63 000	140 157
500	70 593	70 000	140 593
600	58 342	84 000	142 342
700	47 257	98 000	145 257
800	37 339	112 000	149 339
900	28 587	126 000	154 587

Boldface indicates the optimal values.

annuitized marginal cost of transmission investment is 140 $/(MW × km × year). Table 8.4 shows these values together with the annuitized cost of transmission investments and their sum, which is the total annual transmission cost. We only consider the variable part of the cost of transmission investments and calculate it using Equation (8.13). These results show that a transmission capacity of 400 MW is optimum because it minimizes the total cost of transmission.

8.4.5.2 Recovery of Variable Transmission Investment Costs

Let us now examine the effect that a transmission capacity of 400 MW has on electrical energy markets in Borduria and Syldavia.

During off-peak periods, a 400 MW interconnection capacity does not limit the power flow between the two countries. The two markets thus operate as a single market. Generators in Borduria and Syldavia produce 500 and 100 MW, respectively. Since there is only 150 MW of load in Borduria, 350 MW flow on the interconnection to Syldavia. The marginal generation costs and hence the prices in Borduria and Syldavia are identical at 15.00 $/MWh. Therefore, during off-peak conditions, the short-run marginal value of transmission is zero. The congestion surplus or transmission revenue is thus also zero.

During peak periods, Bordurian generators produce only 1300 MW because the local load is 900 MW and the transmission capacity is limited to 400 MW. The generators in Syldavia produce 2300 MW. Because of the transmission congestion, prices in the Bordurian and Syldavian markets are set by the local marginal cost of generation at 23.00 and 59.00 $/MWh. The short-run value of transmission is thus 36.00 $/MWh.

During peak load condition, the hourly congestion surplus is:

$$CS_{\text{hourly}} = 400 \cdot 36 = 14400\,\$/\text{h} \tag{8.32}$$

If we assume that the transmission owner collects this surplus, its annual revenue is equal to this value multiplied by the number of on-peak hours:

$$CS_{\text{annual}} = 14400 \cdot 3889 = 56\,000\,000\,\$/\text{year} \tag{8.33}$$

This amount is equal to the annuitized variable cost of transmission investment:

$$C_V(T) = k \cdot l \cdot T = 140 \cdot 1000 \cdot 400 = 56\,000\,000\,\$/\text{year} \tag{8.34}$$

For the optimal transmission capacity, the revenue earned from the congestion surplus thus covers exactly the variable part of the investment cost. However, it does not cover the fixed part of the transmission investment. Furthermore, this equality holds because we have assumed a constant value for the marginal cost of transmission capacity k. It does not hold if this marginal cost is not constant because of economies of scale.

8.4.6 Revenue Recovery for Suboptimal Transmission Capacity

In practice, the actual transmission capacity rarely coincides with its optimal value. The reasons for this discrepancy are easy to understand if we consider the uncertainties that affect the forecasts of demand and generation prices, the lumpiness of investments in transmission capacity and the legacy of historical investment decisions. Obviously, power system operators run the system on the basis of what the transmission capacity actually is, not on the basis of what an optimization program says it should be. Since the nodal energy prices and the congestion surplus are determined by the actual network, it is important to study how suboptimality affects revenue recovery.

In Example 8.2, the optimal capacity of the interconnection between Borduria and Syldavia was found to be 800 MW. Let us calculate the revenue and cost that would result if the transmission line were built with a capacity of 900 MW. Since this capacity is available, Bordurian generators increase their output to 1400 MW, production in Syldavia drops to 600 and 900 MW flows on the interconnection. Using Equations (8.1) and (8.2), we find that energy prices in Borduria and Syldavia are 24.00 and 25.00 \$/MWh, respectively. The short-run value of transmission thus drops from 4.00 \$/MWh for a capacity of 800 MW to 1.00 \$/MWh for a capacity of 900 MW.

The hourly congestion surplus and the annual revenue are:

$$CS_{\text{hourly}} = 900 \cdot 1 = 900\,\$/\text{h} \tag{8.35}$$

$$CS_{\text{annual}} = 900 \cdot 8760 = 7\,884\,000\,\$/\text{year} \tag{8.36}$$

On the other hand, the annuitized investment cost amounts to:

$$C_V(T) = k \cdot l \cdot T = 35 \cdot 1000 \cdot 900 = 31\,500\,000\,\$/\text{year} \tag{8.37}$$

The revenue generated by the congestion surplus is smaller than it was for the optimal transmission capacity and is not sufficient to cover the cost of this overinvested transmission system.

Let us now examine the case of underinvestment. If the transmission capacity is only 700 MW, the flow on the interconnection is limited to this value. Generators in Borduria produce only 1200 MW (500 MW of local load and 700 MW transmitted to Syldavia) at a price of 22.00 \$/MWh. Syldavian producers generate 800 MW at a price of 29.00 \$/MWh to satisfy the remainder of the 1500 MW Syldavian load. This 7.00 \$/MWh price differential creates a congestion surplus of:

$$CS_{\text{hourly}} = 700 \cdot 7 = 4900\,\$/\text{h} \tag{8.38}$$

Over 1 year this will generate a revenue of:

$$CS_{\text{annual}} = 4900 \cdot 8760 = 42\,924\,000\,\$/\text{year} \tag{8.39}$$

On the other hand, the annuitized cost of investment for a 700 MW interconnection is:

$$C_V(T) = k \cdot l \cdot T = 35 \cdot 1000 \cdot 700 = 24\,500\,000\,\$/\text{year} \tag{8.40}$$

In this case, the income generated by short-run marginal pricing of transmission is thus larger than the cost of building the transmission line. In other words, keeping the transmission capacity below the optimal value increases the revenue collected.

Let us now consider the situation of Example 8.3 where the interconnection between Borduria and Syldavia has a transmission capacity of 500 MW. During off-peak periods, this overinvestment has no effect because even the optimal capacity does not constrain the power flow. The short-run marginal value of transmission and the transmission revenue remain at zero. On the other hand, during peak periods, the system operator makes use of all the 500 MW capacity of the interconnection. Bordurian generators can then produce 1400 MW, while production in Syldavia is only 2200 MW. Equations (8.1) and (8.2) show that the energy prices in Borduria and Syldavia are 24.00 and 57.00 \$/MWh, respectively. The short-run value of transmission is thus 33.00 \$/MWh, instead of 36.00 \$/MWh for a 400 MW transmission capacity.

The congestion surplus collected during hours of peak load is:

$$CS_{\text{hourly}} = 500 \cdot 33 = 16\,500\,\$/\text{h} \tag{8.41}$$

Given the duration of the on-peak period, the annual revenue is:

$$CS_{\text{annual}} = 16\,500 \cdot 3889 = 64\,168\,500\,\$/\text{year} \tag{8.42}$$

On the other hand, the annuitized investment cost amounts to:

$$C_V(T) = k \cdot l \cdot T = 140 \cdot 1000 \cdot 500 = 70\,000\,000\,\$/\text{year} \tag{8.43}$$

The revenue generated by the congestion surplus is larger than it was for the optimal transmission capacity, but is not sufficient to cover the cost of the overinvested transmission system.

Let us now turn our attention to the case of underinvestment. If the transmission capacity is only 300 MW, the flow on the interconnection is limited not only during peak load conditions, but also during the off-peak period.

During the off-peak period, generators in Borduria produce 450 MW (150 MW of local load and 300 MW exported to Syldavia) at a price of 14.50 \$/MWh. Syldavian producers generate 150 MW at a price of 16.00 \$/MWh to satisfy the remainder of the 450 MW Syldavian load. This 1.50 \$/MWh price differential creates a congestion surplus of:

$$CS_{\text{hourly}} = 300 \cdot 1.50 = 450\,\$/\text{h} \tag{8.44}$$

Over the 4871 off-peak hours, \$2 191 950 of congestion revenue is thus collected.

During the peak load period, Bordurian generators produce 1200 MW, out of which 300 MW are transmitted through the interconnection, leaving Syldavian generators to produce 2400 MW. The marginal prices in Borduria and Syldavia are therefore 22.00 and 51.00 \$/MWh, respectively. The hourly congestion surplus amounts to:

$$CS_{\text{hourly}} = 300 \cdot (61.00 - 22.00) = 11\,700\,\$/\text{h} \tag{8.45}$$

Given that peak load conditions span 3889 h, \$45 501 300 is generated in congestion surplus. Considering both the off-peak and on-peak periods, the annual congestion revenue reaches \$47 693 250. On the other hand, the annuitized cost of investment for a 300 MW interconnection is:

$$C_V(T) = k \cdot l \cdot T = 140 \cdot 1000 \cdot 300 = 42\,000\,000\,\$/\text{year} \tag{8.46}$$

In this case, the income generated by short-run marginal pricing of transmission is larger than the cost of building the transmission network. In other words, keeping the transmission capacity below the optimal value increases the revenue collected because congestion is more frequent.

8.4.7 Economies of Scale

We have assumed so far that the cost of investments in transmission equipment is proportional to the power transmitted. However, a significant part of this cost is fixed, i.e. independent of the transmission capacity. Let us remove this simplifying assumption and reconsider the interconnection between Borduria and Syldavia, taking into account component C_F of the total cost C_T of building the line:

$$C_T(T) = C_F + C_V(T) \tag{8.47}$$

The magnitude of the fixed cost does not affect the capacity of the circuit to be built because once we have decided to proceed with a transmission expansion project, we are committed to pay the fixed cost, independently of what decision is made about the capacity. To illustrate this counterintuitive effect, let us go back to Example 8.2 and assume that the fixed cost of the line is 20 000 \$/(km × year). When we add this cost to the variable investment cost of the 1000-km-long Borduria–Syldavia interconnection, it simply shifts the total cost curve of Figure 8.6 upward and does not affect the location of its minimum.

Let us assume that the interconnection has been built with the optimal capacity and that all this capacity is made available. As we saw in the previous section, the pattern of nodal prices is then such that the revenues derived from the price differentials cover exactly the variable part of the cost of building the transmission line. On the other hand, congestion revenues do not cover the fixed component of the cost of building the interconnection.

Hogan (1999) suggested that one way of recovering this shortfall would be to restrict the capacity that is made available. Let us calculate the short-run transmission revenue that its owner would collect if it made available to the system operator only 650 MW of transmission capacity, instead of offering the full 800 MW. The flow between Borduria and Syldavia is then 650 MW. Bordurian generators reduce their output to 1150 MW while production in Syldavia increases to 850 MW. Using Equations (8.1) and (8.2), we find that energy prices in Borduria and Syldavia are 21.50 and 30.00 \$/MWh, respectively. The short-run value of transmission thus increases from 4.00 to 8.50 \$/MWh.

The hourly and annual congestion surpluses are:

$$CS_{hourly} = 650 \cdot 8.5 = 5525\,\$/\text{h} \tag{8.48}$$

$$CS_{annual} = 5525 \cdot 8760 = 48\,399\,000\,\$/\text{year} \tag{8.49}$$

On the other hand, the annuitized investment cost amounts to:

$$C_T(T) = C_F + k \cdot l \cdot T = 20\,000\,000 + 35 \cdot 1000 \cdot 800 = 48\,032\,000\,\$/\text{year} \tag{8.50}$$

In this case, withholding 150 MW of transmission capacity generates enough additional revenue to cover both fixed and variable costs.

Withholding some transmission capacity creates a larger price differential and increases the value of transmission. Network users therefore may be willing to pay more to buy financial transmission rights from the owner of this new line, thereby providing this owner with an opportunity to not only cover their cost but also to make a profit.

Example 8.4

Let us revisit Example 8.3. Table 8.5 illustrates the effect of the annuitized fixed cost on an annual basis and shows that the optimal capacity of the interconnection remains 400 MW, independently of the fixed cost.

Let us assume that the interconnection has been built with the optimal capacity but that some of this capacity is withheld with an eye toward generating enough revenue to recover the fixed cost. Tables 8.6 and 8.7 show how withholding capacity affects the congestion revenue during the off-peak and on-peak periods.

Table 8.5 Annual cost of constraints, annuitized cost of transmission investments (including both fixed and variable costs), and total annual cost of transmission as a function of the capacity of the Borduria–Syldavia interconnection.

Interconnection capacity (MW)	Annual constraint cost (k$/year)	Annuitized fixed investment cost (k$/year)	Annuitized variable investment cost (k$/year)	Annuitized investment cost (k$/year)	Total annual transmission cost (k$/year)
100	135 835	20 000	14 000	34 000	169 835
200	115 993	20 000	28 000	48 000	163 993
300	98 780	20 000	42 000	62 000	160 780
350	91 159	20 000	49 000	69 000	160 159
400	**84 012**	**20 000**	**56 000**	**76 000**	**160 012**
450	77 157	20 000	63 000	83 000	160 157
500	70 593	20 000	70 000	90 000	160 593
600	58 342	20 000	84 000	104 000	162 342
700	47 257	20 000	98 000	118 000	165 257
800	37 339	20 000	112 000	132 000	169 339
900	28 587	20 000	126 000	146 000	174 587

Boldface indicates the optimal values.

Table 8.6 Congestion surplus as a function of the available transmission capacity during the off-peak period.

Available capacity (MW)	Generation in Borduria (MW)	Generation in Syldavia (MW)	Marginal cost in Borduria ($/MWh)	Marginal cost in Syldavia ($/MWh)	Hourly surplus ($/h)	Annual surplus ($/year)
100	250	350	12.5	20	750	3 653 250
200	350	250	13.5	18	900	4 383 900
300	450	150	14.5	16	450	2 191 950

Table 8.7 Congestion surplus as a function of the available transmission capacity during the on-peak period.

Available capacity (MW)	Generation in Borduria (MW)	Generation in Syldavia (MW)	Marginal cost in Borduria ($/MWh)	Marginal cost in Syldavia ($/MWh)	Hourly surplus ($/h)	Annual surplus ($/year)
100	1 000	2 600	20	65	4 500	17 500 500
200	1 100	2 500	21	63	8 400	32 667 600
300	1 200	2 400	22	61	11 700	45 501 300

During the off-peak period, reducing the available capacity from 400 to 200 MW increases the transmission revenue from 0 to 4 383 900 $/year. Further reductions in capacity decrease the congestion revenue. On the other hand, during the on-peak period, withholding any capacity reduces the revenue. This difference stems from the fact that, as illustrated in Figure 8.4, the congestion revenue is a quadratic function of the transmission capacity. For the on-peak period, the range of capacities considered in Tables 8.6 and 8.7 is located to the left of the maximum while for the off-peak period, it spans the maximum. Given that the overall contribution of the on-peak period is much greater than the contribution of the off-peak period, it is not possible to increase the short-term transmission revenues by withholding transmission capacity.

Whenever we assess the effect of fixed costs, we must consider what happens if we decide not to build the transmission line. In this case, the total investment cost would be zero. For the conditions of Example 8.3, the cost of constraints would then be at its maximum, i.e. 158 304 000 $/year. By comparison, building the optimal transmission capacity (i.e. 400 MW) would result in a total cost of 160 012 000 $/year. Under these conditions, building a transmission line would therefore not be justified.

8.4.8 Transmission Expansion in a Meshed Network

We must now explore the effect that Kirchhoff's voltage law has on the value of transmission and the recovery of investments in transmission capacity. To illustrate this issue, we will use the three-bus system shown in Figure 8.10. We consider the effect of changes in demand by assuming that each year can be divided into two demand periods.

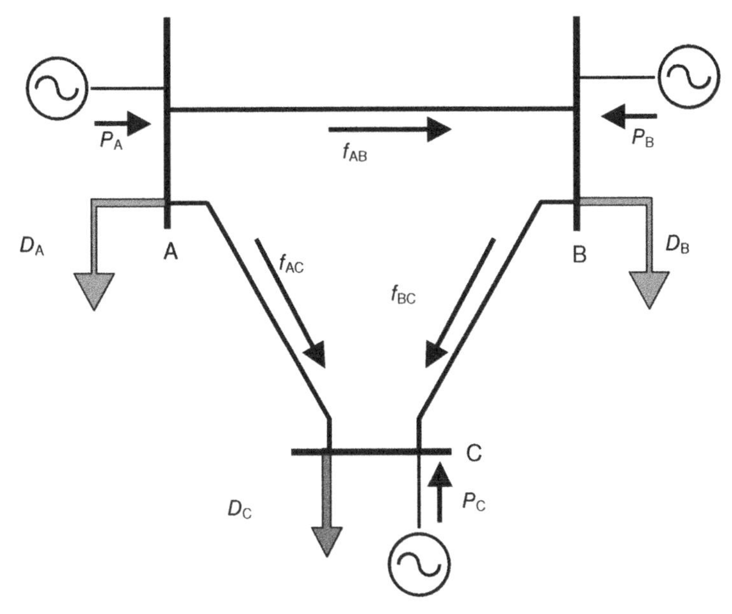

Figure 8.10 Three-bus system used to illustrate the effect of Kirchhoff's voltage law on the value of transmission and the recovery of transmission investments.

Table 8.8 shows the duration of each period and the load at each bus. Note that, unlike in the previous two-bus example, the load profile does not follow the same pattern at all buses. Table 8.9 shows that the marginal cost of generation at each bus increases linearly with output. Once again, we assume that there are enough competitors to ensure that the price of energy at each bus is equal to its marginal cost.

Table 8.8 Variation of the load with time for the three-bus system.

	Period 1	Period 2
Duration (h)	2 190	6 570
Load at bus A (MW)	0	0
Load at bus B (MW)	10 000	5 000
Load at bus C (MW)	2 500	10 000

Table 8.9 Marginal costs of electrical energy for the three-bus system.

Bus	Capacity (MW)	Marginal cost ($/MWh)
A	5000	$0.003\,P_A + 2$
B	7000	$0.003\,P_B + 1.35$
C	8000	$0.003\,P_C + 1.75$

The annuitized transmission investment cost of a transmission line c is proportional to the circuit capacity (T_c) and its length (l_c):

$$\Omega_c(T_c) = k_c \cdot l_c \cdot T_c \tag{8.51}$$

where the marginal annuitized investment cost of the circuit per unit length k_c is 50 \$/(MW × km × year). For the sake of simplicity, we assume that all lines in our three-bus system have the same 600 km length and thus the same reactance.

We want to determine the capacities of the transmission lines that minimize the sum of the operating and investment costs for this network. This minimization must be done over the expected life of the system. Since in this example we assume that the load pattern repeats itself year after year, we can carry out this optimization over an equivalent hour. This is achieved by multiplying the operating cost for each load period by its duration ($\tau_1 = 2190$ h and $\tau_2 = 6570$ h) and dividing by the number of hours in a year ($\tau_0 = 8760$ h). The objective function of this optimization problem is thus:

$$\min_{T_{AB}, T_{AC} T_{BC}} \left[\sum_{t=1}^{t=2} \frac{\tau_t}{\tau_0} \left(\sum_{i \in \{A,B,C\}} a_i P_{it} + \frac{1}{2} b_i P_{it}^2 \right) + \frac{k_{AB} \cdot l_{AB} \cdot T_{AB}}{\tau_0} \right.$$
$$\left. + \frac{k_{AC} \cdot l_{AC} \cdot T_{AC}}{\tau_0} + \frac{k_{BC} \cdot l_{BC} \cdot T_{BC}}{\tau_0} \right] \tag{8.52}$$

This minimization is subject to the following constraints imposed by KCL and KVL for demand period 1:

$$\begin{aligned} f_{AB1} + f_{AC1} - P_{A1} + D_{A1} &= 0 \\ -f_{AB1} + f_{BC1} - P_{B1} + D_{B1} &= 0 \\ -f_{AC1} - f_{BC1} - P_{C1} + D_{C1} &= 0 \\ -f_{AB1} + f_{AC1} - f_{BC1} &= 0 \end{aligned} \tag{8.53}$$

and for demand period 2:

$$\begin{aligned} f_{AB2} + f_{AC2} - P_{A2} + D_{A2} &= 0 \\ -f_{AB2} + f_{BC2} - P_{B2} + D_{B2} &= 0 \\ -f_{AC2} - f_{BC2} - P_{C2} + D_{C2} &= 0 \\ -f_{AB2} + f_{AC2} - f_{BC2} &= 0 \end{aligned} \tag{8.54}$$

Furthermore, the line flows during each period must remain below the (as yet unknown) capacity of the corresponding lines:

$$\begin{aligned} |f_{AB1}|, |f_{AB2}| &\le T_{AB} \\ |f_{AC1}|, |f_{AC2}| &\le T_{AC} \\ |f_{BC1}|, |f_{BC2}| &\le T_{BC} \end{aligned} \tag{8.55}$$

Finally, during each demand period, the output of the generators connected to each bus must be less than the installed generation capacity:

$$\begin{aligned} P_{A1} &\le P_A^{max}; P_{A2} \le P_A^{max} \\ P_{B1} &\le P_B^{max}; P_{B2} \le P_B^{max} \\ P_{C1} &\le P_C^{max}; P_{C2} \le P_C^{max} \end{aligned} \tag{8.56}$$

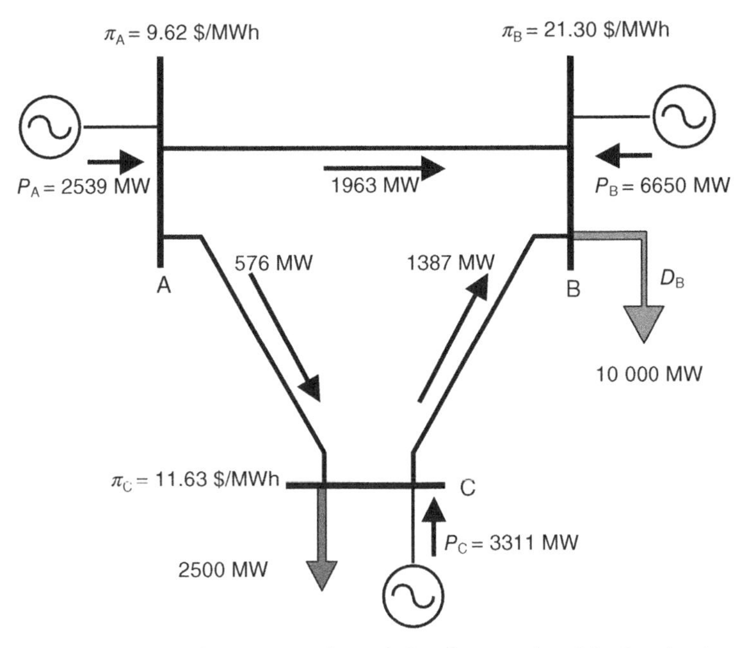

Figure 8.11 Optimal generation dispatch, line flows, and nodal prices for demand period 1.

This quadratic optimization problem is too complex for a manual solution but can be solved numerically using a spreadsheet. Figures 8.11 and 8.12 illustrate the optimal generation dispatch, the line flows, and the nodal prices for the two demand periods. Table 8.10 shows the detail of the operating costs. Since the duration of period 1 represents 25% of the total and the duration of period 2 the remaining 75%, the costs

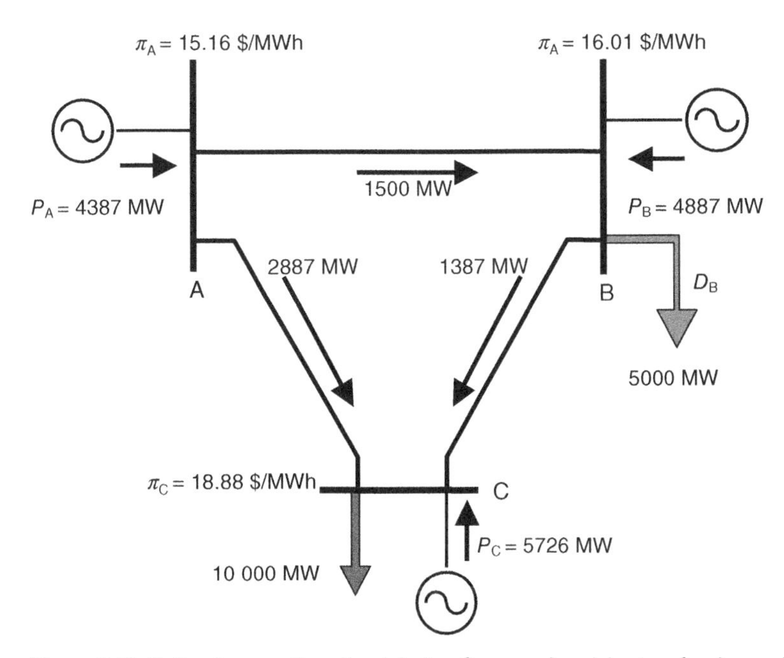

Figure 8.12 Optimal generation dispatch, line flows, and nodal prices for demand period 2.

Table 8.10 Optimal hourly operating cost for the three-bus system.

Bus	Period 1 ($/0.25 h)	Period 2 ($/0.75 h)	Cost for an equivalent hour ($/h)	Annual cost ($/year)
A	3 687	28 233	31 920	279 619 200
B	18 827	31 817	50 644	443 641 440
C	5 519	44 184	49 703	435 398 280
Total	28 033	104 234	132 267	1 158 658 920

for each period are expressed on $/0.25 h and $/0.75 h, respectively. The operating cost for an equivalent hour is then obtained by adding the cost for each period. The annual cost is obtained by multiplying the cost for an equivalent hour by the number of hours in a year.

Table 8.11 shows the flows in each transmission line as well as its optimal capacity and the corresponding hourly and annual investment costs. The flow in each line reaches its maximum (and hence sets the capacity) in one of the periods. Since we minimized the sum of the operating and investment costs, the whole capacity of the line should indeed be fully utilized during at least one period. In this particular case, the flow between buses B and C fully utilizes the capacity of that line during both periods, but in opposite directions.

Table 8.12 summarizes the nodal prices and payments for each period. Negative quantities represent payments to generators, while positive quantities denote payments

Table 8.11 Optimal line capacities and investment costs for the three-bus system.

Line	Flow in period 1 (MW)	Flow in period 2 (MW)	Optimal capacity (MW)	Hourly investment cost ($/h)	Annual investment cost ($/year)
A–B	1 963	1 500	1 963	6 723	58 891 939
A–C	576	2 887	2 887	9 887	86 612 631
B–C	−1 387	1 387	1 387	4 750	41 612 636
Total				21 360	187 117 206

Table 8.12 Nodal prices and revenues for the three-bus system.

Bus	Nodal prices		Revenues		
	Period 1 ($/MWh)	Period 2 ($/MWh)	Period 1 ($/0.25 h)	Period 2 ($/0.75 h)	Equivalent hour ($/h)
A	9.62	15.16	−6 105	−49 885	−55 990
B	21.30	16.01	17 839	1 356	19 195
C	11.63	18.88	−2 359	60 514	58 155
Total			9 375	11 985	21 360

from loads. The payments for each period are proportional to their duration and the payments for an equivalent hour are a weighted average of the payments for each period. The grand total (in the bottom right-hand corner) represents the total congestion surplus that would be collected during an equivalent hour. This quantity is exactly equal to the total hourly investment cost given in Table 8.11. This equivalence demonstrates that, in the absence of fixed costs, short-run marginal pricing generates a sufficient amount of revenue to cover the cost of transmission investments.

Table 8.13 provides the information needed to calculate the revenue "earned" by each line during each period and over the equivalent hour. As we discussed in Chapter 5, differences in nodal prices arise between two buses even when the line connecting these two buses is not congested. For example, during period 1, the flow in the line between buses A and C is 576 MW, well below its 2887 MW capacity. However, congestion in lines A–B and B–C creates a 2.01 \$/MWh price differential between nodes A and C. The flow in that line thus generates a revenue of:

$$R_{AC,1} = 576 \times 2.01 \times 0.25 = 289\,\$/0.25\,\text{h} \tag{8.57}$$

During period 2, when the flow in this line is equal to its capacity, it generates a revenue of:

$$R_{AC,2} = 2887 \times 3.72 \times 0.75 = 8055\,\$/0.75\,\text{h} \tag{8.58}$$

The revenue "collected" by this line during an equivalent hour is thus 8344 \$/h. It is not equal to the 9887 \$/h hourly cost of this line given in Table 8.11. Similarly, the 1500 MW flow in line A–B during period 2 is less than its 1963 MW capacity. However, the price differentials across that line generate some revenue. These results demonstrate that hourly SRMC revenues associated with individual lines do not match their hourly investment cost. However, Table 8.13 also shows that the total congestion surplus recovers exactly the investment cost of this transmission network. This result is not a coincidence and holds for all networks, no matter how complex. If the entire network is owned by the same entity, this cross-subsidization between lines is not a problem. On the other hand, it is not clear how FTRs could be sold to the network users under these conditions. For example, let us suppose that lines A–B and B–C belong to the incumbent utility while line A–C has been developed by a merchant transmission company. If revenues are allocated on the basis of nodal price differentials, the owner of line A–C would recover only 8344 \$/h instead of its cost of 9887 \$/h. On the other hand, the

Table 8.13 Congestion revenues and investment costs for each line of the three-bus system.

Line	Period 1			Period 2				
	Δ Price (\$/MWh)	Flow (MW)	Revenue (\$/0.25 h)	Δ Price (\$/MWh)	Flow (MW)	Revenue (\$/0.75 h)	Total revenue (\$/h)	Investment costs (\$/h)
A–B	11.68	1 963	5 732	0.85	1 500	956	6 688	6 723
A–C	2.01	576	289	3.72	2 887	8 055	8 344	9 887
B–C	−9.67	−1 387	3 353	2.86	1 387	2 975	6 339	4 750
Total			9 374			11 986	21 360	21 360

incumbent utility would recover more than its cost. It is also not clear on what basis negotiation between users and network owners about the purchase of FTRS could proceed.

8.4.9 Concept of Reference Network

In the examples that we discussed in the previous sections, we determined the optimal capacity of a new transmission line by minimizing the sum of the operating cost and the cost of investments in transmission. Maintaining that balance for the system as a whole is a major challenge for regulatory authorities in a competitive environment because generation and transmission operate as separate entities. If we assume that the transmission network operates as a monopoly, the regulator needs to devise a set of incentives that encourages the right level of transmission investments. To achieve this goal, the regulator needs a way to measure the overall performance of the system. This can be achieved using a reference network.

In its simplest form, a reference network is topologically identical to the existing network and generators and loads are unchanged. On the other hand, each transmission line has the optimal capacity determined as in the examples above. One important difference, however, is that instead of optimizing the capacity of one or a few new lines, the procedure is applied to the whole transmission system, including both new and existing lines.

A reference network is thus a network against which the real one can be objectively compared. Optimal investment and congestion costs can be quantified and compared with those of the real system. Furthermore, by comparing the capacities of individual lines in the reference network and the real network, the need for new investment can be identified and stranded investments can be detected. A comparison of optimal and actual operating costs can also be performed. Finally, differences between actual and reference network operations and investments could be used as a measure of the performance of a transmission company and used by the regulator to set financial incentives.

The concept of a reference network has a long history and a solid foundation in economic theory. See, for example, Boiteux (1949), Nelson (1967), and Farmer et al. (1995).

In this section, we present a general formulation of the transmission expansion problem for pricing and regulatory purposes. This involves determining an optimally designed transmission network. Determining such a reference network requires the solution of a type of security-constrained optimal power flow problem. In its simplest form, this problem can be formulated using a conventional DC optimal power flow. The objective of this optimization is to minimize the sum of the annual generation cost and the annuitized cost of transmission. This optimization is constrained by Kirchhoff's current and voltage laws as well as the limits on system components. It must cover several demand levels using a yearly load-duration curve as described earlier. Finally, it must also take into account credible outages of transmission and generation facilities.

8.4.9.1 Notations

In order to state the problem mathematically, we need to introduce the following notations:

np Number of demand periods
nb Number of buses
ng Number of generators

nl Number of lines
nc Number of contingencies
τ_p Duration of demand period p
D_p Nodal demand vector for period p
C_g Operating cost of generator g
P_{pg} Output of generator g during demand period p
P_p Vector of nodal generations for demand period p
$P^{\max}$ Vector of maximum nodal generations
$P^{\min}$ Vector of minimum nodal generations
A^0 Node–branch incidence matrix for the intact system
A^c Node–branch incidence matrix for contingency c
H^0 Sensitivity matrix for the intact system
H^c Sensitivity matrix for contingency c
k_b Annuitized investment cost for line b in \$/(MW × km × year)
l_b Length of line b (km)
T_b Capacity of line b
T Vector of line capacities
F_p^0 Vector of line flows for the intact system during period p
F_p^c Vector of line flows for contingency c during period p.

The sensitivity matrix **H** that relates injections and power flows is defined as follows:

$$[\mathbf{H}] = [Y_\mathrm{d}] \cdot [A^\mathrm{T}] \cdot \begin{bmatrix} 0 & 0 \\ 0 & [Y_\mathrm{bus}^\mathrm{r}]^{-1} \end{bmatrix} \tag{8.59}$$

where Y_d is the diagonal matrix of branch admittances and $Y_\mathrm{bus}^\mathrm{r}$ is obtained from the system admittance matrix Y_bus by removing the row and the column corresponding to the slack bus to make it nonsingular. The elements of the sensitivity matrix **H** are called sensitivity factors:

$$h_{kn} = \frac{\Delta F_k}{\Delta P_n} \tag{8.60}$$

This sensitivity factor relates the change in the power flow in branch k to an increase in injection at node n. In the conventional DC power flow model, these sensitivity factors depend on the topology and reactances of the network but not on the loading conditions. Hence, for a network with a fixed topology, the sensitivity factors are constant and are evaluated without considering generation and demand.

Wood and Wollenberg (1996) show that if branch k connects buses a and b, the sensitivity factors relating the flow in that branch to the injection at bus n can be calculated as follows:

$$h_{kn} = \frac{\Delta F_k}{\Delta P_n} = \frac{1}{x_{ab}}(X_{an} - X_{bn}) \tag{8.61}$$

where X_{an} and X_{bn} are elements of the inverse of the reduced admittance matrix $Y_\mathrm{bus}^\mathrm{r}$. Although the values of the sensitivity factors depend on the choice of reference node, the result of the optimization problem is indifferent to this choice.

8.4.9.2 Problem Formulation

The objective function of this optimization problem can be expressed as follows:

$$\underset{P_{pg},T_b}{\text{Min}}\left(\sum_{p=1}^{np}\tau_p\sum_{g=1}^{ng}C_gP_{pg}+\sum_{b=1}^{nl}k_bl_bT_b\right) \tag{8.62}$$

Since this problem covers several demand periods over a year, it must satisfy the power flow equations for the intact system and the line capacity limits for each demand period. Using a DC power flow formulation neglecting losses, these constraints are:

$$A^0F_p^0 - P_p + D_p = 0 \tag{8.63}$$

$$F_p^0 = H^0(P_p - D_p) \tag{8.64}$$

$$F_p^0 - T \leq 0 \tag{8.65}$$

$$-F_p^0 - T \leq 0 \quad p = 1, np \tag{8.66}$$

Equation (8.63) is a nodal balance constraint derived from Kirchhoff's current law, which requires that the total power flowing into a node must be equal to the total power flowing out of the node. Constraint (8.64) relates flows and injections on the basis of Kirchhoff's voltage law. The last two equations represent thermal constraints on the line flows. All these constraints must also be satisfied for each contingency during each demand period:

$$A^cF_p^c - P_p + D_p = 0 \tag{8.67}$$

$$F_p^c = H^c(P_p - D_p) \tag{8.68}$$

$$F_p^c - T \leq 0 \tag{8.69}$$

$$-F_p^c - T \leq 0 \quad p = 1, np; \;\; c = 1, nc \tag{8.70}$$

Finally, the optimization must respect the limits on the output of the generators:

$$P^{\min} \leq P_p \leq P^{\max} \quad p = 1, np \tag{8.71}$$

Since the object of the optimization is to determine the optimal thermal capacity of the lines, this variable can take any positive value:

$$0 \leq T \leq \infty \tag{8.72}$$

8.4.9.3 Implementation

The above model calculates for each demand period the vector of generation dispatch P_p, the vector of line flows F_p^0, and the vector of optimal capacities T. All other parameters in the above equations are either specified or determined from the network topology and data. Since we have assumed constant generation marginal costs the optimization problem is linear. However, because of its size, this problem is usually not solved in its original form. Instead, it is solved using the iterative algorithm shown in Figure 8.13. At the start of each iteration, we calculate a generation dispatch for each demand period and a capacity for each line such that the demand is met and the transmission constraints are satisfied. Note that at the beginning of the process there are no transmission constraints. The feasibility of this dispatch is then evaluated by performing a power flow analysis for all

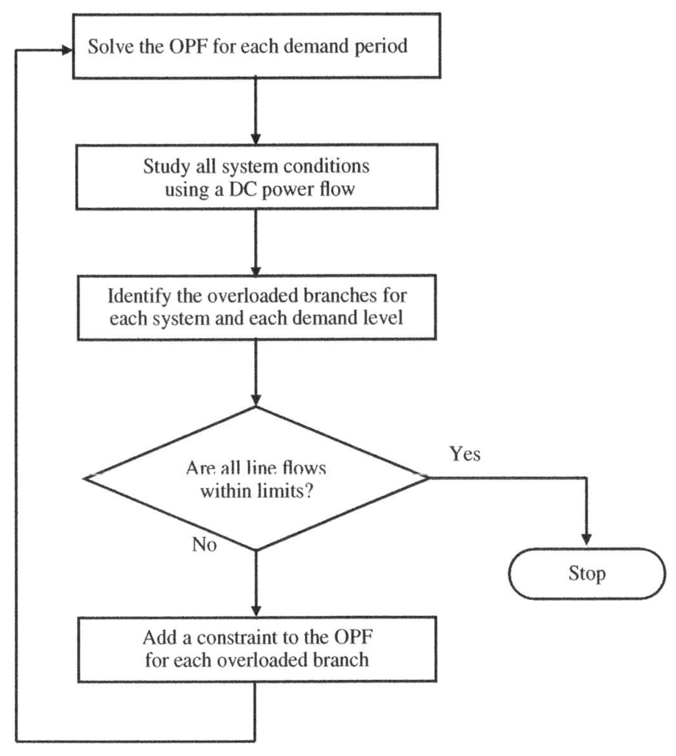

Figure 8.13 Flowchart of the security-constrained OPF problem used to determine the reference network.

contingent networks and each demand period. If any of the line flows is greater than the proposed capacity of the line, a constraint is created and inserted in the OPF at the next iteration. For example, if line b is overloaded, the following constraint is added to the problem:

$$-T_b \leq f_b^{ps} + \sum_{k=1}^{nb} h_{kb}^S \cdot \left(P_k^p - P_k^{p0}\right) \leq T_b \tag{8.73}$$

where S represents the network topology for both the intact and contingent conditions and h_{kb}^S are the corresponding sensitivity factors. This process is repeated until all line overloads are eliminated.

The nodal prices are then calculated as follows:

$$\pi_j^p = \pi^p + \sum_{s=1}^{nc} \sum_{b=1}^{nl} h_{jb}^s \cdot \mu_b^{ps} \tag{8.74}$$

where $\pi^p$is the Lagrange multiplier associated with the load balance constraint for demand period p in the intact network. This quantity is frequently called the system marginal cost. The variables μ_b^{ps} are the Lagrange multipliers associated with the transmission constraints (8.73) that are generated in the iterative process.

Example 8.5

This optimization procedure has been applied to the IEEE 24 bus Reliability Test System (RTS) depicted in Figure 8.14. For the details of this network, see IEEE (1979). Figure 8.15 shows that, except for a small number of lines, line flows are well below 50% of the optimal capacity even during the period of maximum demand. This observation confirms the importance of taking security into consideration when designing and pricing a transmission network.

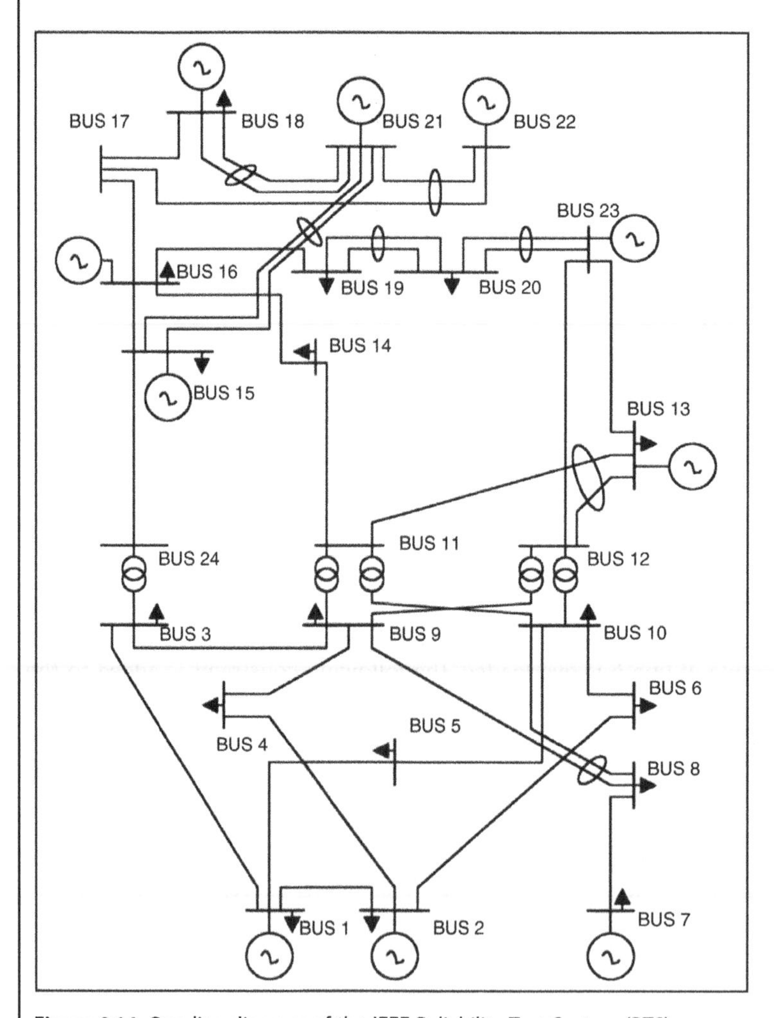

Figure 8.14 One-line diagram of the IEEE Reliability Test System (RTS).

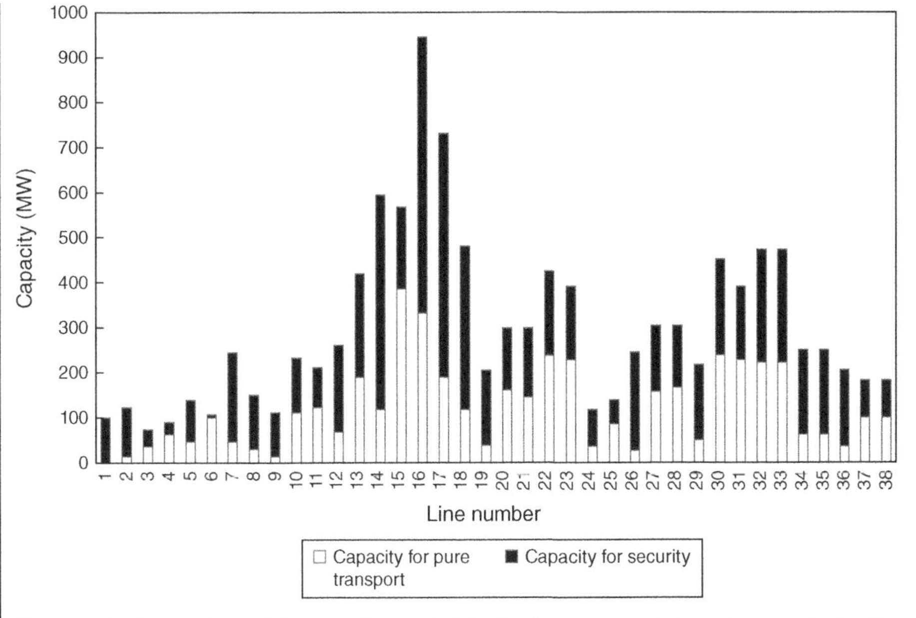

Figure 8.15 Comparison of the capacity needed for the intact network (pure transport) with the capacity needed to ensure security during the maximum demand period for the IEEE RTS.

8.4.9.4 Considering Other Factors

This basic algorithm for constructing the reference network becomes considerably more complex if we want to optimize not only the capacity of the lines but also the network topology and the choice of transmission voltage levels, or if we want to deal with load growth, economies of scale, new transmission technologies such as FACTS, distributed generation, demand-side management, losses, reactive power, network stability constraints, and generation reserve. The appropriate degree of complexity depends on the intended application and the specific system. However, it is important to bear in mind that the purpose of a reference network is not to replace the detailed technical design of the transmission network, but to support decisions regarding regulation, investments, and pricing.

8.5 Other Sources of Value

While the most obvious value of the transmission network stems from its ability to move electrical energy from one region to another, it provides other benefits that can help justify the construction of new lines or an increase in the existing capacity. In the following sections, we discuss how transmission creates value by making it possible to share generation reserves, balancing capacity, and generation capacity margin.

8.5.1 Sharing Reserve

In Section 6.4.3, we argued that some of the available transmission capacity ought to be allocated to the provision of reserve generation capacity and that this allocation was likely to grow as the proportion of stochastic generation from renewable sources increases. This issue should also be considered when decisions about transmission capacity expansions are made because increased operating reserve requirements may justify additional network investments. Currently, and as we discussed in the previous sections, decisions about transmission capacity balance the cost of congestion against the cost of investments that would make possible more efficient energy transfers. Because the need to allocate transmission capacity for reserve is not explicitly taken into account, this may result in underinvestment in transmission capacity, which may, in turn, hamper the ability or increase the cost of integrating renewable generation sources in the system.

Example 8.6

In Example 6.9, we investigated how the existing capacity of the Borduria–Syldavia interconnection should be allocated between energy and reserve. Let us revisit this example from the perspective of a transmission planner who tries to determine how much capacity should be built considering the needs to transmit energy during normal operation and reserve during a contingency. Figure 8.16 illustrates this system. The only difference with the system of Example 6.9 is that the transmission capacity is now a variable T.

As before, our objective is to minimize the sum of the cost of energy and reserve in both countries and the cost of building the interconnection:

$$\text{Min}\{C_B(P_B, R_B) + C_S(P_S, R_S) + C_T(T)\} \tag{8.75}$$

where P_B and P_S represent the power produced in Borduria and Syldavia, respectively, and R_B and R_S the reserve procured in each country. The costs of energy and reserve in Borduria and Syldavia $C_B(P_B, R_B)$ and $C_S(P_S, R_S)$ are the same as those in Example 6.9:

$$C_B(P_B, R_B) = 10P_B + \frac{0.01}{2}P_B^2 + R_B + \frac{0.001}{2}R_B^2 + 0.001P_BR_B \tag{8.76}$$

$$C_S(P_S, R_S) = 13P_S + \frac{0.02}{2}P_S^2 + 5R_S + \frac{0.019}{2}R_S^2 + 0.001P_BR_B \tag{8.77}$$

The investment cost of transmission is assumed to be:

$$C_T(T) = c_T T = 4T \tag{8.78}$$

This minimization is subject to the following constraints:

Power balance in Borduria:

$$P_B - F^E = 500 \quad (\pi_B^E) \tag{8.79}$$

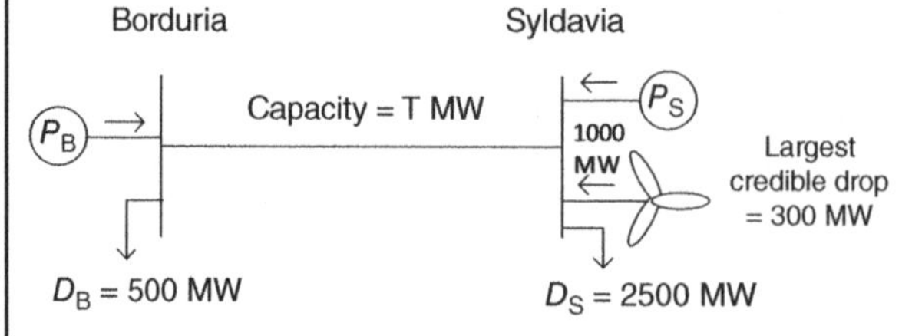

Figure 8.16 Modified Borduria–Syldavia system.

Power balance in Syldavia:

$$P_S + F^E = 2500 - 1000 \quad (\pi_S^E) \tag{8.80}$$

Reserve requirement in Borduria:

$$R_B - F^R = R_B^R \quad (\pi_B^R) \tag{8.81}$$

Reserve requirement in Syldavia:

$$R_S + F^R = R_S^R \quad (\pi_S^R) \tag{8.82}$$

Total transmission capacity limit:

$$F^E + F^R \leq T \quad (\pi^T) \tag{8.83}$$

F^E and F^R denote the capacities allocated for the transfer of energy and reserve. The reserve requirements in Borduria (R_B^R) and Syldavia (R_S^R) can be satisfied either locally (R_B, R_S) or from reserve located in the other country if the transmission capacity allocated for reserve (F^R) is sufficient.

These constraints are identical to those of Example 6.9, except for the last one where the fixed transmission capacity (800 MW) has been replaced by the optimal transmission capacity T that we wish to determine.

Solving this optimization problem gives the following results:

$$\begin{aligned} P_B &= 1293\text{ MW} \quad & P_S &= 1293\text{ MW} \\ R_B &= 256\text{ MW} \quad & R_S &= 44\text{ MW} \\ F^E &= 793\text{ MW} \quad & F^R &= 256\text{ MW} \\ T &= 1049\text{ MW} \end{aligned}$$

Considering the need to provide transmission capacity for reserve thus increases the optimal capacity from 800 (as calculated in Section 8.4.4) to 1049 MW. 793 MW of this transmission capacity is used for the transmission of energy, while the remaining 256 MW is kept idle to transfer reserve when needed.

Assuming that the prices of energy and reserve in both countries are given by the partial derivatives of the cost functions, we get:

$$\begin{aligned} \pi_B^E &= 23.18\,\$/\text{MWh} \quad & \pi_S^E &= 27.18\,\$/\text{MWh} \\ \pi_B^R &= 2.55\,\$/\text{MWh} \quad & \pi_S^R &= 6.55\,\$/\text{MWh} \end{aligned}$$

The most significant effect of providing extra transmission capacity for reserve is to decrease the cost of reserve in Syldavia compared to what it was in Example 6.9.

Note that the price differentials between the two countries for energy and reserve are both 4 $/MWh, which is the value that we have assumed for the LRMC of transmission expansion in Equation 8.78. The marginal revenue from energy is thus equal to the marginal revenue from reserve, and a revenue stream combining both is sufficient to recover the investment cost associated with the optimum transmission capacity.

This example shows that the ability to access remote, low-cost reserves increases the value of transmission capacity and should be incorporated in network expansion planning models. However, this will happen in a market environment only if transmission

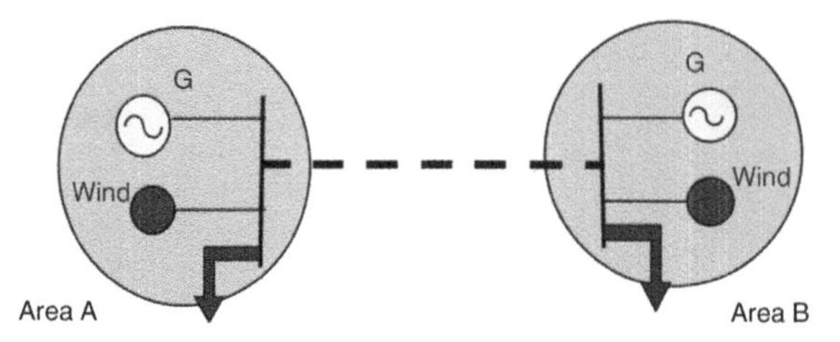

Figure 8.17 Interconnecting two identical areas for the exchange of balancing services.

owners are remunerated for providing capacity headroom that can be used when needed to deliver reserves.

8.5.2 Sharing Balancing Capacity

Connecting two systems via a transmission line gives them the opportunity to share the amount of generation capacity that they need to maintain a balance between load and generation. When stochastic renewable generation becomes a significant portion of the installed generation capacity, sharing balancing resources enhances the value of transmission capacity. To illustrate this observation, let us consider the construction of a transmission line between areas A and B shown in Figure 8.17. These two areas are identical, i.e. they have exactly the same generation mixes and demand profiles. Consequently, prices for energy and reserve are the same in both areas. This assumption helps us demonstrate unambiguously the value of sharing balancing capacity, because it voids any opportunity for locational arbitrage. A transmission line between these two areas therefore provides no value in terms of trading energy or reserve. However, each area is subject to random fluctuations in load, in solar and wind generation and is affected by random generation outages. Operators on both sides therefore need flexibility resources to deal with these random imbalances. Exchanges of short-term balancing services reduce the total operating costs in both systems. These cost reductions represent a benefit from the existence of the transmission line.

Example 8.7

Table 8.14 summarizes the characteristics of the conventional generation plants installed in areas A and B. Both areas have an annual energy requirement of 600 TWh and a 95 GW peak demand. Two levels of wind generation are considered: 15% and 30%. The error on the 4-h-ahead wind generation forecast is assumed to be 10%. Forecasting errors in the two areas are assumed uncorrelated.

Table 8.14 Characteristics of the conventional plants installed in areas A and B.

	Nuclear	CCGT	OCGT
Capacity (GW)	15	60	30
Marginal cost ($/MWh)	7	70	140
No-load cost (k$/h)	0.3	11	34
Startup cost (k$/startup)	100	42	20

Using a stochastic scheduling model (Teng and Strbac, 2017), we can calculate the annual cost of providing balancing resources. When each area is operated independently (i.e. without the benefit of an interconnection to the other area), Figure 8.18 shows that for a 15% wind penetration the annual cost of balancing is about 0.78 b$, while for a 30% penetration it is 2.0 b$.

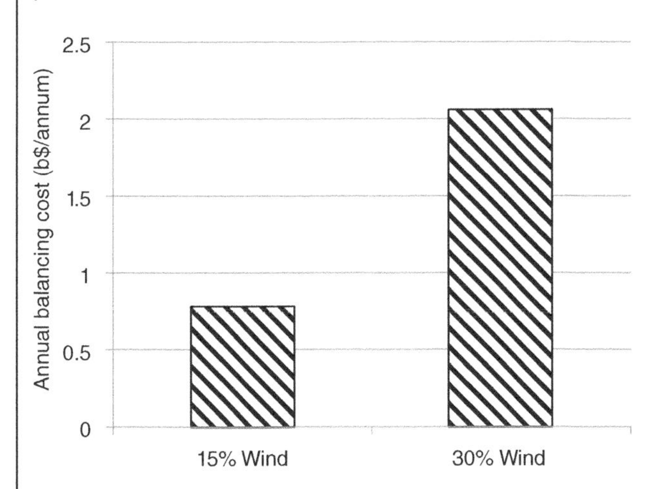

Figure 8.18 Annual cost of balancing services.

Figure 8.19 shows the annual operational cost savings that would accrue if a 5 GW interconnection were built between the two areas. As one would expect, the savings increase significantly with the proportion of wind generation.

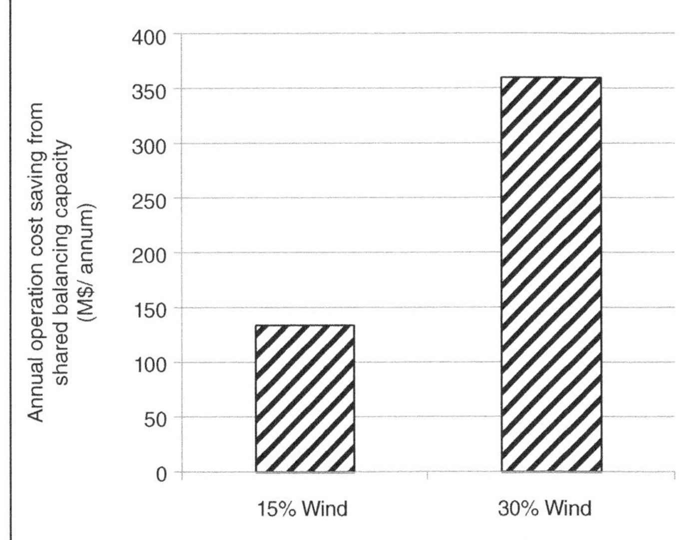

Figure 8.19 Annual savings in the operating cost made possible by the sharing of balancing services between the two identical systems.

Using typical values of annuitized fixed and variable costs of installing undersea cables (Konstantelos et al. 2017), which are around 56 k$/km and 96 k$/GW km, building a

500 km interconnection would cost about 260 M$/year but would bring a net benefit of nearly 100 M$/year for a 30% wind penetration from enabling an exchange of balancing services between the two areas. On the other hand, building such an interconnection would not be cost-effective for a 15% wind penetration because the reduction in the operating cost would be smaller than the annuitized investment cost.

8.5.3 Sharing Generation Capacity Margin

As we discussed in Chapter 7, the total generation capacity installed in a system must exceed the expected peak load by a certain margin because some of the generating units may be unavailable due to a failure or the need to carry out maintenance. In this section, we show that building or expanding interconnections between power systems makes it possible to reduce the size of this generation capacity margins while remaining in compliance with the operational reliability standards. This sharing of generation capacity margin is possible because outages of generating units are stochastically independent events and because interconnecting the power systems enlarges the set of generators contributing to the total installed capacity.

Example 8.8

Let us suppose that the power system of Syldavia has a peak load of 500 MW and that the only generating units that can be built have a capacity of 60 MW and an availability of 90%. The Syldavian operational reliability criterion specifies that the probability of not enough generation being available to meet the peak demand should not exceed 10%, i.e. that the loss of load probability (LoLP) should be less than 10%.

To determine the LoLP under these conditions, we need to construct the Capacity Outage Probability Table[1] shown in Table 8.15. Because we assume that all generating units have the same capacity, each line of this table corresponds to a state where n units are in service. Each line shows the corresponding available generation capacity, whether this capacity is sufficient to meet the load, and the probability that this number of units will be in service. This probability is computed using the binomial distribution formula:

$$\Pr(n) = \binom{N}{n} p^n (1-p)^{N-n} \tag{8.84}$$

where p is the probability that a particular unit is available and N is the total number of units. In Table 8.15, we have set $N = 11$. Since all units are assumed to have an availability of 90%, $p = 0.9$. The probability of each state is thus calculated as follows:

$$\Pr(n) = \binom{11}{n} (0.9)^n (0.1)^{11-n} \tag{8.85}$$

If 8 or fewer generating units are in service, the available generation capacity is less than the peak load. Summing the probabilities of the states where $n \leq 8$ thus gives us an LoLP of 0.09, we can therefore conclude that this system must have at least 11 generating units of 60 MW capacity to satisfy the operational reliability criterion.

1 See Billinton and Allan (1996) for a detailed discussion of how to construct such a table and of the calculation of LoLP.

Table 8.15 Capacity Outage Probability Table for a peak load of 500 MW with 11 generating units of 60 MW capacity and an availability of 90%.

Number of available units n	Available generation capacity (MW)	Available generation capacity – peak load (MW)	Probability of having n units available
11	660	160	0.313 811
10	600	100	0.383 546
9	540	40	0.213 081
8	480	−20	0.071 027
7	420	−80	0.015 784
6	360	−140	0.002 455
5	300	−200	0.000 273
4	240	−260	2.17E−05
3	180	−320	1.20E−06
2	120	−380	4.45E−08
1	60	−440	9.90E−10
0	0	−500	1.00E−11

Example 8.9

The Bordurian power system is identical to the Syldavian system described in the previous example. If these two systems are not interconnected, each of them must have at least 11 generating units of 60 MW capacity to satisfy their operational reliability criterion. Let us show that if an interconnection with an 80 MW capacity is built between these two systems, each of them needs to have only ten 60 MW generating units to meet its peak load of 500 MW with an LoLP of 0.1.

Table 8.16 shows the Capacity Outage Probability Table for the Syldavian power system when this system has only ten 60 MW generating units. Since we assume that the Bordurian system has exactly the same characteristics, it has an identical table. If the Syldavian system relies only on its own resources, 9 or 10 generating units must be in service to meet the peak load. The probability of not meeting the peak load is then:

$$\text{LoLP} = 1 - [\Pr(n = 10) + \Pr(n = 9)] = 1 - (0.3487 + 0.3874) = 0.2369 \tag{8.86}$$

Since this value is greater than 0.1, the operational reliability criterion is not satisfied.

On the other hand, if an 80 MW interconnection with the Bordurian system has been built, Syldavia would be able to meet its peak load with only 8 or 7 of its own units, as long as Borduria has enough spare generation capacity. To calculate the probability that Syldavia would be able to meet its peak load, we must therefore consider the availability of both the Syldavian and the Bordurian units. Here are the possibilities:

- If only 7 units are available in Syldavia, 80 MW must be imported from Borduria. This is possible only if Borduria has at least 80 MW of spare capacity, i.e. if all 10 Bordurian units are in service.

Table 8.16 Capacity Outage Probability Table for a peak load of 500 MW with 10 generating units of 60 MW capacity and an availability of 90%.

Number of available units n	Available generation capacity (MW)	Available generation – peak load (MW)	Probability of having n units available
10	600	100	0.3487
9	540	40	0.3874
8	480	−20	0.1937
7	420	−80	0.0574
6	360	−140	0.0112
5	300	−200	1.49E−03
4	240	−260	1.38E−04
3	180	−320	8.75E−06
2	120	−380	3.65E−07
1	60	−440	9.00E−09
0	0	−500	1.00E−10

- If only 8 units are available in Syldavia, 20 MW must be imported from Borduria. This is possible only if Borduria has at least 20 MW of spare capacity, i.e. if 9 or 10 Bordurian units are in service.
- If 9 or 10 units are available in Syldavia, it does not need to import power to meet its peak load and the availability of Bordurian units is irrelevant.

If we denote by n_S and n_B the number of units available, respectively, in Syldavia and Borduria, and use the probability values from Table 8.16 for both Syldavia and Borduria, the probability that Syldavia will be able to meet its peak load is:

$$\Pr(n_S = 10) + \Pr(n_S = 9) + \Pr(n_S = 8) \times \{\Pr(n_B = 10) + \Pr(n_B = 9)\} + \Pr(n_S = 7) \times \Pr(n_B = 10) = 0.899 \tag{8.87}$$

which gives an LoLP sufficiently close enough to 0.1 for us to declare that it satisfies the operational reliability criterion. Building an 80 MW interconnection therefore avoids the need to build two 60 MW generating plants, one in Syldavia and one in Borduria. Note that in this calculation we neglect the probability that the interconnection might not be available.

8.6 Decentralized Transmission Expansion

8.6.1 Concept

Our discussion of the cost-based and the value-based approaches assumed that decisions about transmission expansion aimed to minimize the sum of the investment and operational cost over the entire system. In other words, we adopted a centralized planning perspective. A radically different, market-based approach has been proposed and occasionally implemented. See, for example, Joskow and Tirole (2005). The idea is that stakeholders or group of stakeholders who would benefit from an expansion of the transmission network should be allowed to invest, collect revenues, and profit from such projects. These self-interested stakeholders could include consumers, generators, as well as merchant transmission companies, i.e. companies whose purpose is to profit from investments in transmission. While this decentralized approach is in line with the overall deregulation of liberalization of the electricity supply industry, questions remain about its ability to deliver a transmission network that maximizes the global welfare.

To model this decentralized approach, we must recognize that each entity considering an investment in transmission behaves independently and strategically, i.e. each of them tries to achieve profits beyond what it would obtain under a centralized approach. However, each of these entities must also take into account the decisions that competing entities could make. Every new transmission asset indeed affects the operation of the system and hence the locational marginal prices. The revenues and profits stemming from different investments in transmission are thus interdependent. Capturing the interactions between these entities and their strategic behavior therefore requires a model based on game theory. In this framework, the regulator determines a final expansion plan that reconciles the interests of different entities. This plan is a Nash equilibrium, i.e. a planning solution where none of the entities can increase its profits by unilaterally modifying its decisions.

8.6.2 Illustration on a Two-bus System

Let us study how decentralized transmission expansion might work on the two-bus Borduria–Syldavia system illustrated in Figure 8.2. As in the previous examples, the operating cost of the generators in Borduria and Syldavia are respectively:

$$C_B = 10P_B + 0.005P_B^2 \ \$/\text{h} \tag{8.88}$$

$$C_S = 13P_S + 0.01P_S^2 \ \$/\text{h} \tag{8.89}$$

where P_B and P_S denote the power outputs of the generators in Borduria and Syldavia, respectively. The inelastic demands in Borduria and Syldavia are:

$$D_B = 500\,\text{MW} \tag{8.90}$$

$$D_S = 1500\,\text{MW} \tag{8.91}$$

We assume that there is initially no interconnection between Borduria and Syldavia and that the hourly long-run marginal cost of building transmission capacity between these two countries is:

$$c_T = 4\ \$/\text{MWh} \tag{8.92}$$

In theory, the generators and consumers in both countries, as well as merchant transmission companies could all consider investing in the construction of a transmission line between the two countries. However, to keep things simple, we assume that only two entities could invest in an interconnection between the two countries: a consortium of the generation companies of Borduria and a consortium of the generation companies of Syldavia.

Since these generating companies sell the energy that they produce at the LMP in effect in their country, they collect the following revenues:

$$\Gamma_B = \pi_B P_B \tag{8.93}$$

$$\Gamma_S = \pi_S P_S \tag{8.94}$$

If the Bordurian generators build a transmission capacity F_B and the Syldavian generators a transmission capacity F_s, they would be entitled, respectively, to the following congestion surpluses:

$$CS_B = (\pi_S - \pi_B) F_B \tag{8.95}$$

$$CS_S = (\pi_S - \pi_B) F_S \tag{8.96}$$

These companies would incur the operating costs given by Equations (8.88) and (8.89) as well as the following transmission investment costs:

$$IC_B = c_T F_B \tag{8.97}$$

$$IC_S = c_T F_S \tag{8.98}$$

Their profits are then the difference between these two streams of revenues and these two types of costs:

$$\Omega_B = \Gamma_B + CS_B - C_B - IC_B \tag{8.99}$$

$$\Omega_S = \Gamma_S + CS_S - C_S - IC_S \tag{8.100}$$

These profits depend on the power outputs of the generators, the LMPs, and the transmission capacities that they build. Since generation in Borduria is cheaper than in Syldavia, the construction of a transmission line will create a power flow from Borduria to Syldavia. Intuition suggests that this power flow will use the total capacity of the interconnection, which is the sum of the capacities built by the two consortia. The production of the Bordurian generators is then equal to the demand in Borduria plus this total transmission capacity:

$$P_B = D_B + (F_B + F_S) \tag{8.101}$$

The production of the Syldavian generators is then equal to the load in Syldavia minus the import from Borduria:

$$P_S = D_S - (F_B + F_S) \tag{8.102}$$

Since we assume that the markets in both countries are perfectly competitive, the LMP in each country is equal to the marginal cost of the local generation. Using Equations (8.88) and (8.89), we get:

$$\pi_B = \frac{dC_B}{dP_B} = 10 + 0.01 P_B = 10 + 0.01(D_B + F_B + F_S) \tag{8.103}$$

$$\pi_S = \frac{dC_S}{dP_S} = 13 + 0.02 P_S = 13 + 0.02(D_S - F_B - F_s) \tag{8.104}$$

Combining Equations (8.88)–(8.104), we can express the profits of the generation companies in Borduria and in Syldavia as a function of the transmission capacities built by both consortia:

$$\Omega_B = -\frac{F_B^2}{40} - \frac{F_B F_S}{50} + 29F_B + \frac{F_S^2}{200} + 5F_S + 1250 \tag{8.105}$$

$$\Omega_S = \frac{F_B^2}{100} - \frac{F_B F_S}{100} - 30F_B - \frac{F_S^2}{50} - 6F_S + 22\,500 \tag{8.106}$$

Each of these consortia will choose to invest in the transmission capacity that maximizes its profit, subject to the constraint that this capacity must be greater than or equal to zero:

$$F_B \geq 0 \tag{8.107}$$

$$F_S \geq 0 \tag{8.108}$$

The solution of these two profit maximization problems, which are coupled through the F_B and F_S decision variables, is the Nash equilibrium of our game theoretic model of decentralized transmission expansion.

To compute this equilibrium, we express these profit maximization problems as an equivalent negative profit minimization problem and construct their Lagrangian functions:

$$\mathcal{L}_B = -\Omega_B - \lambda_B F_B \tag{8.109}$$

$$\mathcal{L}_S = -\Omega_S - \lambda_S F_S \tag{8.110}$$

where λ_B and λ_S are the nonnegative Lagrangian multipliers associated with constraints (8.107) and (8.108).

Taking into account the fact that Bordurian generators can only set F_B and Syldavian generators F_S, the Karush–Kuhn–Tucker (KKT) optimality conditions for these coupled optimization problems are:

$$\frac{\partial \mathcal{L}_B}{\partial F_B} = 0 \Rightarrow \frac{F_B}{20} + \frac{F_S}{50} - \lambda_B - 29 = 0 \tag{8.111}$$

$$\frac{\partial \mathcal{L}_S}{\partial F_S} = 0 \Rightarrow \frac{F_B}{100} + \frac{F_S}{25} - \lambda_S + 6 = 0 \tag{8.112}$$

$$\lambda_B \frac{\partial \mathcal{L}_B}{\partial \lambda_B} = 0 \Rightarrow \lambda_B F_B = 0 \tag{8.113}$$

$$\lambda_S \frac{\partial \mathcal{L}_S}{\partial \lambda_S} = 0 \Rightarrow \lambda_S F_S = 0 \tag{8.114}$$

Solving the set of Equations (8.111)–(8.114) leads to the Nash equilibrium shown in Table 8.17.

As one might have guessed, the Syldavian generators would not invest anything in transmission capacity because, as we discussed in Chapter 5, an interconnection with

Table 8.17 Optimal decentralized transmission investment for the Borduria/Syldavia system.

F_B (MW)	F_S (MW)	λ_B (\$/MWh)	λ_S (\$/MWh)
580	0	0	11.80

Table 8.18 Comparison between the centralized and decentralized transmission expansion approaches.

	Centralized approach	Decentralized approach
Optimal transmission capacity (MW)	800	580
Profits of the Bordurian generators ($/h)	8 450	9 660
Profits of the Syldavian generators ($/h)	4 900	8 464
Total cost ($/h)	38 650	39 376

Borduria decreases their profits. Such an interconnection not only depresses their LMP but also reduces the amount of energy that they produce. In fact, the value of λ_S shows that every megawatt-hour transmitted over the interconnection costs them $11.80. On the other hand, the Bordurian generators would like to build a 580 MW interconnection because it would give them the opportunity not only to increase their production but also to collect the congestion surplus on the energy being exported to Syldavia.

The 580 MW transmission capacity that results from this decentralized approach is smaller than the 800 MW capacity that is optimal from a centralized perspective for this system under the same conditions (Section 8.4.4). Building more than 580 MW of transmission capacity is not in the best interests of the Bordurian generators because it would reduce the price differential between the two ends of the interconnection and hence decrease the congestion surplus.

Table 8.18 compares the consequences of the two transmission expansion approaches. Because the decentralized approach leads to a smaller transmission capacity, it yields larger profits for the generators not only in Borduria but also in Syldavia. On the other hand, the total cost (i.e. the sum of the transmission investment and generation operating costs) increases.

Considering a larger system and more self-interested participants in the decentralized planning process requires the use of more advanced game theoretic and equilibrium programming approaches, such as the ones presented in Shrestha and Fonseka (2007) or Fan et al. (2016).

8.7 Non-wires Alternatives for Transmission Expansion

The examples used in this chapter may give the impression that transmission expansion is synonymous with building new transmission lines or upgrading existing ones. However, environmental considerations often make it very difficult to increase transmission capacity in this manner and the cost of building or upgrading lines can be very high. Transmission planners therefore increasingly consider what is called “non-wires alternatives.”

In the traditional centralized approach, projections about load growth and generation expansion are used to identify when the flow on critical lines are likely to exceed their ratings during peak load periods. While expanding the transmission capacity has value under a variety of loading conditions, investment decisions tend to be driven by the need to reliably meet the peak load. Non-wires solutions often provide a cheaper and environmentally friendlier way of handling peak load conditions. These solutions include the following:

- Activating strategically located demand response during peak load conditions to keep the flow in the critical lines at a safe level.
- Charging storage devices located upstream of critical lines and discharging storage devices located downstream of these lines when the flows reach operational reliability limits.
- Remedial action schemes that relieve overloads and other stability problems caused by unplanned outages.
- Energy efficiency programs that slow the growth in the peak load.
- Distributed generation in load centers to reduce the loading on the transmission network.

Non-wires solutions can defer the need to build or upgrade transmission lines for at least a few years. They can also reduce the risk of large transmission investments getting stranded because of erroneous forecasts. On the other hand, their implementation requires an integrated resource planning process and a regulatory regime that incentivizes the provision of such solutions. See Poudineh and Jamasb (2014) for a discussion of these issues.

8.8 Problems

8.1 Summarize the regulatory process used for transmission expansion in your region or country or in another area for which you have access to sufficient information.

8.2 Identify the method used to allocate the cost of transmission investments in your region or country or in another area for which you have access to sufficient information.

8.3 Consider the two-bus power system shown in Figure P8.1. Assume that the demand is constant and insensitive to price, that energy is sold at its marginal cost of production and that there are no limits on the output of the generators. What is the maximum price that could be charged for transmission if the marginal costs of generation are as follows?

$$MC_A = 25\ \$/\text{MWh}$$
$$MC_B = 17\ \$/\text{MWh}$$

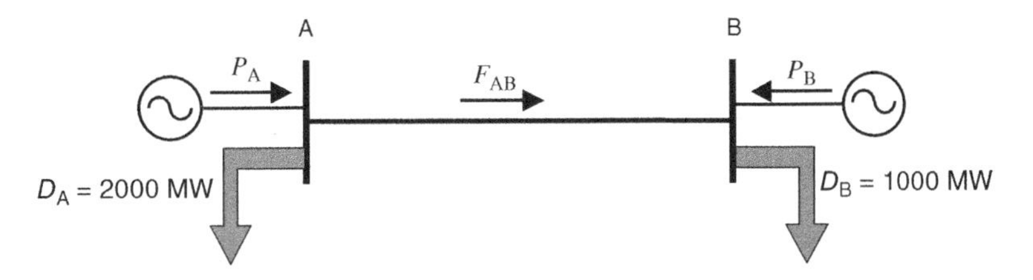

Figure P8.1 Two-bus power system for Problems 8.3–8.9.

8.4 Consider the two-bus power system shown in Figure P8.1. Assume that the demand is constant and insensitive to price, that energy is sold at its marginal cost of production and that there are no limits on the output of the generators. The marginal cost of production of the generators connected to buses A and B are given, respectively, by the following expressions:

$$MC_A = 20 + 0.03P_A \ (\$/\text{MWh})$$
$$MC_B = 15 + 0.02P_B \ (\$/\text{MWh})$$

Plot the marginal value of transmission as a function of the capacity of the transmission line connecting buses A and B.

8.5 Determine the transmission demand function for the system of Problem 8.4.

8.6 Calculate the hourly long-range marginal cost of the transmission line of Problem 8.4 assuming that the line is 500 km long and the amortized variable cost of building the line is 210 \$/(MW × km × year).

8.7 Determine the optimal capacity of the transmission line of Problems 8.4–8.6, assuming the loading conditions shown in Figure P8.1.

8.8 Determine the optimal capacity of the transmission line of Problems 8.4–8.6, for the three-part load-duration curves summarized in the following table. Assume that the periods of high, medium, and low load coincide at both buses.

Period	Load at A (MW)	Load at B (MW)	Duration (h)
High	4000	2000	1000
Medium	2200	1100	5000
Low	1000	500	2760

Compare the amount of congestion revenue collected annually for this optimal transmission capacity with the annuitized cost of building the transmission line.

8.9 Calculate the amount of congestion revenue collected annually for a transmission capacity 33.3% higher and 33.3% lower than the optimal transmission capacity calculated in Problem 8.8. Compare these values to the annuitized cost of building the transmission line.

References

Billinton, R. and Allan, R.N. (1996). *Reliability Evaluation of Power Systems*, 2e. Plenum Press.

Boiteux, M. (1949). La tarification des demandes en pointe: application de la théorie de la vente au coût marginal. *Rev. Gén. Electr.* 58: 321–340.

Fan, Y., Papadaskalopoulos, D., and Strbac, G. (2016). A game theoretic modeling framework for decentralized transmission planning. *2016 Power Systems Computation Conference (PSCC)*, Genoa, pp. 1–7.

Farmer, E.D., Cory, B.J., and Perera, B.L.P.P. (1995). Optimal pricing of transmission and distribution Services in Electricity Supply. *IEE Proc. Generat. Transm. Distrib.* 142 (1).

Hogan, W.W. (1999). Market-based transmission investments and competitive electricity markets. https://sites.hks.harvard.edu/fs/whogan/trans_mkt_design_040403.pdf.

IEEE Power Engineering Society Subcommittee on the Application of Probabilistic Methods (1979). A reliability test system. *IEEE Trans. Power App. Syst.* PAS-98 (6).

Joskow, P. and Tirole, J. (2005). Merchant transmission investment, J. *Industr. Econ.* 53 (2): 233–264.

Konstantelos, I., Moreno, R., and Strbac, G. (2017). Coordination and uncertainty in strategic network investment: case on the north seas grid. *Energy Econ.* 64: 131–148.

Marangon Lima, J.W. (1996). Allocation of transmission fixed charges: an overview. *IEEE Trans. Power Syst.* 11 (3): 1409–1418.

Nelson, J.R. (1967). *Marginal Cost Pricing in Practice.* Prentice-Hall.

Poudineh, R. and Jamasb, T. (2014). Distributed generation, storage, demand response and energy efficiency as alternatives to grid capacity enhancement. *Energy Policy* 67: 222–231.

Shrestha, G.B. and Fonseka, P.A.J. (Sep. 2007). Optimal transmission expansion under different market structures. *IET Gen. Transm. Distr.* 1 (5): 697–706.

Teng, F. and Strbac, G. (2017). Full stochastic scheduling for low-carbon electricity systems. *IEEE Trans. Autom. Sci. Eng.* 14: 461–470.

Wood, A.J. and Wollenberg, B.F. (1996). *Power Generation, Operation and Control,* 2e. Wiley.

Further Reading

Hogan, W.W. (2003). Transmission Market Design. www.ksg.harvard.edu/whogan.

Woolf, F. (2003). *Global Transmission Expansion: Recipes for Success.* PennWell.

Index

Printed and bound by CPI Group (UK) Ltd, Croydon, CR0 4YY
23/04/2025
14660952-0003